A Historical Approach to

Materials Under Irradiation

SCIENCES

Materials Science, Field Director – Jean-Pierre Chevalier

Metallurgy, Subject Head – Jean-Pierre Chevalier

A Historical Approach to Materials Under Irradiation

Serge Bouffard

iSTE

WILEY

First published 2025 in Great Britain and the United States by ISTE Ltd and John Wiley & Sons, Inc.

ISTE Ltd
27-37 St George's Road
London SW19 4EU
UK

www.iste.co.uk

John Wiley & Sons, Inc.
111 River Street
Hoboken, NJ 07030
USA

www.wiley.com

Library of Congress Control Number: 2024950249

British Library Cataloguing-in-Publication Data
A CIP record for this book is available from the British Library
ISBN 978-1-78945-221-1

ERC code:
PE4 Physical and Analytical Chemical Sciences
PE4_14 Radiation and Nuclear chemistry
PE5 Synthetic Chemistry and Materials
PE5_1 Structural properties of materials

Contents

Preface xi

Chapter 1. Preliminary Remarks 1

1.1. References 4

Chapter 2. Prerequisites for the Irradiation of Materials 5

2.1. Materials and purity, an old story 5
2.1.1. Materials and disorder in ancient history 5
2.2. Discovering the particles behind irradiation 14
2.2.1. High-voltage generation 15
2.2.2. Vacuum control 19
2.2.3. Cathode rays 24
2.2.4. Discovery of X-rays 27
2.2.5. Discovery of the electron 28
2.2.6. Discovery of radioactivity 29
2.3. First irradiation experiments 30
2.3.1. Pleochroic halos 34
2.4. Secondary effects of radiation 35
2.5. Chapter 2 references 38
2.5.1. A brief chronology 38
2.5.2. Biographies of some of the chapter's personalities 39
2.6. References 40

Chapter 3. Particle Transport 47

3.1. It all started with collision experiments 47
3.2. Slowing down in the matter 50
3.3. Particle stopping power 60
3.3.1. Electronic stopping power 60
3.3.2. Nuclear stopping power 64
3.4. Particle range 67
3.5. Transport simulation 68
3.5.1. First simulations 70
3.6. Channeling effects 74
3.7. Chapter 3 references 79
3.7.1. Short chronology 79
3.7.2. Biographies of some of the chapter's personalities 79
3.8. References 81

Chapter 4. First Notions of Defects 89

4.1. First observations of defects 89
4.1.1. Photographic processes 89
4.1.2. First experiments: an approach guided by sight 92
4.1.3. Defects, a useful concept for diffusion 96
4.2. Notions of defects 98
4.3. Chapter 4 references 100
4.3.1. Biographies of some of the chapter's personalities 100
4.4. References 103

Chapter 5. Defect Creation Mechanisms 107

5.1. Production of defects by irradiation 109
5.1.1. Creation of defects by electronic excitations and ionizations 109
5.1.2. Models for the creation of defects by elastic collisions 114
5.2. Determination of threshold displacement energy 120
5.2.1. Threshold displacement energy mapping 122
5.3. Numerical simulations 124
5.3.1. Creation and stability of point defects 124
5.3.2. Thermal spike 126
5.4. Irradiation-induced sputtering 128
5.4.1. Metal sputtering 129
5.4.2. Uranium sputtering 130

5.5. Chapter 5 references 132
5.5.1. Biographies of some of the chapter's personalities 132
5.6. References 134

Chapter 6. Metals Under Irradiation. 139

6.1. Notions shared with other disciplines 141
6.1.1. Self-diffusion in metals 141
6.1.2. Cold metalworking 142
6.1.3. Dislocation theory 145
6.2. Creation of defects in metals by irradiation 146
6.2.1. Irradiation of pure metals 147
6.2.2. Irradiation of ordered alloys. 149
6.3. Displacement threshold. 151
6.4. Description of defects. 154
6.4.1. Experimental observations of point defects. 155
6.5. Defect annealing. 159
6.6. Chapter 6 references 167
6.6.1. Biographies of some of the chapter's personalities 167
6.7. References 168

Chapter 7. Semiconductors Under Irradiation 175

7.1. First irradiation of semiconductors 176
7.2. Defect generation and counting 181
7.2.1. Determining the displacement threshold 181
7.2.2. High-energy deposits. 183
7.2.3. Description of defects 184
7.3. Diffusion in semiconductors 187
7.3.1. Smart Cut process. 188
7.4. Chapter 7 references 189
7.4.1. Laboratories and personalities in this chapter 189
7.5. References 191

Chapter 8. Iono-covalent Insulators Under Irradiation. 195

8.1. Iono-covalent materials under irradiation 195
8.1.1. Defects in iono-covalent materials 197
8.1.2. Threshold displacement energy in inorganic insulators. 200
8.1.3. Phase transformation under irradiation 202

8.2. Biographies of some of the chapter's personalities 203
8.3. References 203

Chapter 9. Polymers Under Irradiation 207

9.1. First irradiations of polymers 207
9.2. Research into degradation mechanisms 211
9.3. Radio-oxidation of polymers 217
9.4. Research and development, an active field 218
9.5. Chapter 9 references 219
9.5.1. Biographies of some of the chapter's personalities 219
9.6. References 219

Chapter 10. Radiolysis of Liquids 223

10.1. Upstream of the notion of radiolysis 223
10.2. Activated water 227
10.3. Free radicals 228
10.4. Solvated electrons 229
10.4.1. Solvated electrons, an old story 230
10.5. Effects of the spatial structure of energy deposits 234
10.6. Radiolysis yields 237
10.7. Chapter 10 references 237
10.7.1. Biographies of some of the chapter's personalities 237
10.8. References 239

Chapter 11. Irradiation Tools 243

11.1. Accelerators 244
11.1.1. Radio-frequency cavity accelerators 244
11.1.2. Electrostatic accelerators 249
11.1.3. Tandem electrostatic accelerators 252
11.1.4. Pulsed electron accelerators 254
11.2. Nuclear reactors 255
11.3. Recent developments 260
11.4. Chapter 11 references 260
11.4.1. Biographies of some of the chapter's personalities 260
11.5 References 262

Chapter 12. Irradiation Applications 267

12.1. Medical applications. 269
12.1.1. Radiography 269
12.1.2. Radiotherapies 271
12.1.3. Nuclear medicine 273
12.1.4. Radiosterilization 274
12.2. Food processing 274
12.3. Polymer irradiation applications. 277
12.4. Semiconductor doping 278
12.4.1. Doping by implantation 279
12.4.2. Transmutation doping. 281
12.5. Radiation resistance of electronic components 282
12.6. Ion track technology. 283
12.7. Cultural and historical heritage materials 287
12.8. References. 289

Conclusions 293

C.1. An active community. 293
C.2. Future prospects. 295
C.3. References 297

Index 299

Preface

The history of science generally focuses on major scientific breakthroughs or personalities who left their mark. When it comes to the history of research on materials under irradiation, there are very few documents providing a historical overview of this research. A few short texts do, however, provide a starting point for developing this history. These include articles by Seitz (1952), Dienes (1953), Billington and Crawford (1961), Seeger (1980), Quéré (1993) and Ewing (1994). Research into materials under irradiation is closely linked to related fields, which have been the subject of numerous reports. These include, on the one hand, the discoveries of X-rays and radioactivity (Pais 1986; Fernandez 2006) and, on the other hand, the development of materials science and solid state physics, with names like F. Seitz, F. Bloch, N.F. Mott, R.E. Peierls and J. Bardeen. These fields are well documented; see, for example, Mott's (1980) book on the beginnings of solid state physics. In addition, the "irradiation of materials" community has benefited greatly from its proximity to the "dislocation" community (F.R.N. Nabarro, W.G. Burger, F.C. Frank, A.H. Cottrell and J. Friedel). Particle transport in matter, a more specialized field than material damage, has also been the subject of reviews; see, for example, the introduction to the book "*The Stopping and Range of Ions in Matter*" (Ziegler et al. 1985).

The aim of this book is therefore to provide researchers, engineers and students with a historical approach to materials under irradiation, presenting the appearance and development of the various concepts, and the experiments that led to their emergence. These theories and experiments are presented only in a broad outline, and readers that would like to know all their subtleties will need to study the original texts. It should also be noted that the lack of information on most of the actors involved makes it difficult to flesh out this story.

This book is a project which, from the initial research of articles to the printed book, has taken more than 10 years to complete, albeit on a part-time basis. Having worked on a corpus of some 1,690 articles, it was reasonable to bring it to a close when some aspects were only imperfectly covered. Moreover, I am an experimental physicist, and experiments are certainly presented in greater detail than theoretical approaches. Finally, as Jonah writes in *A Short History of the Radiation Chemistry of Water*:

> There will certainly be omissions, misattributions and misunderstandings of the past. This is unavoidable because first, I wasn't there and second, I have been influenced by my surroundings and the people that I have known (Jonah 1995).

This book would not exist without the people who guided my first steps in research, Gérard Maeder, professor at ENSAM (École Nationale Supérieure des Arts et Métiers), and Libéro Zuppiroli, my thesis supervisor. Nor would it have been possible without the colleagues I enjoyed working with, Florence Rullier-Albenque, and László Forró at SESI[1] (Section d'Études des Solides Irradiés, CEA Fontenay-aux-Roses), my colleagues at CIRIL[2] (Centre Interdisciplinaire de Recherche avec les Ions Lourds, Caen) and, especially, Emmanuel Balanzat. I would like to thank the CEA and CNRS for their documentation service, which enabled me to find practically all the documents online. Thanks also to Chantal Brassy of CIMAP in Caen, who provided me with many articles.

Serge Bouffard
December 2024

References

Billington, D.S. and Crawford, J.H. (1961). *Radiation Damage in Solids*. Princeton University Press, Princeton, NJ.

Dienes, G.J. (1953). Radiation effects in solids. *Annual Review of Nuclear Science*, 2, 187–220.

Ewing, R.C. (1994). The metamict state: 1993 – The centennial. *Nuclear Instruments & Methods in Physics Research Section B: Beam Interactions with Materials and Atoms*, 91, 22–29.

1 Now LSI (Laboratoire des Solides Irradiés).

2 Now known as CIMAP (Centre de recherche sur les Ions, les Matériaux et la Photonique).

Fernandez, B. (2006). *De l'atome au noyau : une approche historique de la physique atomique et de la physique nucléaire.* Éditions Ellipses, Paris.

Jonah, C.D. (1995). A short history of the radiation chemistry of water. *Radiation Research,* 144(2), 141–147.

Mott, N.F. (1980). The beginning of solid state physics. *Proceedings of the Royal Society of London. Series A, Mathematical and Physical Sciences,* 371(1744).

Pais, A. (1986). *Inward Bound: Of Matter and Forces in the Physical World.* Clarendon Press, Oxford.

Quéré, Y. (1993). Radiation effects in solids: A brief history. *Solid State Phenomena,* 30(31), 1–6.

Seeger, A.K. (1980). Some recollections of the radiation damage work of the 1950s. *Proceedings of the Royal Society of London. Series A, Mathematical and Physical Sciences,* 371, 165–172.

Seitz, F. (1952). Radiation effects in solids. *Physics Today,* 5(6), 6–9.

Ziegler, J.F., Biersack, J.P., Littmark, U. (1985). *The Stopping and Range of Ions in Solids.* Pergamon Press, New York.

1

Preliminary Remarks

The essential difference between a physics or chemistry book and a historical study lies in the accuracy and completeness of the sources. Any physics or chemistry article usually begins with a short introduction describing the background to the research, without necessarily going back to the original work. Furthermore, it is questionable whether all the articles cited have actually been read. For example, nearly all theses on material irradiation refer to an article by Kinchin and Pease for the creation of defects, more out of tradition than necessity. On the other hand, a historical approach must necessarily go back to the emergence of the idea, the concept, etc., which poses a number of problems for an amateur historian who, in order to be as close as possible to the truth, comes up against several difficulties. In the 19th century, articles contained few references to previous works and ideas were often germinated collectively, hence their appearance in several laboratories. Also, many early articles were written in the authors' own languages (in German, or Swedish, for example; fortunately, translation software makes it possible to understand them). Moreover, during the Second World War, a time of intense discoveries in the field of irradiation, advances were only disseminated in confidential internal reports. Added to this was the difficulty of obtaining certain reports, articles and books. Despite all this, writing this book required me to consult some 1,700 documents and translate around a hundred from German, only to cite 620 of them in this book.

The number of articles collected is nevertheless sufficient to extract some information on the evolution of the subject, bearing in mind that this corpus is dependent on the way in which I have traced the history of the concepts. The more difficult it is to find the source, the more articles there will be on the subject. Nevertheless, Figure 1.1 shows two key dates for the irradiation of materials: around 1900, when the discoveries of X-rays and radioactivity opened up the field of irradiation, and around 1940, when the Manhattan Project and the first nuclear reactors created a strong demand for scientific and technological data. The decline in the number of articles from the 1960s onwards does not correspond to a reduction in activity, but to the fact that the main concepts were well known.

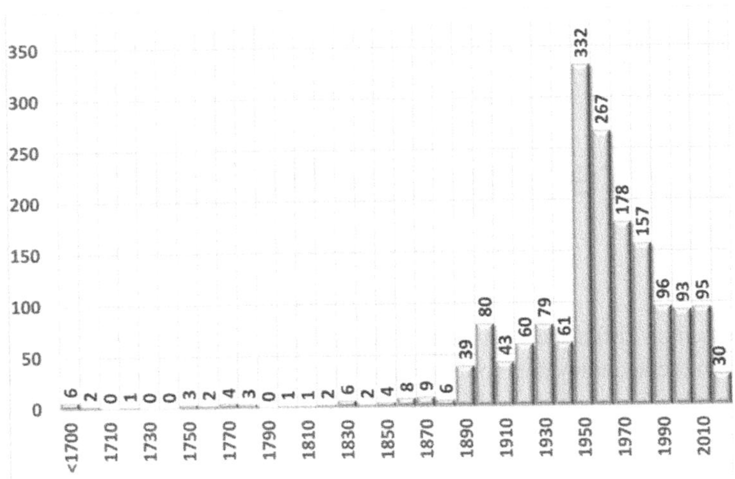

Figure 1.1. *Calendar distribution of publications consulted, highlighting the two key dates (1900 and 1950) in research on the irradiation of materials. For a color version of this figure, see www.iste.co.uk/bouffard/irradiation.zip*

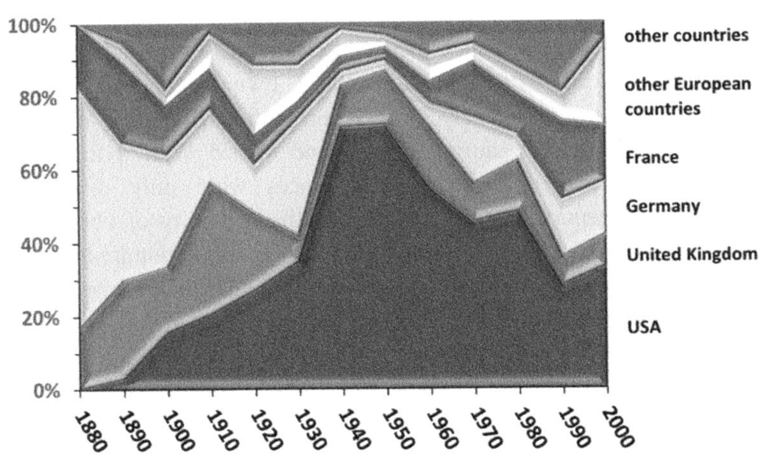

Figure 1.2. *Trends in the geographical origin of publications. The rise of the United States can be seen as early as the 1920s, while the Second World War considerably reduced the importance of Germany, the United Kingdom and France. For a color version of this figure, see www.iste.co.uk/bouffard/irradiation.zip*

Figure 1.2 shows the geographical distribution of those involved in this research. In the 19th century, advances came first from Germany, then from the UK and France. By the 1910s, Germany and the UK were competing with the United States. By the end of the Second World War, the situation had changed radically. The United States accounted for around two-thirds of all publications, while German research had virtually disappeared, although it would regain significant influence from 1975 onwards. The importance of France from the 1960s onwards is certainly overestimated, thanks to a better knowledge of the actors, most of whom I have come across during my career. Other countries that have contributed to our understanding of the effects of irradiation include Japan, Canada, Hungary and Russia, followed by Poland, Sweden, Holland, Australia, Spain and many others.

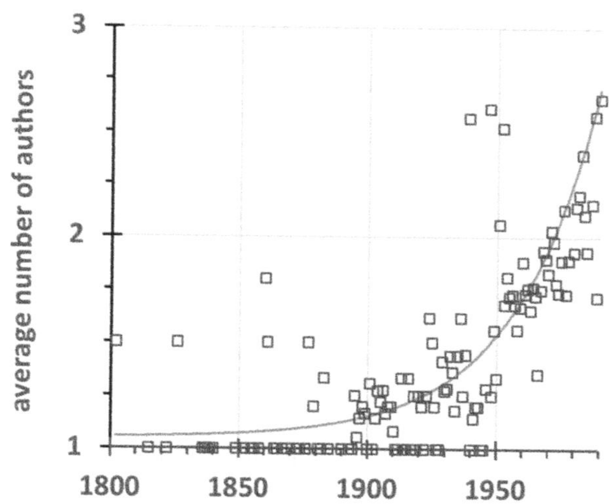

Figure 1.3. *Evolution of the average number of authors per publication, calculated on the basis of 1,690 references grouped together for this work. Before 1900, the vast majority of publications were signed by a single author. The solid line is a guide for the eye, adjusted for exponential variation*

Figure 1.3 shows the average number of authors per article. Until 1900, this number remained constant at around 1.1 authors per article, then rose steadily to reach three in 2000. This evolution is consistent with the evolution of research organization described by Krzysztof Szymborski in "*The Physics of Imperfect Crystals – A Social History*" (Szymborski 1984). He divides this history into three periods. First, research was carried out by independent researchers, often from the aristocracy. The notion of a research laboratory did not exist. The signatures on articles were simply "W.C. Röntgen", "*Note de H. Becquerel*" or "by William

Crookes, Fellow of the Royal Society "or, more precisely, "by J.J. Thomson, M.A., F.R.S., Cavendish Professor of Experimental Physics, Cambridge". The number of people who made a major contribution in this period remained limited to around 30, including E. Goldstein, C. Doelter, J.J. Thomson, E. Rutherford, W.H. Bragg, P. Curie, J. Frenkel and K. Przibram.

Next came the formation of schools of thought around a single personality, usually at a university. Examples include the Cavendish Laboratory at Cambridge University, home to eight Nobel Prize winners including J.J. Thomson, E. Rutherford and W.L. Bragg; the Physical Institute at Göttingen University, with N. Born, R.W. Pohl and J. Frank; the Institut für Radiumforschung, Wien, with K. Przibram, the Leningrad Institute of Physics and Technology with A. Joffe and J. Frenkel and many others.

Finally, the last stage developed by Szymborski is entitled "emergence of speciality". In the case of irradiated materials, this stage appeared during the Second World War with the Metallurgical Laboratory at the University of Chicago and the creation of nuclear research centers: Oak Ridge National Laboratory (1943) in Tennessee, Brookhaven National Laboratory (1947) in New York State (USA), Chalk River Nuclear Laboratories (1942) in Ontario (Canada), Atomic Energy Research Establishment (1945) in Harwell, Oxfordshire (United Kingdom), Centre d'Études Nucléaires de Fontenay aux Roses (1946) and Saclay (1947) (France), Kernforschungsanlage Jülich GmbH (1956) and Forschungszentrum Dresden–Rossendorf (1956) (Germany).

Before delving into the history of materials under irradiation, we will take a look at what led to the emergence of this discipline: the notions of impure or disordered materials, instrumental developments in high-voltage and vacuum techniques and, of course, the discoveries of X-rays, electrons and radioactivity.

1.1. References

Szymborski K. (1984). The physics of imperfect crystals – A social history. *Historical Studies in the Physical Sciences*, 14(2), 317–355.

2

Prerequisites for the Irradiation of Materials

By modifying the properties of certain materials, natural irradiation has long posed a problem for geologists. Why do materials with the same chemical composition and the same angles between their faces have different properties, such as the more or less birefringent nature of crystals? We now know that the origin lies in self-irradiation due to the presence of actinides. Thus, the notions of perfect/imperfect, pure/impure, ordered/disordered have been part of the preoccupations of different civilizations from very early on.

2.1. Materials and purity, an old story

2.1.1. *Materials and disorder in ancient history*

It was in a still young universe that the first particle–material interactions took place, but it was not until 700 million to one billion years later that the first generations of end-of-life stars dispersed the products of their nucleosynthesis, creating elements heavier than iron in supernova explosions. The first materials were then formed by the condensation of these atoms. Cosmic radiation and the radioactivity of nucleosynthesis elements were intense sources of irradiation. Much later, around a young star, planet Earth had the brilliance of having the right initial mixture and a temperature compatible with carbon chemistry, providing a favorable environment for the emergence of life. Here again, irradiation certainly helped organic chemistry to explore the possibilities. The appearance of life on Earth is estimated at 3.5 billion years ago in an environment that was far more radioactive than it is today. The ratio $^{235}U/^{238}U$ then exceeded 11% (currently 0.7%). Note that 1.5 billion years later, this ratio was still close to 3.5%, enabling the divergence of the natural reactors at Oklo

(Gabon), albeit under very specific conditions of uranium concentration and presence of water (Bodu et al. 1972; Neuilly et al. 1972). They operated for several hundred thousand years.

The first hominids appeared in the Paleolithic era, some 3.3 million years ago. Then, as evolution did its work, knowledge of materials and, above all, the ability to modify them enabled hominids to distinguish themselves from other living beings. The first manufactured mineral was probably flint[1] or obsidian[2], whose conchoidal fractures, obtained by percussion, form sharp edges. Using an empirical approach, prehistoric man selected flint and the properties of the nanometric texture of chalcedony, combined with impurities. Much more recently, metals took social organization a step further, copper and gold at first, but their mediocre mechanical properties practically limited their use to objects of worship and social distinction. Alloy synthesis was the great breakthrough between the 5th and 3rd millennium before the current era (BCE). In particular, bronzes, binary copper–tin alloys, demonstrate very good mechanical strength and have a lower melting point than copper. They can therefore be processed even with rudimentary furnaces. When working with alloys, the aim is to master the microstructure. Improvements in furnaces enabled higher temperatures to be reached and the development of iron metallurgy (around 1,100 BCE). The great advantage of iron over bronze is its greater mechanical strength, while at the same time being ductile when hot, so it can be worked by hammering. Furthermore, mankind learned fairly quickly how to modify the mechanical properties of iron by adding carbon. These latest technological developments profoundly altered the organization of society, whose economy developed through the specialization of production and workplaces, increased agricultural yields, the emergence of merchants, etc.

In the last millennium BCE, we were a long way from materials under irradiation. Nevertheless, in Greco-Roman antiquity, the notions of disorder, defect, amorphous, etc., without being clearly named, were underlying in mineralogical works. Thus, the Athenian philosopher Theophrastus (371–288 BCE) in his treatise on stones, a veritable inventory of these materials and their origins, often attributed the qualifiers “true”, “bad”, “indefinite” and “pure” to minerals. For example, he wrote: “There are still in the fossil kingdom certain remarkable Earths which are formed in a purer and more homogeneous manner than the others” (Theophrastus, 1754 [4th century BCE], author’s translation). Theophrastus was a disciple and

1 Flint is a hard rock composed of one form of silica, chalcedony (microcrystalline intergrowth of quartz and a quartz polymorph, moganite).

2 Obsidian, rarer than flint, is a glassy volcanic rock containing 70–75% silica, the remainder being composed of MgO and Fe_3O_4.

friend of Aristotle. On Aristotle's death, he succeeded him as head of the Lyceum, the philosophical school founded around 335 BCE by Aristotle.

Figure 2.1. *Statue of Theophrastus (Palermo Botanical Gardens). We owe this Athenian philosopher a veritable inventory of stones and their origins*

In Rome, three centuries later, Pliny the Elder's *Naturalis Historiæ* included book XXXVII[3] on precious stones – what we would today call an encyclopedia. In addition to describing Roman society's relationship with materials, paragraphs were devoted to their properties. In the case of diamonds, for example, he wrote: "All these diamonds are tested on the anvil, and they resist blows so well that the iron rebounds and the anvil itself cracks. Indeed, their hardness is incredible: moreover, they triumph over the action of fire and never heat up; it is this indomitable strength that has given them the name they bear in Greek[4]" (Pliny the Elder 1830 [79 BCE], author's translation). Their properties are somewhat idealized, and I would not advise anyone who owns a diamond to test its resistance to hammer and fire. But he also indicated that, because of its great hardness, diamond dust could be used to cut and polish other precious stones.

During the Middle Ages, scientific work continued, combining philosophy, theology, astronomy and alchemy. In the 13th century, we find a remarkable figure, the Dominican Albertus Magnus (Bavarian from the Holy Roman Empire), also known as Albert de Lauingen. He taught in Rome and Paris, where he was so

3 Book XXXIII on the nature of metals is devoted to gold and silver.

4 *α δαμας*: that cannot be tamed or broken (Dictionnaire Grec – Français, A. Bailly).

successful that he was obliged to hold his lectures in the open air on a square that came to be known as Place Maître-Albert (this square still exists today as Place Maubert, an alteration of the original name). His output was enormous: more than 70 authentic works on theology, Aristotle, animals, plants, women, meteorology and minerals. This work on minerals was produced between 1248 and 1262, and was printed in Latin in 1569 (Magnus 1569); and, more recently, translated into English with commentary (Wickoff 1967). The originality of this work lies in the fact that, unlike what had been or would be published in subsequent centuries, his work was not a simple compilation of the various minerals, but an attempt to explain their properties. Thus, we find chapters entitled "The cause of the different colors of gemstones", "The cause of differences in the hardness of stones" and "The cause of the porosity and compactness of stones, as well as of their heaviness and lightness". Nevertheless, the content often remains at the philosophical or assertive level: "Let us say, therefore, that the general cause of hardness is dryness".

Figure 2.2. *Albertus Magnus, author of "De mineralibus", a compilation on materials and their properties. Image: Posthumous fresco by Tommaso da Modena (1352, Treviso). For a color version of this figure, see www.iste.co.uk/bouffard/irradiation.zip*

The Renaissance, with its questions about the world around it and major discoveries such as printing, paved the way for the scientific revolutions of modern times. Descriptions of materials became increasingly precise. Examples include *De re metallica* (Agricola 1556), an impressive compilation of 16th-century knowledge on the metallurgy of various metals, published in 1556 by Georgius de Agricola (real name Georg Bauer, a Saxon from the Holy Roman Empire).

A few centuries later, Baron Axel Fredrik Cronstedt, a Swedish chemistry enthusiast, discovered nickel and recognized stilbite as belonging to a new class of minerals: zeolites. In 1758, he published an essay on a system of the mineral kingdom, *Försök till Mineralogiens eller mineral-Rikets upställing* (Cronstedt 1758).

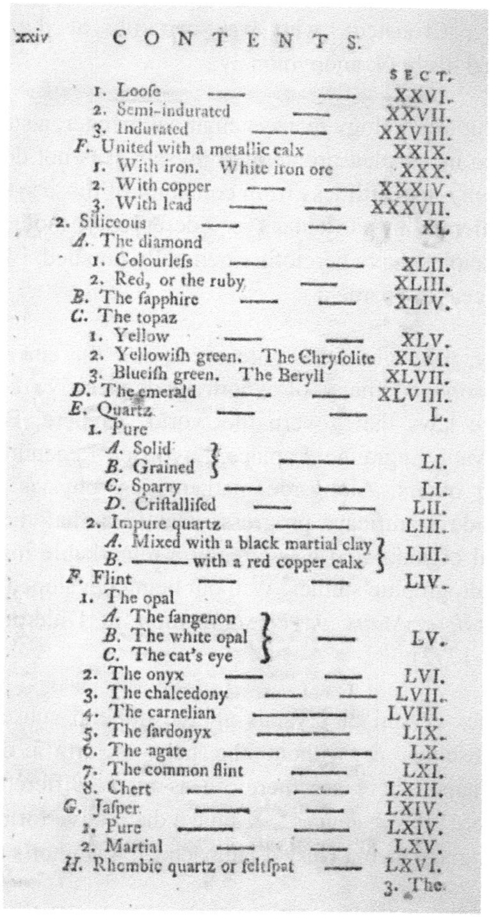

xxiv CONTENTS.	SECT.
1. Loose —— ——	XXVI.
2. Semi-indurated ——	XXVII.
3. Indurated ——	XXVIII.
F. United with a metallic calx	XXIX.
1. With iron. White iron ore	XXX.
2. With copper —— ——	XXXIV.
3. With lead —— ——	XXXVII.
2. Siliceous —— ——	XL.
A. The diamond	
1. Colourless —— ——	XLII.
2. Red, or the ruby ——	XLIII.
B. The sapphire —— ——	XLIV.
C. The topaz	
1. Yellow —— ——	XLV.
2. Yellowish green. The Chrysolite	XLVI.
3. Blueish green. The Beryll	XLVII.
D. The emerald —— ——	XLVIII.
E. Quartz —— ——	L.
1. Pure	
A. Solid } *B.* Grained } ——	LI.
C. Sparry —— ——	LI.
D. Cristallised —— ——	LII.
2. Impure quartz ——	LIII.
A. Mixed with a black martial clay } *B.* —— with a red copper calx }	LIII.
F. Flint —— ——	LIV.
1. The opal	
A. The sangenon } *B.* The white opal } *C.* The cat's eye } ——	LV.
2. The onyx —— ——	LVI.
3. The chalcedony —— ——	LVII.
4. The carnelian ——	LVIII.
5. The sardonyx ——	LIX.
6. The agate —— ——	LX.
7. The common flint ——	LXI.
8. Chert —— ——	LXIII.
G. Jasper —— ——	LXIV.
1. Pure —— ——	LXIV.
2. Martial —— ——	LXV.
H. Rhombic quartz or feltspat ——	LXVI.
	3. The

Figure 2.3. *Partial reproduction of the table of contents of A.F. Cronstedt's essay on a mineral kingdom system (Cronstedt 1758) as translated by G. von Engeström (Von Engestrom 1770). For a color version of this figure, see www.iste.co.uk/bouffard/irradiation.zip*

Published in Swedish, it was translated into English 12 years later (Von Engestrom 1770). Materials belonging to the mineral kingdom are divided into four categories: earths, flammables, salts and metals. A.F. Cronstedt classified around 360 minerals and, for each one, gave its origin and main properties, appearance, hardness, brittleness, fusion, chemical attack, etc. In the table of contents (Figure 2.3), he considers two main categories of quartz, pure and impure, with subdivisions based mainly on visual appearance, color, transparency, inclusions and origin. He also classifies diamonds in the "Siliceous" category and rubies in the "Diamond" category. Cronstedt, who was certainly at the origin of modern mineralogy, published his book anonymously:

> [...] for treating Mineralogy in a systematical manner; a study to which I have with so much pleasure applied myself. It is not done from the desire of novelty, and still less from contempt of those systems, which Swedish gentlemen in particular, very deservedly, though chiefly on the same principles, have heretofore generally pursued. I have thought proper to conceal my name...

The 18th century, the Age of Enlightenment, saw the emergence of numerous scientists and philosophers, many of whom made decisive contributions to our understanding of the laws that govern the world: Ampère, Bernoulli, Coulomb, Euler, Franklin, Galvani, Lagrange, Laplace, Lavoisier, Legendre, Newton, Venturi, Volta, Watt, among others. Alongside mathematics, physics and chemistry and mineralogy also made significant progress. The idea that the shape of crystals reflects their internal organization was already a remarkable insight that paved the way for future crystallographic studies. Without being developed, this idea is present in the article *Crystal, crystaux ou crystallisation* in Diderot and d'Alembert's *Encyclopédie* (1751–1772):

> [...] one names crystal or crystals all the mineral substances which take of themselves & without the help of art, a constant and determined shape: there are therefore as many different species of crystals as there are substances... & take a distinctive form by which it is easy to recognize them (Thiry d'Holbach 1754, author's translation).

In the same spirit, Abbé Jean-Baptiste Romé de l'Isle went even further, describing the geometric shape of a very large number of materials (Romé de l'Isle 1783). The example of rock crystal (quartz) is shown in Figure 2.4: "The primitive figure of the rock crystal is a dodecahedron, formed by two hexahedral pyramids with isosceles triangular planes, joined base to base...". He then gives possible variants, introducing truncations. In his discussion of the results, he demonstrates a very direct critical spirit:

> [...] I confess that these experiments will always leave something to be desired, as long as we have not succeeded in regenerating quartz or rock crystal, by directly combining the principles of which it is said to be composed. This is also what Messrs Achard & Bergman claim to

> have done; but as the experiments of the former have hitherto been repeated without success, & as the unique experiment of the latter leaves doubts as to the nature of the crystals which were its product, it is appropriate to suspend judgment, until one or other of these skilful chemists has put the seal of evidence on his discovery (Volume 2, p. 55, author's translation).

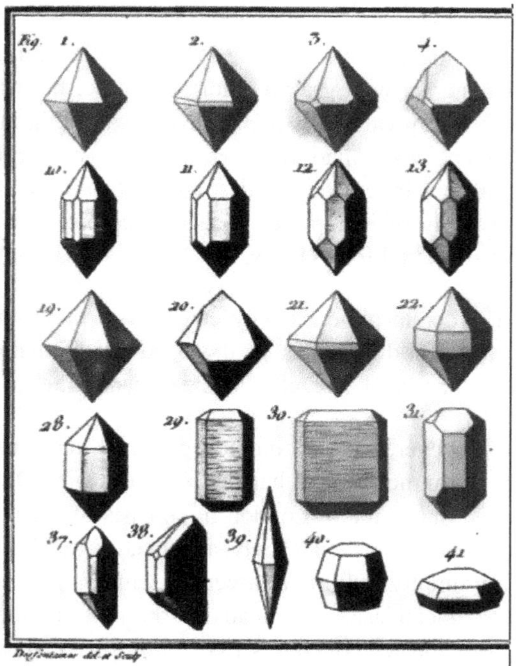

Figure 2.4. *Reproduction of a plate showing the different dodecahedron shapes with truncated planes after J.B. Romé de l'Isle (Romé de l'Isle 1783)*

Many geologists were also interested in the systematic identification and classification of minerals. However, these considerations on the internal structure of minerals were slow to spread among geologists and mineralogists. Thus, in the 1763 edition of his comprehensive *dictionnaire oryctologique*[5], Swiss geologist Elie Bertrand describes minerals as formed from primitive principles, "Earths and sands serve to form ROCKS...". However, in discussing the difference between pyrite and iron marcasite, he refers to the internal structure of these materials: "Amorphous pyrite is according to them [the alchemists] the marcasite of iron" and further, in a

5 Mineral science.

contradictory manner, "We believe to reserve the word marcasite to designate an angular, crystallized, faceted form of pyrite" (Bertrand 1763, p. 329, author's translation). Finally, we should mention the work of Abraham Gottlob Werner, a Saxon from the Holy Roman Empire. He developed a methodological system that turned geology into a systematic science. In 1774, he published the first modern textbook on descriptive mineralogy: *Von den äusserlichen Kennzeichen der Fossilien* (On the external characteristics of minerals) (Werner 1774).

At the beginning of the 19th century, the Frenchman René Just Haüy described the organization of minerals with greater rigor. To do so, he measured the angles between crystal faces with great precision. He hypothesized that minerals are made up of elementary molecules bound together by affinity. Elementary molecules assemble into integral molecules, the smallest element that preserves the nature of the mineral. The assembly of integrating molecules composes a regular body, the crystal, and gives it its external form (Haüy 1822). He listed 1,040 crystalline forms.

The notion of disorder in minerals only emerged in the 19th century, initially through the observation of differences in the behavior of materials of the same composition. In 1815, Jöns Jacob Berzelius wrote of gadolinite[6] of different origins:

> All of them, however, have a common property that I have not yet seen described in mineralogical handbooks, and which Dr. Wollaston recently pointed out to me in London: it consists of the fact that, if a piece of gadolinite is heated slowly and evenly by the flame of a blowtorch, it promptly reddens at a certain temperature, as if it were on fire, and the incandescence spreads all the more rapidly the more evenly it has been heated (Berzelius 1815, author's translation)[7].

The phenomenon reported to Berzelius by Wollaston[8] corresponds to the release of energy during the recrystallization of amorphous gadolinite. This explanation would not be known until many years later (Gibson and Ehlmann 1970).

---

6 Gadolinite or ytterbite, silicate with formula $(Ce,La,Nd,Y)_2FeBe_2Si_2O_{10}$.

7 The original article was published in French in the Journal de Chimie et de Physique in September 1816.

8 A few words about the two protagonists of this discovery: the British scientist William Hyde Wollaston was very eclectic in his work. He discovered rhodium and palladium and made his fortune with his invention for extracting platinum from its host mineral. The Swedish Jöns Jacob Berzelius can be considered one of the founders of modern chemistry, along with J. Dalton and A. Lavoisier. Professor of chemistry and pharmacy at the Karolinska Institute (Sweden), he identified cerium, selenium, silicon and thorium, and helped discover lithium and vanadium.

Although Alfred Des Cloiseaux and Alexis Damour were not concerned with defects or amorphization, they did show that materials of the same chemical composition can have different properties (Des Cloizeaux and Damour 1860). Their study focused on several materials, including gadolinite. The latter material, of fifteen different origins for this study, apparently formed a homogeneous group, but some behaved like a birefringent crystal, while others appeared to be composed, in variable proportions, of a birefringent component embedded in a mono-refractive material. After the same observations on allanite[9], they concluded:

> We can see from our observations on Allanites that some samples behave in polarized light like a birefringent crystallized body with two axes, while others do not enjoy double refraction. The latter must be regarded as amorphous substances that have taken the place of epidote crystals whose shape is very close to that of Allanite...

Further, they noted that

> [...] Allanites and orthites, whether or not they possess double refraction, always offer the same external form (Des Cloizeaux and Damour 1860).

The polarization of light has been known since the early 19th century. In 1860, no one could have imagined the origin of these differences. This French-language paper on the analysis of 44 origins or types of material represents a very extensive work on minerals, mostly from private collections (Adam, Damour) or donations (Berzelius, Prince Napoleon). Curiously, while this study was signed with both names, some paragraphs are written in the first person singular, in which case Des Cloizeaux is the author.

In a presentation to the Académie des Sciences de Paris (Paris Academy of Science) on Monday January 18, 1864, M.A. Damour discussed the density of zircons[10], which, despite identical composition, varies from 4.04 to 4.67 g/cm^3. By heating these zircons to a high temperature (red-white), whatever the initial density, it converges to a single value. He then hypothesized that zircons have two allotropic states[11] (Damour 1864).

In 1890, W. Petersson published a long article summarizing gadolinite research and presenting new chemical analyses (Peterson 1890). It includes a description of

9 Allanite, sorosilicate with formula $(Ce,Ca,Y,La)_2(Al,Fe^{3+})(SiO_4)_3(OH)$ and may contain significant amounts of Th, U, Zr, Ba, etc.

10 Zircon, silicate with the formula $ZrSiO_4$.

11 Allotropy is the ability of a material to exist in several different crystalline forms.

Berzelius and Wollaston's experiment, followed by an attempted explanation taken from a publication by Scheerer (1840), which assumes that gadolinite solidifies from molten magma. When cooling is very slow, the molecules choose the loosest system, whereas rapid cooling results in a different molecular arrangement. Three years later, in his classification of minerals, Norwegian geologist Waldemar Christofer Broegger differentiated between crystalline and amorphous materials and classified the latter into two categories: porodine amorphous formed by slow cooling of gelatinous substances and hyaline amorphous formed by rapid cooling from the molten state. However, he was obliged to add a third group, materials which, while having a shape reminiscent of crystals, have isotropic optical properties, a lower density than normal, conchoidal-type fractures. He proposed to call this category of materials metamict. He first used the word in an article on amorphous materials in the Danish encyclopedia *Salmonsens Store Illustrere Konversations Lexicon* (Broegger 1893). W.C. Broegger postulated that metamict minerals existed as perfect crystals, but were probably disordered by external agents because their structure is relatively unstable:

> Many of these minerals are even known only as amorphous, but from their original formation they have retained their outer crystalline boundaries (e.g. euxenite, thorite, etc.); others are amorphous in some deposits and crystalline in others unchanged. In the case of gadolinite, it has been proven that the amorphous gadolinite, when heated to annealing, resumes its original crystal structure and becomes birefringent (translation from Danish by the author).

Swedish mineralogist Axel Hamberg was probably the first to suggest that metamictization is caused by α particles emitted by the radionuclides uranium and thorium included in minerals, but this was not until 1914 (Hamberg 1914). Rodney Ewing's article *The Metamict State: 1993 – The Centennial* is a very good introduction to the evolution of knowledge about the metamict state (Ewing 1994).

2.2. Discovering the particles behind irradiation

The notion of disordered materials therefore predates the first studies of intentionally irradiated materials. It was only with the discovery of energetic particles, the observation of their effects on materials and the development of structural analysis tools that real progress was made in understanding the phenomena observed. In this context, the discoveries of X-rays and natural radioactivity were major events. However, before describing the first irradiation experiments, let us take a brief look back at the succession of conceptual and experimental advances that made these discoveries possible. The key element in the case of X-rays is the discovery of cathode radiation, which required knowledge of high-voltage generation and the notion of vacuum. We will take a look at the history of these two techniques, starting with electricity.

2.2.1. *High-voltage generation*

The word electricity comes from the name of a fossilized oleoresin: amber named in Greek *ἤλεκτρον* (elektron). Its electrostatic properties were observed by Thales of Miletus around 600 BCE and succinctly described by Theophrastus a century later (Theophrastus 1754 [4th century BCE]).

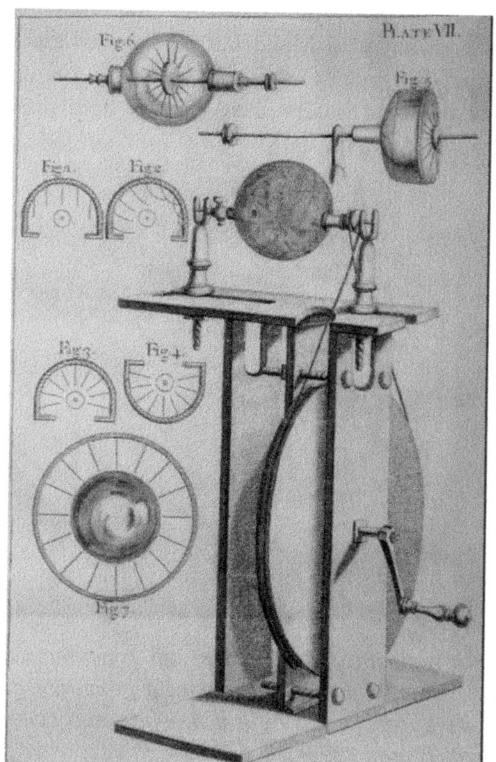

Figure 2.5. *F. Hauksbee's electric machine. The glass sphere, partially emptied of air and set in rotation, is electrified by friction. Fine wires indicate the orientation of the field lines (Hauksbee 1709). For a color version of this figure, see www.iste.co.uk/bouffard/irradiation.zip*

In 1775, in his history of electricity, Joseph Priestley credited Guilielmi Gilberti (William Gilbert) with inventing modern electricity, even though he lived in the late 16th–early 17th century (Priestley 1775a). The London physicist carried out numerous experiments on magnets and electrifiable materials. He showed that amber's property of attracting light objects when rubbed could be generalized to many materials (diamond, sapphire, beryl and glass). He wrote that the electrical effect was stronger when the air was dry, which seems logical, but more unlikely when the wind blows from the east or north. In 1600, he published his observations in what may be considered the first book on experimental physics: *De magnete* (Gilberti 1600).

The first voltage generators were electrostatic machines whose electrification was achieved by friction (triboelectricity). In 1660, the Prussian Otto von Guericke, fortification engineer and mayor of Magdeburg, generated electricity by rubbing a rotating ball of sulfur. He mainly studied the attraction/repulsion phenomena between electrified bodies, and observed an electric light that he likened to that emitted when sugar is crushed in the dark.

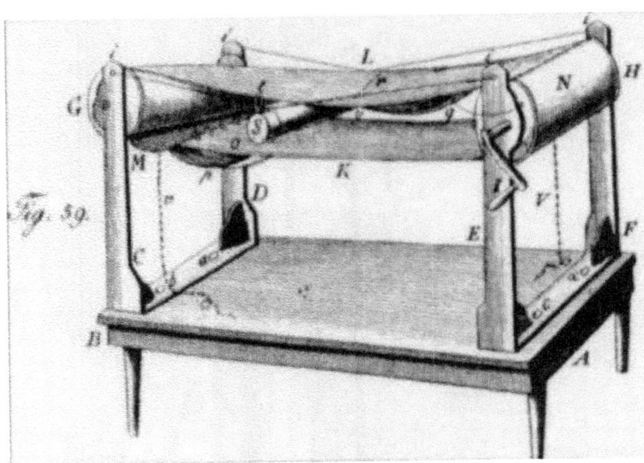

Figure 2.6. *N. Rouland's electrostatic machine, an improved version of the one by C.L. Walckiers de Saint Amand. In this machine, a precursor of the Van de Graaff generators, charges were transported by a silk ribbon and collected by cushions to generate a high voltage (Brenni 1999)*

The first real electric machine was built by Francis Hauksbee (see section 2.5.2). Seeking to produce light by friction, he designed a device consisting of a glass sphere rotating around an axis, in which a partial vacuum could be created (Figure 2.5). When the sphere was set in motion, friction with a hand electrified the sphere, and he observed the electric field lines using fine wires running along these field lines (Hauksbee 1709). During the 18th century, machines made progress, notably by increasing the size of glass spheres or cylinders, improving glass quality and, above all, electrical machines themselves. The latter were capable of producing sparks over distances of around 18 inches, that is, for dry air, a voltage of the order of 450 kV! For more details on this exciting period, see Joseph Priestley's *The History and Present State of Electricity*, published in 1775 (Priestley 1775a).

In 1784, Charles-Louis Walckiers de St Amand developed an electric machine in which charges were carried by an endless silk ribbon. The charges were deposited by cushions covered with cat skins or wool. The machine had impressive dimensions, with a silk taffeta ribbon five feet wide by 25 feet long (1.5×7.6 m^2). It was presented to the Royal Academy of Sciences in Paris (Walckiers de St Amand 1784). From its very first tests, this generator became one of the best machines based on glass loading capability. It was the forerunner of the Van de Graaff generator. However, this machine had the disadvantage of its size, so it was not developed further, despite a few proposals for vertical versions, notably in Germany, by Gottlieb Christian Bohnenberger. It is interesting to note that it was not until 1800 that the electric battery was invented by Alessendro Volta. In 1872, following the same idea as C.L. Walckiers de Saint Amand, Augusto Righi designed a machine in Bologna to create a strong electric field by adding small electric charges. An endless Indian rubber tube was mounted on two pulleys. Rings of copper wire wound at regular intervals on the tube transported the charges deposited near the earthed pulley. They were discharged by means of a contact spring inside a sphere which acquires a very high charge, even though the initial charge in the influence body was very low. The principle of today's Pelletron closely resembles this device. The imagination of physicists and engineers has produced a large number of different devices, all based on the transport of charges deposited by influence. John Gray's book *(Electrical Influence Machines: A Full Account of Their Historical Development, and Modern Forms, With Instructions for Making Them)* gives a fairly complete description on pages 188–190 (Gray 1890).

Michael Faraday's discovery of electromagnetic induction in 1831 opened up a new way of creating high voltages: induction coils. This type of electrical transformer produces high-voltage pulses from low-voltage direct current. The direct current in the primary coil is periodically interrupted by a vibrating mechanical contact to create the flux changes required to induce a voltage in the secondary coil. This device was invented simultaneously in 1836 by American physicist Charles Grafton Page (1837) and Irish scientist Nicholas Callan (1836). This question of anteriority of discovery seems particularly important to C.G. Page. The tone is set by the title of his book *History of Induction. The American Claim to the Induction Coil and its Electrostatic Development*, in which he devotes 46 pages to proving its anteriority:

> Prof. Callan's description of this coil is found in the London and Edinburg Philosophical Magazine for December, 1836. This communication is dated August 23, 1836. Prof. Page's communication to Sill's Journal was dated May 12th , 1836, and its postscript June 8th , 1836. It did not appear in that Journal until the January No. for 1837; [...] These dates and peculiar facts settle the question of priority in favour of Prof. Page, beyond controversy (Page 1867).

Sill's Journal, in which Page publishes, is short for Silliman's Journal, the name often given in the early days to *The American Journal of Science and Arts*, which Benjamin Silliman founded in 1818.

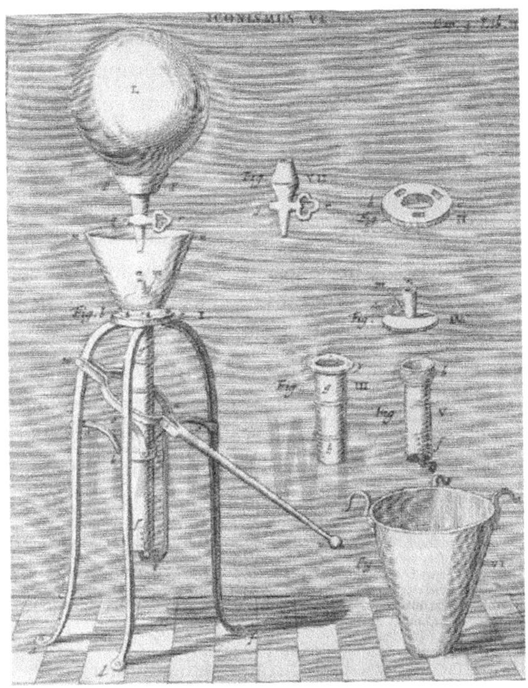

Figure 2.7. *Otto von Guericke's air pump. Its principle is similar to that of a bicycle pump, in which the valve would have been replaced by tap I, opened and closed to the rhythm of the piston's movement (de Guericke 1672). For a color version of this figure, see www.iste.co.uk/bouffard/irradiation.zip*

In 1851, Heinrich Daniel Ruhmkorff, a German-born Parisian, perfected the induction coil by improving the insulation of the secondary windings and segmenting the soft-iron core to reduce eddy currents. This generator, known as the

Ruhmkorff coil, was capable of producing electric arcs 30 cm long. It was used in the experiments that led to the discovery of X-rays and electrons. To reach higher voltages, several devices were then proposed, some of which were never developed, such as the Kelvin water dropper imagined by William Thomson (Lord Kelvin) in 1867 (Thomson 1867). This device consisted of two electrically isolated reservoirs into which positively charged droplets fell in one and negatively charged droplets in the other. Later, in the 20th century, two devices made it possible to reach very high voltages, the Cockroft–Walton generator in 1930. Powered by a low alternating voltage, this generator delivers a very high, quasi-continuous voltage. Its creators, John Douglas Cockroft and Ernest Walton, were both physicists at the Cavendish Laboratory (Cockcroft and Walton 1930). At virtually the same time, Robert Van de Graaff, Karl Taylor Compton and Lester Clare Van Atta published a paper on an electrostatic generator based on the charging of a sphere by a belt carrying charges (Van de Graaff et al. 1933). Technological developments in these two ways of obtaining high voltages are described in section 11.1.2 (Chapter 11).

However, we have strayed from the subject of this chapter: the technical developments that preceded Röntgen's decisive experiment.

2.2.2. *Vacuum control*

Mastery of the vacuum, however imperfect, was the second essential element in the discovery of X-rays. The existence of the vacuum is an age-old question which, from Aristotle to Descartes, remained a philosophical one. For Aristotle, the very existence of the vacuum called into question his entire coherent construction of a full, spherical and immobile universe. In particular, for Aristotle, the falling speed was inversely proportional to the density of the surrounding environment. In a vacuum, velocity becomes infinite (Aristotle 1802 [between 335-323 BCE]). This conception of the universe gave rise to the principle that "nature abhors a vacuum". In contrast, some 2,000 years later, Galileo did not accept Aristotle's physical doctrine as true, but subjected it to an experiment. In fact, he used a thought experiment to describe the fall of bodies in a vacuum. The first real experimental breakthrough came in 1644, when Evangelista Torricelli demonstrated atmospheric pressure. His invention of the mercury-tube barometer followed directly from this, as did his demonstration that the space above the mercury column is empty.

To take this a step further, what was missing was the tool needed to master vacuuming. It was Otto von Guericke[12] who, around 1647, invented the first air pump, similar in principle to a bicycle pump with the equivalent of a non-return

12 The same von Guericke studied frictional charge generation.

valve mounted upside down. He used it to empty a glass balloon in which he carried out numerous experiments (Figure 2.7). He showed that the remaining air did not accumulate like liquid at the bottom of the balloon, and that light propagates in a vacuum, but sound does not. More spectacular were his so-called Magdeburg hemispheres experiments, demonstrating the force of atmospheric pressure: with the two hemispheres next to each other and in a vacuum, four horses could not separate them. First performed in Magdeburg in 1654, the experiment was repeated many times, notably at the court of Elector Friedrich Wilhelm in 1663.

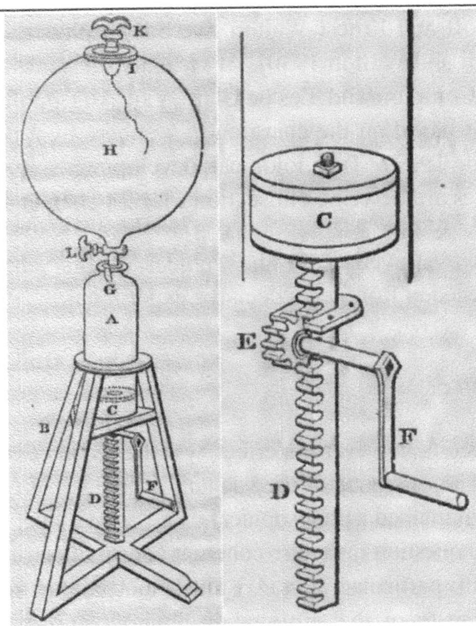

The pump has a single brass cylinder approximately 14 inches long and 3 inches in internal diameter. It rests on a sturdy wooden tripod. The piston rod is a rack driven by a toothed wheel. A hole drilled in the side of the upper end of the cylinder is fitted with a G plug, which can be removed or pushed in by hand.

Operating principle: the above figures show the arrangement of parts when the pump has completed a pumping cycle. The cylinder plug G is closed, the reservoir valve L is open and closed, G is open, and the piston is raised. G is then closed, L opened again, the piston lowered and so on (according to Wilson (1849)).

Box 2.1. *R. Boyle's first pump*

This pump was improved by Robert Boyle in 1659 (actually by Robert Hooke, his assistant). It enabled him to carry out numerous experiments on the properties of air, and it was thus that he demonstrated that the volume of a gas varies as the inverse of its pressure: Boyle's law (or Boyle–Mariotte's law in continental Europe). This pump was particularly tricky to build, especially when it came to sealing the moving parts. It was also capricious and very expensive. Between 1660 and 1670, only six pumps were built for London, Oxford, Cambridge (Great Britain), The Hague (Netherlands) and Paris (France). Studies on the effects of high electrical voltage on rarefied air did not begin until much later. Francis Hauksbee, whom we have just encountered for his electric machines, paid tribute to R. Boyle for his invention of the air pump:

> The Honourable and moſt Excellent Mr. Boyle, by great Variety of Experiments, in almoſt every part of Philoſophy, gave much Light into the Cauſes and Operations of Nature; and particularly by the Invention of that moſt Uſeful Inſtrument the Air-Pump (Preface of Hauksbee (1709)).

F. Hauksbee was not content with this tribute: he built an excellent double-piston air pump that rapidly achieved a pressure of 0.75 inches of mercury, or 26 mbar (Figure 2.8).

In fact, the authorship of R. Boyle's improvements was disputed, notably by a certain George Wilson, who published a history of air pumps in England:

> [...] I have had occasion to study with some attention, the treatises in which he[13] describes the air-pumps which were first employed in England, and the steps by which they were improved. I have been led in consequence to notice, that, in the hands of recent writers, the whole history of the English air-pump has been allowed to fall into great confusion, so that the steps by which the instrument was improved, the periods at which those improvements were made, and the parties by whom they were effected, are all more or less confounded with each other or misstated. The authors of those errors, moreover, are parties justly held in the highest esteem among men of science, as the late Professor Robison, Dr Thomas Young, and Professor Baden Powell. They have been implicitly followed, as might be expected, by writers of less note, and errors have been multiplied, as the erroneous account has been repeated... (Wilson 1849).

---

13 R. Boyle.

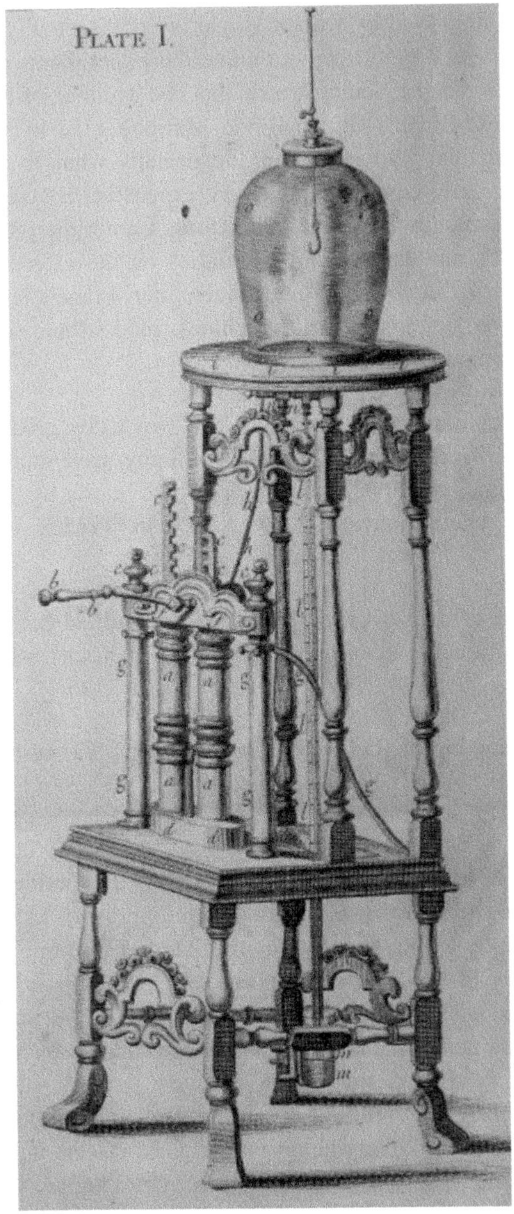

Figure 2.8. *The magnificent vacuum pump used by F. Hauksbee. This pump features two pistons operating in phase opposition, giving it a significant gain in pumping speed (Hauksbee 1709). For a color version of this figure, see www.iste.co.uk/bouffard/irradiation.zip*

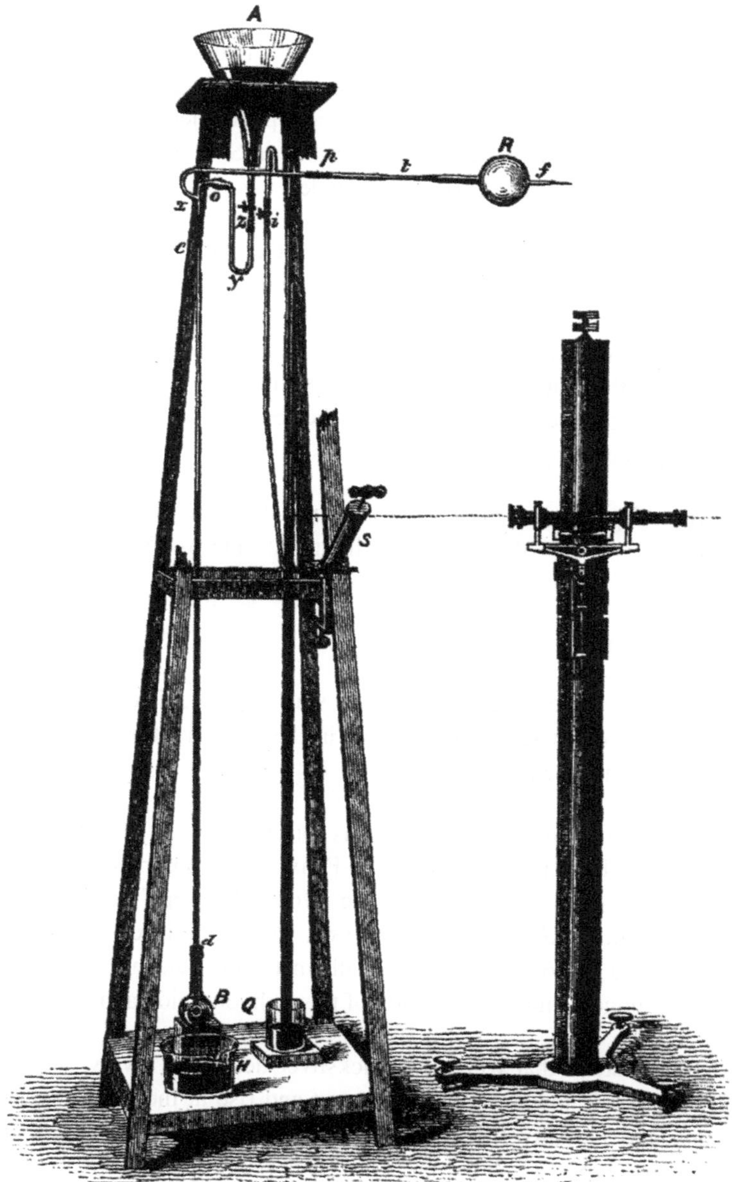

Figure 2.9. *The mercury pump designed by Hermann Sprengel. As the mercury flows from reservoir A to reservoirs B and H, it "sucks" the air from sphere R. To obtain a "good" vacuum, the operation must be repeated several times and 10–15 lbs of mercury (4.5–6.8 kg) is required (Sprengel 1865)*

The next major breakthrough came in 1865 with the development of the mercury pump by Hermann Sprenger. The invention of this British researcher of German origin made it possible to reach much lower pressures (theoretically down to 10^{-5} mbar or 1 mPa) and thus gain access to new phenomena such as the ballistic transport of electrons. Its operation, well described in Sprengel (1865), was similar to that of the water tube (Figure 2.9). He concluded:

> The main fact which I have established in this paper may be shortly stated to be that, if a liquid be allowed to run down a tube, to the upper part of which a receiver is attached by means of a lateral tube, and if the height at which the receiver is attached be not less than that of the column of the liquid which can be supported by the atmospheric pressure, a vacuum will be formed in the receiver, minus the tension of the liquid employed.

For the general public, the most important impact of improved pumping was that it enabled the development of incandescent light bulbs (1880–1881).

2.2.3. *Cathode rays*

Mastery of electric charge and vacuum was a prerequisite for studying the luminous effects of discharges. Experiments in this field multiplied and led to two major discoveries: the electron and X-rays.

The first observation of luminescence was fortuitous. In 1675, while moving his barometer at night, French astronomer Jean-Félix Picard observed a glow in space above the mercury when it was agitated. He informed Europe's oldest literary and scientific periodical, the *Journal des Sçavans*[14], which published a short article on the subject (Picard 1676). The Swiss mathematician Johann Bernoulli, then a professor in Groningen, undertook a study of this light emission and concluded that:

> [...] I understood that when the quicksilver descends, a very delicate and very subtle material must come out and rise at the same time to occupy and partially fill the space of the pipe that the quicksilver leaves... (author's translation).

---

14 Created in 1666 under the patronage of Jean-Baptiste Colbert, Louis XIV's Controller General of Finances, its name evolved from *Journal des Sçavans* to *Journal des Savans* in 1791, then to *Journal des Savants* in 1830. The journal still exists today as the biannual journal of the Académie des inscriptions et belles-lettres.

He also showed that the presence of moisture prevents light emission, but that wine spirit[15] exalts it (Bernoulli 1707).

We have already mentioned Francis Hauksbee, the British scientist who designed one of the first electric generators, enabling him to carry out a large number of experiments on electrical attraction and repulsion. With a better grasp of experimental conditions, he resumed his observations of the light emitted by mercury set in motion in an airless enclosure. He observed strange flashes of pale light, which he called "mercurial phosphorus". The text is not always crystal-clear, at least for a non-professional historian of science, as can be seen from this partial conclusion:

> The Refults of all which Obfervations put together, is, That a peculiar Figure and Motion of Parts, as well as a proper Medium for thofe Motion to be perform'd in, are requifite to the Production of the Mercurial Phofphorous.

More interesting was his discovery in 1704 of "electric light" using a device consisting of a sphere rotating around an axis, in which a partial vacuum can be created. When the sphere is set in motion, friction with a hand electrifies the sphere and, under the effect of the electric field, ionizes atoms in the residual gas, creating a strong diffuse glow. Through numerous experiments, he tried to understand its origin, but never theorized that it could come from glass (Hauksbee 1709; Priestley 1775b).

Little progress was made on the subject during the 18th century, but developments in knowledge and techniques led to significant advances in the 1830s. Although it had already been reported in the early 18th century that the spark from an electrostatic generator propagated further in low-pressure air than in air at atmospheric pressure, it was not until the work of Michael Faraday that this research gained momentum. Between 1832 and 1865, Faraday published a series of 13 papers in the *Royal Society's Philosophical Transactions, entitled Experimental Researches in Electricity*. In the thirteenth paper, he describes in detail his observations on glow discharge and dark discharge (Faraday 1839). He attributed the appearance of luminescence to the (electrical) charging of the air, and the dark zone to the fact that air particles may be able to discharge each other without emitting luminescence. The subject made a significant breakthrough thanks to a technical advance by glassmaker Henrich Geissler (German Confederation). His major contribution to the understanding of cathode rays was the development of a new mercury vacuum pump that gave access to much better vacuums, in the millibar range. Under these

15 Wine spirit: ethyl alcohol obtained by distilling wine (ethanol).

conditions, a voltage of a few kilovolts generated by an induction coil between the two electrodes creates a glow throughout the sphere, rather than an electric arc. This is the operating range of fluorescent tubes, commonly known as neon tubes.

In 1858, Prussian Julius Plücker, who worked at the same university as H. Geissler (Rhein-Universität, Bonn), showed that cathode rays are deflected by a magnetic field. He was also one of the first to carry out spectroscopy of the light emitted by Geissler tubes. He showed that the line spectrum is characteristic of the chemical substance. A few years later, William Crookes, using an improved Sprengel vacuum pump, achieved a vacuum in his tubes of up to 1.5 mbar, compared with 25 mbar for Geissler tubes. A detailed description of this pump is given in Gimingham and Crookes (1877). The use of higher voltages with Ruhmkorff coils is also important. These differences radically change the tube's behavior, as electron transport by multiple diffusion becomes ballistic transport between the two electrodes, making it possible to study what the Prussian Eugen Goldstein called "Kathodenlichts" cathode rays in 1876 (Goldstein 1876).

Plücker's pupil Johann Wilhelm Hittorf showed that cathode rays travel in a straight line by inserting a mask – the famous Maltese cross – into the Crookes tube:

> The clearest evidence of the rectilinear propagation of the glow can be seen when it starts from a point cathode and causes the glass surface to fluoresce at a great distance. Under such conditions, if an object stands in the space filled by the glow, it casts a sharp shadow on the fluorescent wall, excluding the cone of light that runs from the cathode as a point around its surface (Hittorf 1869, author's translation) [the original text is in German].

To carry out his rarefied gas spectroscopy studies, he needed a continuous discharge. It took him several years to build the high-voltage generator consisting of 2,400 Bunsen battery cells[16] giving a voltage of almost 5,000 V. His achievements are described in Van der Willigen (1856).

To find out more about the evolution of knowledge before 1881, see Robert Routledge's *A Popular History of Science* (Routledge 1881) and, for the history of luminescence, Edmund Newton Harvey's *A History of Luminescence: From the Earliest Times Until 1900* (Newton Harvey 1957).

---

16 The Bunsen battery, invented in 1841, consists of a zinc anode in dilute sulfuric acid, and a carbon cathode in concentrated nitric acid. The two liquids are separated by a porous ceramic wall.

2.2.4. *Discovery of X-rays*

On November 8, 1895, in an experiment with a Crookes tube, the Prussian Wilhelm Conrad Röntgen observed fluorescence outside the tube with each pulse from his induction coil. He described his discovery less than two months later in the first paragraph of an article (Röntgen 1895):

> 1) If a good-sized induction coil is discharged into a sufficiently evacuated Hittorf vacuum tube or Lenard tube or Crookes tube or similar, the tube being covered with thin black cardboard providing an acceptable seal, and if the tube is placed in a totally dark room, a bright illumination of a paper screen coated with platinum-barium cyanide, placed in the vicinity of the induction coil, is observed at each discharge, the fluorescence thus produced being independent of whether the coated surface of the paper or its other face is directed towards the discharge tube. This fluorescence is visible even if the paper screen is two meters away from the device. It is easy to prove that the cause of the fluorescence is linked to the discharge device and not to any point in the conductive circuit (author's translation) [original text is in German].

This comprehensive article summarizes a series of experiments designed to characterize this new ray, which he proposed to call X-ray in a footnote: "For the sake of brevity, I will use the term 'rays', and to distinguish them from others of the same name, I will call them 'X-rays'". The rest of the article is devoted to the study of the transparency of various materials to these new rays and the effect of thickness, density etc., then their optical behavior. Already in this first article, he points out the possibility of seeing bones: "If the hand is interposed between the discharge tube and the screen, the darker shadow of the bones is seen in the slightly marked shadow of the hand itself". This article was republished in *Annalen der Physik und Chemie* (Röntgen 1898). In 1901, W.C. Röntgen was awarded the first Nobel Prize in Physics for his discovery[17]. Although Röntgen only wrote a second paper on X-rays, the response was enormous, with numerous articles published on the subject. Equally exceptional was the speed with which applications were proposed. Just two weeks after announcing the discovery of X-rays, Otto Walkoff made the first dental X-ray by placing a photographic plate between the teeth and

17 W.C. Röntgen donated the Nobel Prize money to a fund set up to provide scholarships for needy students.

the tongue for a pause of 25 min. The following year, together with Friedrich Oskar Giesel, he founded[18] the first dental X-ray practice.

After that, it only took a few months for mineralogists to see the value of W.C. Röntgen's discovery. Indeed, as early as 1896, V.C. Doelter suggested applications for Röntgen rays in mineralogy, notably to distinguish diamond from its imitations or to characterize inclusions (Doelter 1896). In January 1897, John E. Burbank of Harvard University became interested in the fluorescence of materials subjected to X-rays. A year later, he published a very short article in the American Journal of Science in which he gave some general results, such as "The group of minerals containing Ca were most susceptible to the X-rays" or "The effect of heat on other crystals in some cases causes increased emission of light, in others diminished effect, and in several cases no effect was observed" (Burbank 1898).

The first popular article on X-rays and their applications was published in 1900 in *Popular Sciences Monthly* (Trowbridge 1900).

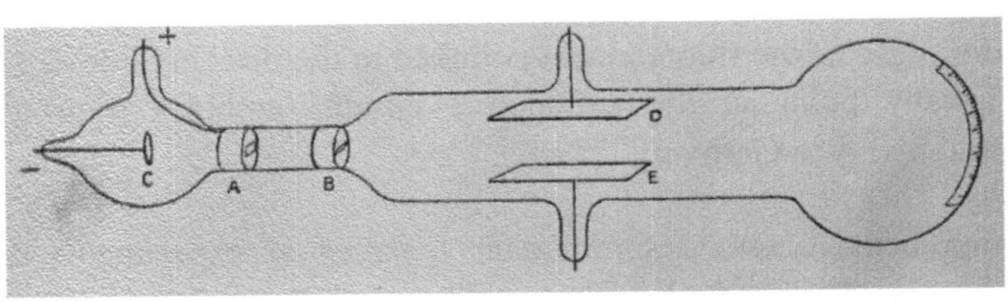

Figure 2.10. *Diagram of the device used by J.J. Thomson to measure the effect of an electrostatic field C cathode, A anode, B slots (Thomson 1897)*

2.2.5. *Discovery of the electron*

In Thomson (1897), the British physicist Joseph John Thomson described a set of experiments that enabled him to determine that cathode rays are negatively charged particles, and not a process in the ether, as physicists such as Goldstein, Hertz and Lenard had claimed. The poor vacuum in the tube makes these experiments tricky, as the residual air becomes conductive under the effect of the cathode rays. From the deflections due to the electrostatic and magnetic fields, he calculated the velocity v and the mass-to-charge ratio (m/e) of the particles emitted from the cathode (mass and charge are designated by these symbols in Thomson's paper). He found that m/e is of the order of 10^{-7} (he did not give the units), which is considerably lower than the same value for the hydrogen ion (10^{-4}). Various considerations led him to believe that this value came from a small m and a large e.

18 F. Geisel died in 1927 of metastatic carcinoma caused by high doses to his hands.

In reality, the difference is equal to the ratio of electron and proton masses. As for the speed, he found it to be in excess of 10^9 cm/s, a remarkable value. Indeed, with the currently accepted electron mass and an acceleration voltage of 1,800 V, we find 2.5×10^9 cm/s. For his discovery of a new particle lighter than ions, Thomson was awarded the Nobel Prize in Physics in 1906. It should be noted that two years earlier, the Frenchman Jean Perrin had found that the charge of the particles making up cathode rays was negative, but he considered these particles to be negative ions (Perrin 1895).

2.2.6. *Discovery of radioactivity*

While the discoveries of X-rays and electrons came from physics, the discovery of natural radioactivity owes a great deal to chemistry for the characterization and separation of elements.

On February 23, 1896, in a note to the Académie des Sciences de Paris, Henri Becquerel described his experiments in which he exposed a photographic plate wrapped in lightproof paper to the combined action of sunlight and a phosphorescent substance, potassium uranium sulfate. After several hours, the silhouette of the phosphorescent substance could be revealed on the photographic plate (Becquerel 1896e). A week later, on March 2, Becquerel returned to the Académie des Sciences with a new note (Becquerel 1896a), in which, as a preamble, he recalled his previous experiments, and in particular the fact that radiation is emitted by phosphorescent matter when exposed to the sun. He then insisted that the observations made previously could be reproduced in the absence of solar excitation. He then described the well-known story of the revelation of films that had been kept in a drawer in contact with uranium salt, the initial experiment having been postponed because of two days without sunlight. His great merit is to have immediately understood that light excitation is not necessary, and to have verified this with a series of experiments. He concluded by pointing out that it was difficult to indicate the nature of the particles emitted, but that he was currently carrying out experiments that should shed some light on the matter. In a short space of time, he returned several times to the Académie des Sciences with an increasingly precise vision of the reality of these invisible radiations:

– March 9,

> [...] invisible radiation [...] has the property of discharging electrified bodies. [...] It is remarkable to note that, since March 3, i.e. after more than one hundred and sixty hours, the intensity of the radiation emitted in the dark has not diminished appreciably.

This last point, and especially the fact that these invisible radiations were emitted only by a few phosphorescent materials, were beginning to raise questions about the link between the two phenomena (Becquerel 1896b).

– On March 23, he reported, among other things, his experiments on the emission of various uranium salts:

> The phenomenon therefore appears to be one of invisible phosphorescence, but does not seem to be related to visible phosphorescence or fluorescence.

His report also mentions other phosphorescent substances:

> Uranium salts emit invisible radiation with remarkable constancy, but the same is not true of other phosphorescent substances (Becquerel 1896d, author's translation).

– On March 30, he showed that the radiation emitted by uranium salts and the radiation from a Crookes tube had characteristics that differentiated them from each other (Becquerel 1896f) on May 18; using metallic uranium, he demonstrated that invisible radiation is emitted by uranium (Becquerel 1896c).

2.3. First irradiation experiments

At the end of the 19th century, everything was available to carry out the first irradiations. The first experiments were observations of gas discharges, the observable being the phosphorescence of the gas, that is, experiments involving molecular collisions or spectroscopy. Prior to this, Edmond Becquerel, father of Henry Becquerel, was one of the first to study the conditions under which mineral phosphorescence occurs. In an 1839 article, he studied the phosphorescence of oyster shells calcined with a pyrophore, calcium sulfate[19]. Having shown that phosphorescence by temperature rise is greater if the material has previously been exposed to sunlight or electric light, he hypothesized that these radiations created a new mode of molecular arrangement, and that the released electricities combine to form the phosphorescent glow. He then stated: "Good conducting materials would never be phosphorescent, as experience shows, because the electricities released by the action of solar rays come together on contact before they have acquired a certain voltage" (Becquerel 1839, author's translation), a way of saying that electronic excitations are ineffective in metals.

19 It does not say why this strange material was chosen.

In 1894, when it was well established that cathode rays produced phosphorescent light in many substances, Eugen Goldstein of Berlin reported for the first time the "permanent" color change of alkali halide crystals irradiated by these cathode rays. In addition to the original color, each material exhibits characteristic colors: a phosphorescence color, a color of the irradiated material and the color after heating. In the case of sodium chloride, he observed that the buff-yellow coloration only affected a thin layer near the surface. If the cathode rays are focused with a magnet, or if the sample is heated, the color changes to dark blue. However, if this blue material is subjected to further "low flux" irradiation, it regains its buff-yellow color (Goldstein 1894). In the years that followed, Goldstein published several articles in which he distinguished between two types of coloration: first-class coloration, which is formed as soon as the material is struck by cathode rays, and second-class coloration, which is formed after long periods of exposure and high-temperature heating (Goldstein 1901). Today, both colorations are attributed to the presence of point defects (colored centers) and clusters of defects. Blue coloration has been extensively studied, as blue rock salt is found in some deposits. Since the search for a dye proved unsuccessful, and the coloration disappeared without loss of mass when the salt was heated, the origin of this color was attributed to the presence of cavities that diffract blue light (Wittjen and Precht 1883).

During these same years, the effects of Becquerel's rays on matter were the subject of study. In their 1903 article in *Science*, George Kunz and Charles Baskerville describe the action of radium, Röntgen rays and ultraviolet light on minerals and precious stones (Kunz and Baskerville 1903). This was a major undertaking, since the study of fluorescence and phosphorescence[20] involved some 13,000 minerals from the Morgan–Tiffany[21] collection at the American Museum of Natural History in New York City. The article remains highly descriptive and led to a proposal for classifying minerals according to their response to different types of radiation. But whatever the publication on the subject, these luminescence phenomena are not the subject of any satisfactory theoretical explanation within the framework of the classical physics known at the time.

20 G. Kunz gave a definition of fluorescence and phosphorescence: fluorescence is a luminosity that lasts only during the direct influence of the exciting agent; phosphorescence is the emission or propagation of ethereal stresses, which affect the optical centers, producing a light that persists after the cause has been removed.

21 Charles Lewis Tiffany was a jeweler with a passion for rare gems. George Kunz's expertise earned him a job at Tiffany & Co. with the mission of finding the finest gems. He identified a new mineral that came to be known as Kunzite. John Pierpont Morgan was a financier and a great collector of colored gems. Kunz catalogued his collection.

German radiochemist Fritz Giesel[22], who purified radium and polonium and co-discovered actinium, listed the effects of radiation on materials: minerals are colored, paper, cellulose, etc., become brown and brittle and eventually disintegrate completely. He also described a physiological effect on the skin, for which he served as a test subject.

> The burn I received on my arm after applying a capsule containing about 0.3 g of radium bromide for two hours about 1 ½ years ago is still visible today, and the hair growth is permanently destroyed (Giesel 1902, author's translation) [original text is in German].

It is difficult to estimate the dose received by the skin without knowing the geometry of the source, but given a few assumptions, for a 0.6-mm-thick source we find a dose of around 6,000 Gy[23] – an enormous dose, given that 30 Gy is enough to induce tissue necrosis.

In Paris, Pierre Curie studied the increase in conductivity of liquid dielectrics under the influence of radium and Röntgen rays. Probably because of the many uncertainties involved, he found no difference between the effects of alphas and X-rays, but it was difficult to justify this non-effect, as he gave no indication of the fluxes or the absorption or loss of energy in the walls. In fact, Henri Becquerel's paper to the Académie des Sciences on February 17, 1902, contained no attempt to explain the phenomenon (Curie 1902). In the same years, August Becker, a young assistant to Philipp Lenard at the University of Kiel, took advantage of the experimental facilities developed by P. Lenard to carry out similar experiments on solid insulators. The device he used was certainly one of the first to study the effects of irradiation in situ (Figure 2.11). To do this, he used cathode rays to irradiate capacitors consisting of two metal plates separated by a solid dielectric, in this case kerosene, whose thickness could be as thin as 11 μm (Becker 1903, 1904). The results lacked precision, but resembled those obtained with modern techniques showing the existence of two types of charge carriers: hot electrons responsible for the rapid response (radiation-induced conductivity (RIC)) and thermalized electrons whose response depend on the trapping–untrapping and recombination phenomena (delayed radiation-induced conductivity (DRIC)).

Recruited in 1899 by Carl Zeiss for his expertise in ultramicroscopic on the coloring of antique glass, Henry Siedentopf hypothesized that rock salt color was

22 His real name was Friedrich Oskar Giesel.

23 Gy: Gray (J/kg), legal unit of dose.

due to the presence of ultramicroscopic metal particles (Siedentopf 1905). He headed the Carl Zeiss Microscopy Department from 1907 to 1938.

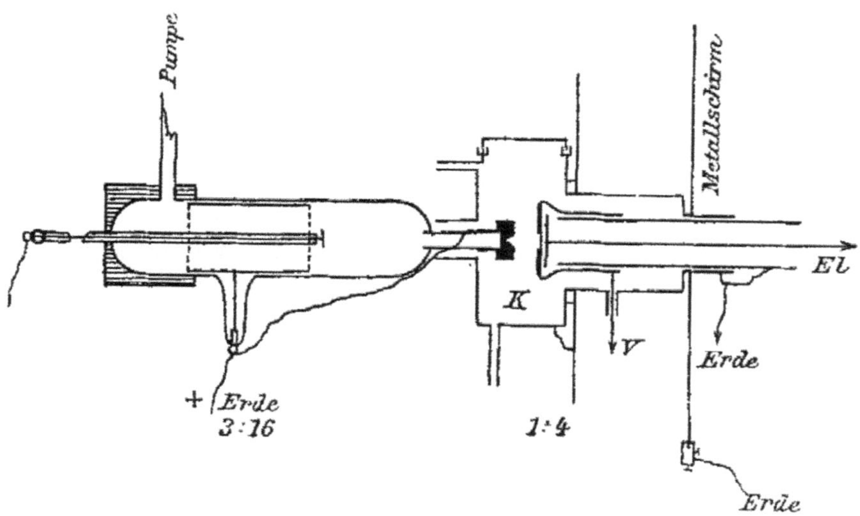

Figure 2.11. *Device used by A. Becker to measure the effect of cathode rays on the conductivity of insulators (Becker 1904)*

Direct access to irradiation changes in insulators led many people to take an interest, hence the large number of publications in this field: for example, von Karl Przibram (Austria) studied radio-photoluminescence and showed that when exposed to beta radiation, fluorite becomes blue and thermo-luminescent. He also reported:

> Some bottles of radium solution were coloured brown where the liquid was in contact with the glass and blue above the surface, the latter colouration being probably due to the radium emanation (Meyer and Przibram 1912).

In 1910, Cornelio Doelter[24] published a book entitled *Das Radium und die Farben* in which he attributed the origin of the colour of minerals to Becquerel's rays (Doelter 1910).

24 His full name was Cornelio August Severinus Doelter y Cisterich.

I will conclude this chapter with a few lines on the observation of effects due to natural irradiation: pleochroic halos.

2.3.1. *Pleochroic halos*

The observation in various crystals such as micas or fluorite of spherical regions, a few tens of micrometer in size and of a different color from the host mineral, intrigued scientists at the end of the 19th century. These zones were called pleochroic halos[25]. Hypotheses on their nature were numerous, but it was not until 1907 that a link between these halos and radioactivity was proposed almost simultaneously by John Joly in Dublin (Joly 1907a) and Otto Mügge in Königsberg (Mügge 1907). In the same publication, O. Mügge presented an experiment in support of his hypothesis. After depositing microscopic grains of radium bromide on lamellas of cordierite[26], he observed concentric halos around the grains after five days. Six months later, in a very short article in *Nature* (Joly 1907b), J. Joly reported that he had tried, unsuccessfully, the same experiment as O. Mügge on biotite[27]. On discovering the latter's paper, he re-examined his preparations and found that they displayed the characteristic brown coloration of pleochroic halos. In 1910, E. Rutherford, in possession of a large quantity of radium emanation, introduced it under pressure into a fine soda-lime glass capillary tube. A month later, the central hole of the capillary was surrounded by a well-defined zone colored red. The thickness of this colored zone was 0.039 mm, in agreement with the 0.041 mm given by a calculation based on the law of absorption of α rays found by W.H. Bragg. This confirms the interpretation of the origin of halos in mica (Rutherford 1910).

Continuing this work, notably in collaboration with E. Rutherford at Cambridge, in 1913, J. Joly estimated, using the radioactive decay of minerals, that the beginning of the Devonian – a geological era situated between the Silurian and Carboniferous – was more than 400 million years old, an age still accepted today (Joly 1922).

After this preamble, which takes us back to the first irradiation experiments, and before moving on to the modern history of materials under irradiation, we will take a brief look at the radiological risks incurred by the players in this history.

---

25 1855 (Des Cloizeaux, in ACP, S. 3, XLX, 290 ds Fonds Barbier) from Greek πλεων "many", and χροιζω "color".

26 Cordierite is a silicate, with the formula $Al_3Mg_2AlSi_5O_{18}$.

27 Biotite is a silicate of the mica family with the ideal formula $K(Mg, Fe)_3(OH,F)_2$.

Figure 2.12. *Rutherford's experiment to verify the origin of pleochroic halos. A capillary with outer diameter EF equal to 0.6 mm and orifice C filled with radium emanium shows a colored zone of thickness 0.039 mm (Rutherford 1910). For a color version of this figure, see www.iste.co.uk/bouffard/irradiation.zip*

2.4. Secondary effects of radiation

We can only imagine the radiation environment in which our pioneers worked. As the voltage of the Crookes tubes increased, the X-rays emitted became harder and harder, increasing the flux through the glass wall. This is how W.C. Röntgen discovered X-rays, but at what cost? Before we became aware of the risks of X-rays, the craze for these rays had its victims, from Mrs. Röntgen for the first X-ray, to the many patients and, above all, operators in Röntgenology centers. As early as July 1896, that is, six months after the discovery of X-rays, dermatitis was observed in a young man who had been used as a subject for commercial demonstrations of lung and heart visualization using X-rays for four weeks, once or twice a day (Lacassagne

1969). In November 1896, at the Sir Patrick Dun Hospital in Dublin, a 35-year-old man with symptoms of kidney stones was exposed to X-rays to determine the presence or absence of stones. The first exposure lasted three hours; however, the photographic plate proved unsatisfactory, and he underwent a second session. The rest of the article by Henry C. Drury, physician assistant, is a description of the patient's clinical condition: nausea, purulent wound... plus worry and anxiety. The author concludes with "I think it right, however, to report the case without any further delay, though incomplete" (Drury 1896).

In terms of radioactivity, the problem affected fewer people. But given the quantity of radioactive material they handled, Marie and Pierre Curie, Henri Becquerel and their collaborators were surely contaminated. In 1901, five years after the discovery of radioactivity, Pierre Curie and André Debierne noted the deplorable state of the air in their laboratory, and pointed out that it contained disseminated radioactive dust:

> The air in the laboratory where we have been working for several years has gradually become more and more conductive; it is no longer possible to have a well-insulated apparatus, and we can only make rough electrometer measurements. This deplorable state does not seem to us to be explained by the direct radiation of radioactive dust scattered around the laboratory; it is probably due in large part to the continuous formation of radioactive gases analogous to those we have just mentioned (Curie and Debierne 1901, author's translation).

Numerous examples of X-ray side effects are described in the article (Lederman 1981; Sansare et al. 2011). Despite knowledge of the biological effects of ionizing radiation and recommendations for protection, it was not until 1928, at the Second International Congress of Radiology, that the mobilization of radiologists led to the creation of the International Commission on Protection against X-rays and radium. However, it was not until after the Second World War that the newly created International Commission on Radiological Protection (ICRP) became a truly international institution, with its affiliation to the World Health Organization (WHO) in 1956, the International Atomic Energy Agency (IAEA) in 1959, the International Organization for Standardization (ISO), the Food and Agriculture Organization (FAO), the United Nations Educational, Scientific, and Cultural Organization (UNESCO), etc. As early as 1955, the ICRP wrote:

> In view of the incomplete evidence on which the values are based, and the knowledge that some effects of radiation are irreversible and may be cumulative, it is strongly recommended that every effort be made

to reduce exposure to all types of ionizing radiation to the lowest possible level. The Commission will revise the proposed permissible doses as new information becomes available (author's translation).

This recommendation prefigured the ALARA (As Low As Reasonably Achievable) principle that was to be issued in the late 1960s. Radiological accidents are well documented when their origin is not discovered late or covered by defense secrecy. Figure 2.13 shows the distribution of the number of accidents between the various fields of radiation use, bearing in mind that the weight of each event can vary from one victim for a handling error to a very large number for an accident like Chernobyl.

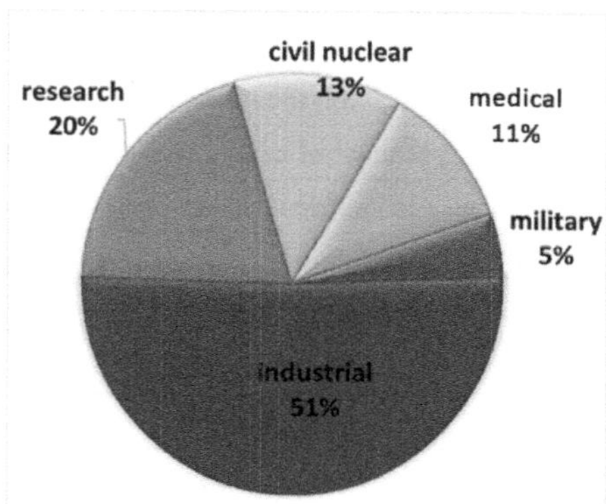

Figure 2.13. *Breakdown of radiological accidents by field of radiation use (Nénot and Gourmelon 2007). For a color version of this figure, see www.iste.co.uk/bouffard/irradiation.zip*

Radioactive sources are the cause of a number of accidents involving the bypassing of radiation protection systems, the failure to remove the source in a safe position and the loss of the source, particularly during the dismantling of an installation. In the nuclear sector, there have been a few criticality accidents, most notably the Chernobyl disaster. Military accidents are more difficult to list. They range from the contamination of a large area in the Urals by the explosion of a tank containing fission fragments, to accidents involving nuclear submarines (two American and seven Soviet submarines carrying nuclear weapons were lost at sea), to the loss of nuclear weapons following the crash of bombers at Palomarès in Spain and Thule in Greenland. In the medical field, it was mainly problems with the calibration of radiotherapy equipment or dose calculations that led to patients being overdosed. Nénot and Gourmelon (2007) describe numerous accidents and their consequences.

2.5. Chapter 2 references

2.5.1. *A brief chronology*

Date	Person	Advances
1647	Guericke	Air pump
1659	Boyle	Air pump
1660	Guericke	Electrostatic machine
1675	Picard	Glow on the surface of mercury
1704	Hauksbee	Electrostatic machine
1709	Hauksbee	Electric light
1727	Schulze	Effect of light on silver salt
1758	Cronsted	Essay on the mineral kingdom
1774	Werner	Manual of descriptive mineralogy
1783	Romé de l'Isle	Geometric shape of crystals
1784	Walkiers de St Amand	Forerunner of Van de Graaffs
1815	Berzelius	Gadolinite
1822	Haüy	Treatise on mineralogy
1824	Niepce	First photograph
1831	Faraday	Electromagnetic induction
1836	Page / Callan	Induction coils
1840	Scheerer	Effect of cooling rate on structure
1851	Ruhmkorff	Ruhmkorff coil
1855	Geissler	Mercury pump
1858	Plücker	Cathode ray deflected by a magnetic field
1860	Des Cloizeaux	Bi and mono refringence
1865	Sprenger	Mercury pump
1870	Crookes	Crookes tube
1876	Goldstein	Cathode rays
1893	Broegger	Metamic state
1894	Goldstein	First irradiation, the salts change color
1895	Röntgen	X-rays (Nobel 1901 Physics)
1896	Becquerel	Natural radioactivity (Nobel Prize 1903 Physics)
1897	Thomson	Electrons (Nobel Prize 1906 Physics)

Table 2.1. *A brief chronology*

2.5.2. *Biographies of some of the chapter's personalities*

Numerous individuals have contributed to the advancement of knowledge in the field of irradiated materials. Below are brief biographies of five of them, arbitrarily chosen.

Francis Hauksbee the Elder (1660 Colchester–1713 London) was an English scientist renowned for his work on electrostatics and, above all, for the quality of the instruments he developed. His double-piston air pump was his most remarkable achievement. The system of two pistons operating alternately produced a better vacuum with less physical effort. For the operating principle, see Figure 3, page 258 in Brundtland (2012). He was the first to observe the luminous emission of mercury when shaken under vacuum in a glass vessel. He called this light mercurial phosphorus, but did not understand its origin. He was Isaac Newton's assistant, then appointed Chief Experimentalist of the Royal Society, of which he was elected a member in 1709. It should be noted that he was the lowest ranking member of this society.

Abbé René Just Haüy (1743 Saint-Just-en-Chaussée–1822 Paris), a member of the Prémontrés religious order, can be considered alongside Jean-Baptiste Romé de l'Isle as the founder of geometric crystallography. He conceived that the crystalline form of a body depends on its chemical composition and results from the stacking of elementary molecules. During the French Revolution, he refused to take the oath to the New Civil Constitution of the Clergy. He was imprisoned, although released thanks to the support of scientists. Although he still refused to take the oath, in 1795 he was appointed curator of mineral collections at the Agence des Mines. In 1801, he published his *Traité de minéralogie* and two volumes of his *Traité de physique*. Both volumes were a great success. Napoleon told him: "Monsieur Haüy, I took your Physique with me to Elba, and re-read it with the greatest interest".

Figure 2.14. *Abbé René Just Haüy. For a color version of this figure, see www.iste.co.uk/bouffard/irradiation.zip*

Glassmaker *Henrich Geissler* (German Confederation) set up his workshop in Bonn in 1852, specializing in the manufacture of physics and chemistry instruments. He was awarded the gold medal at the Paris Universal Exhibition in 1855 for the quality of his glass. His major contribution to the understanding of cathode rays was the development of a new mercury vacuum pump, which gave access to much higher vacuums, in the millibar range.

William Crookes (1832 London–1919 London), chemist and physicist, was best known for the invention of the tube that bears his name: the Crookes tube. He discovered thallium.

Eugen Goldstein (1850 Gleiwitz–1930 Berlin) spent his career at the Potsdam Observatory, where he became interested in electrical discharges under different vacuum conditions. In 1886, he discovered the existence of positive radiation, which he called "kanalstrahlen" (channel rays). He used this name in reference to the arrangement of the electrodes. The electrodes formed a kind of channel through which positive particles could be channeled. But he did not attribute this radiation to positive ions. He also discovered that there are several types of cathode rays.

2.6. References

Agricola, G. (1556). *De re Metallica*, Hoover, H.C., Hoover, L.H. (trans). Dover Publications, New York.

Aristotle (1802) *Physique, tome 2, livre IV, de l'espace, du vide et du temps*, Barthélemy Saint-Hilaire, J. (trans). Librairie Philosophique de Lagrange, Paris [Original published between 335-323 BCE].

Becker, A. (1903). Über die Leitfähigkeit fester Isolatoren unter dem Einfluß von Radiumstrahlen (On the conductivity of solid insulators under the influence of radium rays). *Annalen der Physik*, 317(9), 124–143.

Becker, A. (1904). Über den Einfluß von Kathodenstrahlen auf feste Isolatoren (The influence of cathode rays on solid insulators). *Annalen der Physik*, 318(2), 394–421.

Becquerel, E. (1839). Recherche sur la production de la phosphorescence et sur diverses propriétés de l'étincelle électrique. *Annales de Chimie et de Physique*, 71, 36–48.

Becquerel, H. (1896a). Sur les radiations invisibles émises par les corps phosphorescents. *Comptes Rendus des Séances de l'Académie des Sciences*, CXXII(9), 501–553.

Becquerel, H. (1896b). Sur quelques propriétés nouvelles des radiations invisibles émises par divers corps phosphorescents. *Comptes Rendus des Séances de l'Académie des Sciences*, CXXII(10), 559–564.

Becquerel, H. (1896c). Emission de radiations nouvelles par l'uranium métallique. *Comptes Rendus des Séances de l'Académie des Sciences*, CXXII(20), 1086–1088.

Becquerel, H. (1896d). Sur les radiations invisibles émises par les sels d'uranium. *Comptes Rendus des Séances de l'Académie des Sciences*, CXXII(12), 689–694.

Becquerel, H. (1896e). Sur les radiations émises par phophorescence. *Comptes Rendus des Séances de l'Académie des Sciences*, CXXII(8), 420–421.

Becquerel, H. (1896f). Sur les propriétés différentes radiations invisibles émises par les sels d'uranium, et du rayonnement de la paroi anticathodique d'un tubes de Crookes. *Comptes Rendus des Séances de l'Académie des Sciences*, CXXII(13), 762–767.

Bernoulli, J. (1707). Nouvelle manière de rendre les baromètres lumineux. In *Histoire de l'Académie royale des sciences, avec les mémoires de mathématique & de physique*. Académie Royale des Sciences, Paris.

Bertrand, E. (1763). *Dictionnaire universel des fossiles propres et des fossiles accidentels contenant une description des terres, des sables, des sels, des soufres, des bitumes... (tome 1)*. Louis Chambeau, Avignon.

Berzelius, J. (1815). Bilargor till upplysning af atkilliga i förgaende Afhandling förefallande ämnen. *Afhandl. Fys. Kem. Min.*, 4, 217.

Bodu, R., Bouzigues, H., Morin, N., Pfiffelmann, J.P. (1972). Sur l'existence d'anomalies isotopiques rencontrées dans l'uranium du Gabon. *Comptes Rendus des Séances de l'Académie des Sciences - Série D*, 275, 1731–1733.

Brenni, P. (1999). The Van de Graaff generator. An electrostatic machine for the 20th century. *Bulletin of the Scientific Instrument Society*, 63, 6–13.

Broegger W.C. (1893). Amorf. In *Salmonsens Store Illustrerede Konversationsleksikon*. Brodrene Salmonsen, Norway, 742–743.

Brundtland, T. (2012). Francis Hauksbee and his air pump. *Notes and Records of The Royal Society*, 66(3), 253–272.

Burbank, J.E. (1898). X-rays and mineral phosphorescence. *American Journal of Science*, s4-5(25), 53–54.

Callan, N.J. (1836). A new galvanic battery. *The London and Edinburgh Philosophical Magazine and Journal of Science*, 9, 472–478.

Cockcroft, J.D. and Walton, E.T.S. (1930). Experiments with high velocity positive ions. *Proceedings of the Royal Society of London, Series A Mathematical and Physical Sciences*, 129(811), 477–489.

Cronstedt, A.F. (1758). *Försök till Mineralogiens eller mineral-Rikets upställing*, von Engestrom, G. (trans). Edward and Charles Dilly, London.

Curie, P. (1902). Conductibilité des diélectriques liquides sous l'influence des rayons du radium et des rayons de Röntgen. *Comptes Rendus Hebdomadaires des Séances de l'Académie des Sciences*, 134(7), 420–423.

Curie, P. and Debierne, A. (1901). Sur la radio-activité induite et les gaz activés par le radium. *Comptes Rendus des Séances de l'Académie des Sciences*, CXXXII, 768–770.

Damour, A. (1864). Note sur la densité des zircons. *Comptes Rendus Hebdomadaires des Séances de l'Académie des Sciences*, 58, 154–159.

Des Cloizeaux, A. and Damour, A. (1860). Des propriétés optiques et pyrogénétiques des minéraux connus sous le noms de gadolinites, allanites, othites, euxénite, tyrite, yttrotantalite et fergusonite. *Annales de Chimie et de Physique*, 59(3), 357–379.

Doelter, C. (1896). Influence des rayons de M. Rœntgen sur les minéraux. *Bulletin de la Société Française de Minéralogie*, 19(3–4), 88–90.

Doelter, C. (1910). *Das Radium und die Farben (Radium and the Colours).* Verlag von Theodor Steinkopff, Dresden.

Drury, H.C. (1896). Dermatitis caused by Roentgen X rays. *British Medical Journal*, 2(1871), 1377–1378.

Ewing, R.C. (1994). The metamict state: 1993 – the centennial. *Nuclear Instruments & Methods in Physics Research Section B-Beam Interactions with Materials and Atoms*, 91, 22–29.

Faraday, M. (1839). Glow discharge. In *Experimental Research in Electricity*, Taylor, R. and Taylor, J.E. (eds). University of London, London.

Gibson, S.J. and Ehlmann, A.J. (1970). Annealing characteristics of metamic gadolinite from Rode Ranch Texas. *American Mineralogist*, 55, 288–291.

Giesel, F. (1902). Über Bequerelstrahlen und die radioaktiven Substanzen (About Bequerel rays and radioactive substances). *Zeitschrift für Elektrochemie*, 8, 579–585.

Gilberti, G. (1600). *De magnete, magneticisque corporibus, et de magno magnete tellure; Physiologia nova, plurimis & argumentis, & experimentis demonstrata.* Peter Short, London.

Gimingham, C.H. and Crookes, W. (1877). I. On a new form of the 'sprengel' air-pump and vacuum-tap. *Proceedings of the Royal Society of London*, 25(171–178), 396–402.

Goldstein, E. (1876). Vorläufige Mittheilungen über elektrische Entladungen in verdünnten Gasen (Preliminary reports on electrical discharges in dilute gases). *Monatsberihte der Königlch Preussischen akademie der Wissenschaften zu Berlin*, 38, 279–295.

Goldstein, E. (1894). Über die Einwirkung von Kathodenstrahlen auf einigen Salze (On the action of cathode rays on some salts). *Sitzungsberichte der Königlich Preussischen Akademie der Wissenschaften zu Berlin*, 937–945.

Goldstein, E. (1901). Über Nachfarben und die sie erzeugenden Strahlungen (About secondary colours and the radiations that produce them). *Sitzungsberichte der Königlich Preussischen Akademie der Wissenschaften zu Berlin*, 10, 222–229.

Gray, J. (1890). *Electrical Influence Machines: A Full Account of Their Historical Development, and Modern Forms, with Instructions for Making Them.* Whittaker and Co, London.

de Guericke, O. (1672). *Experimenta Nova (ut vocantur) Magdeburgica de vacuo spatio.* Amstelodami, Joannem Janssonium Waesberge, Amsterdam.

Hamberg, A. (1914). Die radioaktiven Substanzen und die geologische Forschung (Radioactive substances and geological research). *Geologiska Föreningen i Stockholm Förhandlingar*, 36, 31–96.

Hauksbee, F. (1709). *Physico-Mechanical Experiments on Various Subjects Containing an Account of Several Surprinzing Phenomena Touching Light and Electricity. Producible on the Attrition of Bodies with Many Others Remarkabke Appearences Not Before Observ'd. Together with the Explanation of All the Machines.* R. Brugis for the Author, London.

Haüy, R.J. (1822). *Traité de minèralogie (tome 1).* Bachelier, Paris.

Hittorf, J.W. (1869). Ueber die Elektricitätsleitung der Gase (On the electrical conduction of gases). *Annalen der Physik und Chemie*, CXXXVI, 1–31.

Joly, J. (1907a). Pleochroic halos. *The London, Edinburgh and Dublin Philosophical Magazine and Journal of Science (6th series)*, XIII(LXXV), 381–383.

Joly, J. (1907b). Pleochroic halos. *Nature*, 76, 589.

Joly, J. (1922). The age of the Earth. *Nature*, 109(2737), 480–485.

Kunz, G.F. and Baskerville, C. (1903). The action of radium, Roentgen rays and ultra-violet light on minerals and gems. *Science*, 18(468), 769–783.

Lacassagne, A. (1969). Les pionniers de la protection contre les rayonnements ionisants. *Radioprotection*, 3(4), 305–310.

Lederman, M. (1981). The early history of radiotherapy: 1895–1939. *International Journal of Radiation Oncology Biology Physics*, 7(5), 639–648.

Magnus, A. (1569). *De mineralibus et Rebus Metallicis Libri Quinque.* Ioannem Birckmannum and Theodorum Baxmium.

Meyer, S. and Przibram, K. (1912). Influence of radium radiations on glass and certain minerals. *Sitzungsberichte der Akademie der Wissenschaften in Wien, Mathematisch-Naturwissenschaftliche Klasse*, 121(2a), 1413–1418.

Mügge, O. (1907). Radioaktivität als Ursache der pleochroitischen Höfe des Cordierit (Radioactivity as the cause of the pleochroic haloes of cordierite). *Centralblatt für Mineralogie, Geologie und Palaontologie*, 397–399.

Nénot, J.-C. and Gourmelon, P. (2007). Les accidents dus aux rayonnements ionisants – le bilan sur un demi-siècle. Report, IRSN-2007/79.

Neuilly, M., Bussac, J., Frèjacques, C., Nief, G., Vendryes, G., Yvon, J. (1972). Sur l'existence dans un passé reculé d'une réaction en chaîne naturelle de fissions, dans le gisement d'uranium d'Oklo (Gabon). *Comptes Rendus des Séances de l'Académie des Sciences*, 275, 1847–1849.

Newton Harvey, E. (1957). *A History of Luminescence: From the Earliest Times Until 1900.* The American Philosophical Society, Philadelphia.

Page, C.G. (1837). Method of increasing shocks, and experiments, with Prof. Henry's apparatus for obtaining sparks and shocks from the Calorimotor. *The American Journal of Science and Arts*, XXXI, 137–141.

Page, C.G. (1867). *History of Induction: The American Claim to the Induction Coil and Its Electrostatic Developments.* Intelligencer Printing House, Washington, D.C.

Perrin, J. (1895). Nouvelles propriétés des rayons cathodiques. *Comptes Rendus des Séances de l'Académie des Sciences*, CXXI, 1131–1134.

Peterson, W. (1890). Studier öfver gadolinit (Studies on gadolinite). *Geologiska Föreningens i Stockholm Förhandlingar*, 12, 275–347.

Picard, J. (1676). Experience fait à l'Observatoire sur le Barometre simple touchant un nouveau Phenomene qu'on y a découvert. *Journal des Sçavans, de l'An M.DC*, LXXVI, 132.

Pliny the Elder (1830). Livre XXXVII traitant des pierres précieuses (French translation by Emile Littré). In *Naturalis Historiæ*, Firmin-Didot et Cie (ed.). Paris [Original published in 79 BCE].

Priestley, J. (1775a). *The History and Present State of Electricity: With Original Experiments.* London.

Priestley, J. (1775b). The experiments and discoveries of Mr. Hauksbee. In *The History and Present State of Electricity with Original Experiments*. London.

Romé de l'Isle, J.-B. (1783). *Cristallographie ou description des formes propres à tous les corps du règne minéral, dans l'état de combinaison saline, pierreuse ou métallique, avec figures & tableaux synoptiques de tous les cristaux connus.* Imprimerie de Monsieur, Paris.

Röntgen, W.C. (1895). Über eine neue Art von Strahlen (About a new type of rays). *Sitzungsherichte der Physicalisch-Medicinischen Gesellshaft zu Würsburg*, 137, 132–141.

Röntgen, W.C. (1898). Ueber eine neue Art von Strahlen (About a new type of rays). *Annalen der Physik und Chemie*, 64(1), 1–11.

Routledge, R. (1881). *A Popular History of Science.* George Routledge and Sons, London.

Rutherford, E. (1910). Action of the α rays on glass. *The London, Edinburgh and Dublin Philosophical Magazine and Journal of Science (6th series)*, XIX(CIX), 192–194.

Sansare, K., Khanna, V., Karjodkar, F. (2011). Early victims of X-rays: A tribute and current perception. *Dento Maxillo Facial Radiology*, 40(2), 123–125.

Scheerer, T. (1840). Untersuchung des Allanit, Orthit, Cerin und Gadolinit (Investigation of allanite, orthite, cerine and gadolinite). *Annalen der Physik und Chemie*, CXXVII, 407–422.

Siedentopf, H. (1905). Ultramikroskopische Untersuchungen über Steinsalzfärbungen (Ultramicroscopic studies on rock salt colourations). *Physikalische Zeitschrift*, 6(24), 855–866.

Sprengel, H. (1865). III—Researches on the vacuum. *Journal of the Chemical Society, Faraday Transactions*, 18, 9–21.

Theophrastus (1754). *On the Stone*. Translation from the original Greek published by Jean-Thomas Herrisant (Paris) [Original published in 4th century BCE].

Thiry d'Holbach, P. (1754). Crystal, crystaux ou crystallisation. *Encyclopédie de Diderot et d'Alembert*, IV, 523–524.

Thomson, W. (1867). On a self-acting apparatus for multiplying and maintaining electric charges, with applications to the Voltaic theory. *The London, Edinburgh and Dublin Philosophical Magazine and Journal of Science*, 34, 391–396.

Thomson, J.J. (1897). Cathode rays. *The London, Edinburgh and Dublin Philosophical Magazine and Journal of Science*, 44, 293–316.

Trowbridge, J.T. (1900). Latest developments with the X-rays. *Popular Science Monthly*, 56, 659.

Van de Graaff, R.J., Compton, K.T., Van Atta, L.C. (1933). The electrostatic production of high voltage for nuclear investigations. *Physical Review*, 43, 149–164.

Van der Willigen, V.S.M. (1856). Ueber das geschichtete Licht im elektrische Ei (About the layered light in the electric egg). *Annalen der Physik und Chemie*, XCVIII, 494.

Von Engestrom, G. (1770). *Essay Towards a System of Mineralogy: By Axel Fredric Cronstedt*. Edward and Charles Dilly, London.

Werner, A.G. (1774). *Von den Äusserlichen Kennzeichen der Fossilien (From the external characteristics of the fossils)*. Giegfried Lebrecht Crusius, Leipzig.

Wickoff, D. (1967). *Albertus Magnus, Book of Minerals*. Oxford University Press, Clarendon Press, London.

Wilson, G. (1849). On the early history of the air-pump in England. *The Edinburgh New Philosophical Journal*, XLVI, 330–354.

Wittjen, B. and Precht, H. (1883). Zur Kenntnis des blaugefärbten Steinsalzes (On the knowledge of the blue colored rock salt). *Berichte der Deutschen Chemischen Gesellschaft*, 16(1), 1454–1457.

3

Particle Transport

Shortly after the discovery of X-rays (1895) and Becquerel rays (1896), work began on their interaction with matter, their transport or rather their absorption as it was then called. This was a truly scientific approach, despite the absence of any theoretical guide; the experimental approach was built up as discoveries were made.

A few months after his discovery, W.C. Röntgen presented his results on the absorption of X-rays by matter to the Würzburg Physico-medical Society on January 23, 1896 (Röntgen 1895). This was to be his only lecture on the subject. He began by citing the X-ray "permeability" of various materials, such as aluminum, rubber, glass (noting that leaded glass is less permeable), water, air, etc., and even the hand. He concluded that, for the same thickness, permeability was essentially determined by density, but the increase in permeability was greater than the decrease in the product of density × thickness (surface density in g/cm^2). Then, showing the absence of refraction and reflection, he indicated that X-rays could not be concentrated with a lens. He concluded by suggesting that X-rays could be longitudinal oscillations of the ether (sic).

3.1. It all started with collision experiments

The first experiments that were important for understanding the transport of charged particles in matter were actually atomic physics experiments, most of which focused on gas-phase ionization.

In addition to the phosphorescence induced by rays penetrating matter, the modification of gas conductivity was an effect discovered very early on, notably by Joseph John Thomson at Cambridge, initially in the form of observations:

> If, now, the Röntgen rays are directed on to the plate a violent leakage of electricity from the plate occurs, the potential of the quadrants connected with the plate changes, and in a few seconds the spot of light reflected from the mirror of the electrometer is driven off the scale (Thomson 1896).

With the recruitment of Ernest Rutherford, who had just been awarded a grant by the *Royal Commission for the Exhibition of 1851*[1] to continue in England the research he had begun in New Zealand, studies became more systematic. Indeed, E. Rutherford soon began working on gas ionization, and in particular on the recombination of ions produced by Röntgen rays passing through a gas (Rutherford 1897). He used two methods to achieve this. The first involved flowing ionized air at a known velocity and testing its conductivity at different distances. In the second, he applied an electromotive force to the gas at different times after the Röntgen rays had been interrupted, and then he measured the charge collected. This work, along with that of C.T.R. Wilson on the condensation of water vapor, was an important basis for J.J. Thomson's study of the electric charge of ions. Indeed, the current flowing in a gas exposed to the Röntgen rays and subjected to an electromotive force was equal to nev, where e is the charge of the ions, v is the average velocity and n is the number. The velocity v was determined by Rutherford and n was measured using the ancestor of the fog chamber invented by Wilson (1897). Droplet size depended on charge and surface tension. Since direct measurement of droplet size was impossible, droplet radius was determined by observing the speed at which the droplets fell (Thomson 1898). Using this difficult-to-implement protocol, he obtained the following values for ion charge e 7.4×10^{-10} electrostatic units, that is, 2.47×10^{-19} Coulomb. A deviation of 54% from the actual value – which is exact because it is considered to be a universal constant – was already an excellent result, given the difficulty of the measurements.

In 1899, E. Rutherford, then Professor of Physics at McGill University in Montreal, demonstrated experimentally that uranium radiation is complex and composed of at least two types of radiation:

> one that is very readily absorbed, which will be termed for convenience the α radiation, and the other of a more penetrating character, which will be termed the β radiation (Rutherford 1899).

1 Past holders of this prestigious prize include 13 Nobel Prize laureates (Chadwick, Cockcroft, Dirac, Higgs, Walton, etc., and Rutherford) and many others who have become eminent in their field.

Assuming that the intensity of this radiation decreases according to an ordinary absorption law $r = e^{-\lambda d}$, he found that, for aluminum, for example, λ equals 1,600 and 15 for α and β, respectively (Rutherford 1899). The second part of this long article is devoted to the absorption of radiation by gases: the effects of pressure on absorption and discharge velocity, ion velocity, the relationship between the current flowing through the gas and the applied electromotive force, etc.

Quadrant electrometer

At the beginning of the 20th century, electrical charges were measured using quadrant electrometers, including the one designed by Sir W. Thomson. A needle attached to a torsion wire is connected to the charge to be measured, for example, that of the ionization chamber. It carries quadrants A and B and a mirror to measure rotation through a telescope and graduated ruler. Quadrants 1–3 and 2–4 are at opposite potential. Most researchers used the Dolezalek configuration, similar in principle but allowing faster measurements.

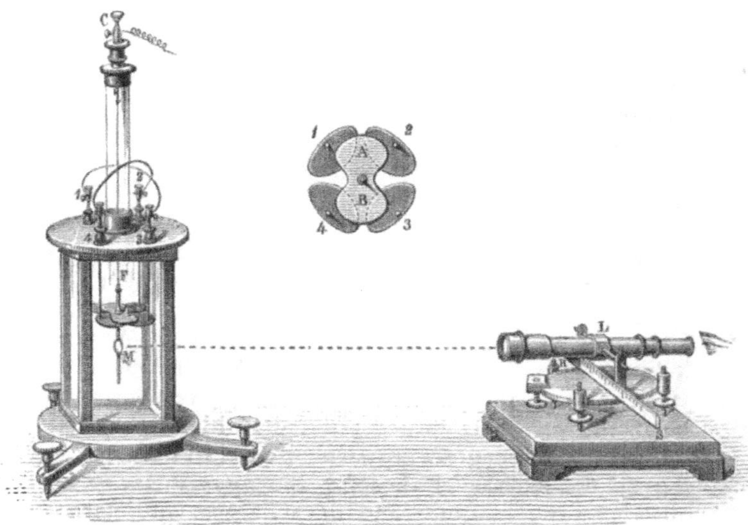

Figure 3.1. *Drawing of Sir W. Thomson's quadrant electrometer (Thompson 1905)*

Two years later, Ernest Rutherford and his physics demonstrator, R.K. McClung, published an article aimed at determining the amount of energy required to create an ion in the gas phase (Rutherford and McClung 1901). After a precise description of the experimental procedure, they found that the energy required to produce an ion is of the order of 2.1×10^{-10} erg or 131 eV (1 erg = 6.242×10^{11} eV). This value, which he found to be only slightly gas dependent, was at least an order of magnitude too high.

Many researchers were interested in these gas ionization experiments, as witnessed by the important series of experiments carried out in Cambridge by Strutt (1901).

3.2. Slowing down in the matter

In 1900, Henri Becquerel presented a report to the Académie des Sciences on Marie Curie's work on the penetration of Becquerel rays that cannot be deflected by a magnetic field (alphas) (Sklodowska-Curie 1900). It states:

> [...] my experiments have shown that, on the contrary, non-deviating rays are all the more absorbable the greater the thickness of matter they have already passed through. This singular law of absorption is contrary to that known for other types of radiation; it is rather reminiscent of the way in which a projectile behaves, losing part of its momentum as it passes through obstacles.

This is the form of the electronic stopping power integral Marie Curie described, but with a rather coarse resolution.

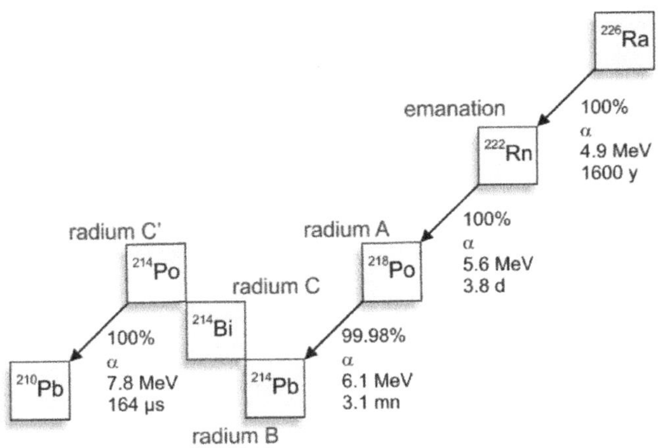

Figure 3.2. *Radioactive decay chain of ^{226}Ra. At the beginning of the 20th century, radium's descendants had not yet been identified; they were called emanation, radium A, B, C, C', etc. For a color version of this figure, see www.iste.co.uk/bouffard/irradiation.zip*

The foundations of our knowledge of particles α were laid in the years 1904–1908, in Australia, Canada, the UK, Germany, France and the United States. This research focused on the different types of alphas emitted by radium, their range through the air, their slowing down in matter, their stopping power as a function of energy and the material they pass through.

At the January 1904 meeting of the *Australasian Association for the Advancement of Sciences*, and in two articles (Bragg 1904; Bragg and Kleeman 1904), William Henry Bragg[2], a professor at the University of Adelaide, described with much greater precision the absorption of β and α radiation as it passes through matter. He asserted that these particles undergo a continuous decrease in velocity due to energy losses in ionization processes, this phenomenon being the only active one for alphas. Thus, each alpha has a defined range through a given medium, the length of which depends on the initial velocity and the medium traversed. On the other hand, β rays, which are high-speed electrons, are liable to be deflected by close collisions with atomic electrons, which explains why the limits of a well-collimated beam quickly become ill-defined. It should be remembered that, at the time, the internal structure of atoms was not known. However, W.H. Bragg indicated that the movement of electrons in atoms should be taken into account, and positive charges are introduced. A simple but ingenious improvement to their device would greatly enhance the quality of their measurements: the ionization chamber received only particles selected by a set of tubes perpendicular to the surface (see inset in Figure 3.3). By varying the air thickness in this device, W.H. Bragg and colleagues measured with good accuracy the range of the various alphas emitted by the thin film (Bragg and Kleeman 1905). Indeed, the decay chain of ^{226}Ra comprises five α emitters, the last one being emitted later. For freshly purified radium, only four alphas of different energies are observed. Table 3.1 shows the excellent agreement between the ranges measured by W.H. Bragg for the four alphas and the SRIM calculations (Ziegler et al. 1985).

This precision enabled them to study the dependence of ionization on the medium through which it passes. Two types of medium were studied: metal foils and gases. These studies concluded that the energy loss of α particles is approximately inversely proportional to the square root of their velocity and proportional to the square root of their atomic mass. They assumed that the energy required to create an ionization does not depend on atomic mass, and that it is the number of ionizations that varies. To explain the square-root law of atomic mass, they assumed that atoms have a surface area proportional to their mass, and that only

2 William Henry Bragg is best known for his work with his son William Lawrence Bragg on the diffraction of X-rays by crystals, which led to the resolution of atomic structures (Bragg's law). They were awarded the Nobel Prize in 1915 for this work.

peripheral collisions can ionize (the circumference is proportional to the square root of the mass). The curves in Figure 3.3 closely resemble the curves representing the evolution of electronic stopping power as a function of ion travel, by swapping the abscissa and ordinate axes. They show an ionization peak (energy loss peak) that we now call the Bragg peak.

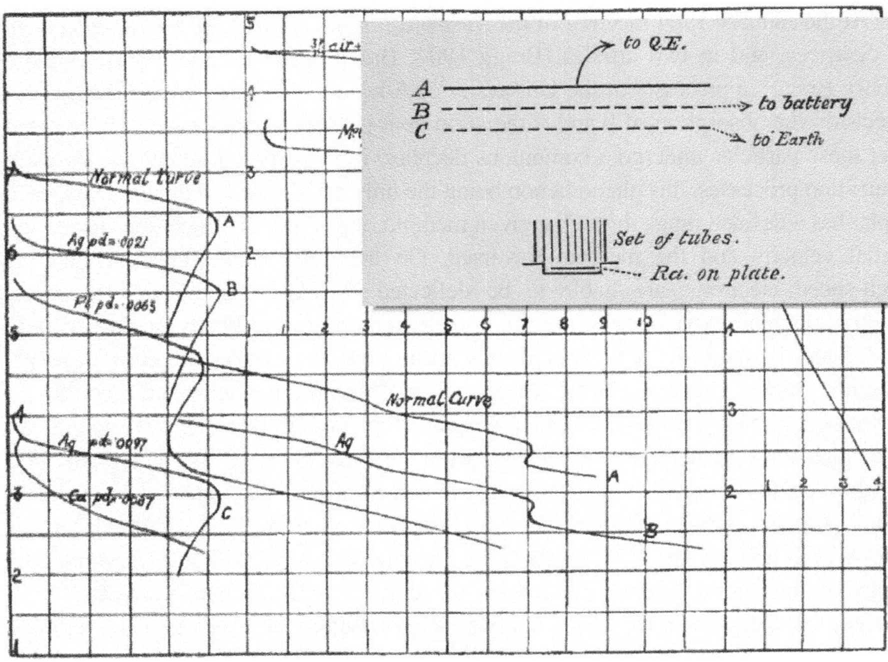

Figure 3.3. *Evolution of ionizations in the air of a radium source for different absorber sheets. Curve A corresponds to the ionization curve in air, and curve B to the same experiment, but with the interposition of a silver foil of surface thickness ρ.d = 0.002 (density x thickness). The unit is not indicated in the article, but if we consider that it is 0.002 lb/in², the thickness of the silver foil is equal to 134 µm (Figure 2 of Bragg and Kleeman (1905)). The inset shows the device used. For a color version of this figure, see www.iste.co.uk/bouffard/irradiation.zip*

The notion of stopping power is defined by W.H. Bragg in the article (Bragg and Kleeman 1905). This notion of stopping power is only indirectly linked to the current definition (energy loss per unit range). It is calculated from the reduction in the ion path in air when a sheet is introduced into the trajectory of α particles. The *"stopping power" is* expressed as follows:

$\frac{"loss\ of\ range"}{e_{sheet}} \frac{d_{air}}{d_{sheet}} \frac{M_{sheet}}{M_{air}}$, where e is the sheet thickness, d is the densities and M is the atomic masses. It is therefore a relative quantity whose behavior is difficult to predict.

Alpha emitting atom (1905 vision)	Measured range (cm)	Alpha emitting atom (modern vision)	Alpha energy (MeV)	Calculated range (cm)	Deviations (%)
Radium	3.50	^{226}Ra	4.78	3.39	3.2
Emanation or radium A	4.23	^{222}Rn	5.49	4.13	2.5
Radium A or emanation	4.83	^{218}Po	6.00	4.78	1
Radium C	7.06	^{214}Po	7.68	7.08	−0.3

Table 3.1. *Comparison of air range measurements (Bragg and Kleeman 1905) to the calculation of these same ranges with the SRIM software (Ziegler et al. 1985)*

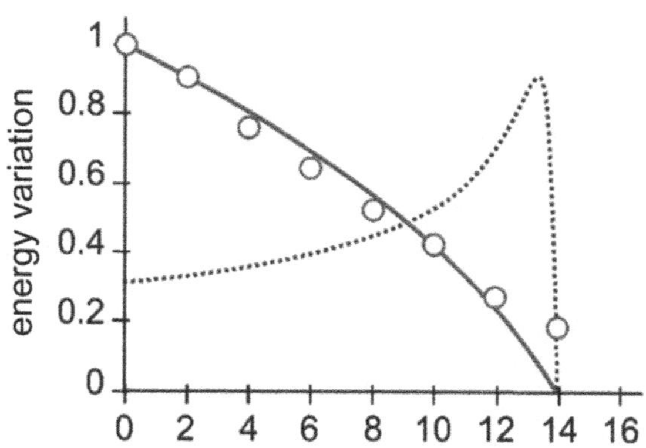

Figure 3.4. *Variation in the energy of α of radium C as a function of the range through aluminum. Rutherford's results are the red dots (Rutherford 1906b). The solid and dashed lines are the energy variation and stopping power, respectively, calculated with SRIM software for 2.9-µm-thick aluminum foils*

In the same years, Ernest Rutherford, then Professor of Physics at McGill University in Montreal, determined the speed v and the charge-to-mass ratio e/m of α particles emitted by radium C (^{214}Po). The source α was a thin wire activated by radium. A slit 2 cm from the source defined the beam that would print a photographic plate 7 cm away. The whole system was in a vacuum. The application of magnetic and electric fields gave access to the quantities v and e/m. E. Rutherford pointed out that the precision of the value of e/m did not allow for ruling definitively on the important question: are α particles accelerated helium atoms? In any case, knowing these two quantities, it was possible to measure the deceleration as it passed through matter. To do this, stacks of 3.1-μm-thick aluminum foil were inserted between the source and the slit. Measurement of the deflection by the magnetic field gave the residual velocity. The evolution of the velocity could be followed up to a thickness of 12 sheets, that is, 37 μm. The α particles then lost 36% of their initial velocity. But the addition of a thirteenth sheet made the beam disappear. The end of the article is devoted to hypotheses to explain this anomaly (Rutherford 1905). Ten months later, in a letter to the editors of Philosophical Magazine, E. Rutherford reported that he had been able to extend his measurements to greater thicknesses and a velocity loss of 57% (Rutherford 1906a). In June 1906, he wrote a third paper based on the 1905 paper, but for a more intense source and optimized geometry (Rutherford 1906b). In it, he plotted the alphas' energy as a function of the number of aluminum foils crossed. The thickness of the foils was not indicated, but their air equivalent was given. The energy decreased linearly with the thickness crossed, that is, the slowing power was independent of energy, an assumption often made at the time. The discrepancies with "real" variations can be seen in Figure 3.4. Only the point for 14 sheets was problematic.

Discovering Rutherford's paper, Henri Becquerel resumed the study he had carried out in 1905 using a procedure similar to Rutherford's, but performed in air. A platinum wire about 0.15 mm in diameter, uniformly excited by radium emanation, served as the source. The detector was a photographic plate inclined in such a way that it was possible to reconstruct the trajectory of alphas subjected to a magnetic field and having (or not) having passed through a 15-μm-thick aluminum foil (Figure 3.5). The magnetic field was reversed every five minutes to increase measurement accuracy. However, H. Becquerel did not really exploit the results; he simply concluded:

> To sum up, the measurements described confirm the existence of a retardation of the α rays when they traverse a sheet of aluminum such as observed by Rutherford (Becquerel 1906).

That same year, B. von Kucera and B. Masek, from the Institute of Physics at the University of Prague, published three papers on radiotellurium radiation (Kucera and Masek 1906a, 1906b, 1906c). The first describes the experimental set-up and ionization curves obtained with polonium, demonstrating the quality of the method.

From their experiments, they concluded that the absorption of α rays depended on their velocity, and increased with velocity. What is more, these α rays were diffusely scattered when they passed through a metal foil. The scattering capacity of metals increased with their atomic weight, as they put it: "perhaps as the square root of the atomic weight".

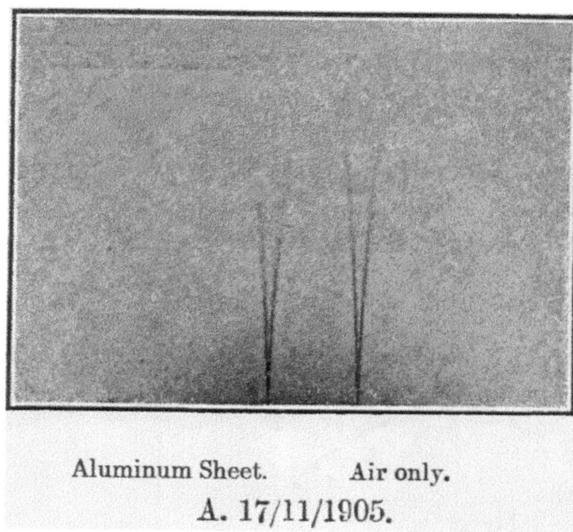

Figure 3.5. *Photographic plate showing the trajectories of α in air alone or behind a 15-μm-thick aluminum foil for two magnetic field orientations. The horizontal distance between the two experimental conditions was due to a translation of the plate (Becquerel 1906)*

In his paper, Bragg (1907) took up the discussion of the variation in "stopping power" on the basis of new measurements. He concluded: "...the stopping-power of a metal increases with the speed of the α particle, and more so the greater the atomic weight of the metal". At present, we know that in the energy range of radium α, stopping power decreases with increasing energy.

In the second edition of "Conduction of electricity through gases", J.J. Thomson describes and explains in Chapter XII, entitled *Rays from Radioactive Substances*, what was known in this field at the beginning of the 20th century (Thomson 1906). Of all the points covered, we will just mention a few relating to the range of particles. J.J. Thompson studied the absorption of alphas, explaining their loss of ionizing power beyond a certain distance by the neutralization of the α particle when their energy drops below a certain value, with a negatively electrified particle attaching itself to the α particle. He describes the electronic capture of ions at low speeds. On the other hand, he attributes the increase in ionization rate as energy decreases (at high speeds) to secondary radiation. Next comes an attempt to calculate the energy loss of α in a gas. For this, he relied on the method developed by James Clerk Maxwell in *The Dynamical Theory of Gases*. With the force between particles proposed by J.C. Maxwell of the form K/r^5 the deceleration is independent of velocity V and $dV/dx \approx 10^{10}$ (no unit is given, but it is certainly in s^{-1}). However, Rutherford's experiment showed that the speed of alphas decreased from 2.6×10^9 to 2×10^9 cm/s in 7 cm of travel in air, that is, $dV/dx \approx 10^8$. A force in K/r^4 makes it possible to reduce the difference. Still in the same paragraph, discussing the absorption of kinetic energy, J.J. Thomson states that it is important to separate the decrease in kinetic energy from angular scattering.

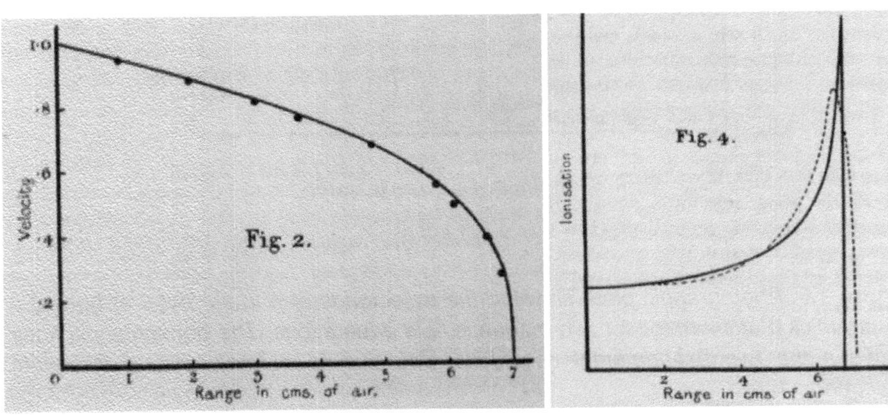

Figure 3.6. *Left: velocity evolution of radium C alphas as a function of their range through the air. Right: evolution of air ionization calculated from velocity evolution measurements (solid line) and measured experimentally (dotted line) (Geiger 1910a)*

An important advance in our understanding of particle transport in matter was made a few years later. In two articles, H. Geiger presented a very convincing measurement of the evolution of alpha velocity as a function of their trajectory and, above all, the link between ionization and slowing down (Geiger 1909, 1910a). To this end, he took up Rutherford's procedure, which involved measuring velocity by deflection in a magnetic field and following its evolution along the range by interposing sheets of different thicknesses. Progress was made on three fronts: (a) a more suitable geometry; (b) the replacement of photographic detection by direct observation of the scintillation of a zinc sulfide screen; and (c) the use of mica foil instead of aluminum foil to slow down the alphas. This last point is probably the most important. Indeed, the quality of measurements close to the stop zone is highly dependent on the dispersion of the thicknesses passed through. This is certainly the reason for the absence of data in this zone for Rutherford's experiment (Rutherford 1906b). Mica sheets, a cleavable material, have a very homogeneous thickness. The slowdown of radium C alphas shown in Figure 3.6 can be represented by a law of the type $v^3 = a(R - x)$, where R is the maximum range. H. Geiger then made the hypothesis, which he described as natural, that the ionization I produced on a given thickness of material is proportional to the energy absorbed by that thickness: $I = b\frac{d(v^2)}{dx} = c(R - x)^{-1/3}$. Ionization therefore diverges as x tends toward the maximum range, as Bragg and Geiger observed in their respective measurements. We can only wonder that W.H. Bragg did not make this connection earlier, since in his 1904 publication, referring to the α and β particles, he wrote: "Both kinds of rays suffer a continual diminution in speed through the expenditure of energy on ionization; but in the case of the α rays this is the only cause to which their so-called 'absorption' is due" (Bragg 1904).

These early results were confirmed and improved upon by the first actors (Bragg, Rutherford, Thomson and Geiger) and by new researchers such as T.S. Taylord (Yale University, New Haven, USA), S.C. Lind (University of Michigan, USA) and N. Bohr (Copenhagen, Denmark and Manchester, UK). By installing a diaphragm in the beam axis, T.S. Taylord demonstrated the role of diffusion in range measurement (Taylot 1909).

Experimental research often gave rise to original instrumental developments. In Cambridge, Richard Whiddington set up a device to measure the transmission laws of cathode rays and to distinguish between dependencies of the type $v_0^4 - v_x^4 = ax$ and $v_0^2 - v_x^2 = ax$ (v_0 being the speed of incident rays, v_x the speed after passing through an absorber of thickness x). The apparatus, described in Figure 3.7, was quite modern in spirit and comparable to accelerator-based atomic or nuclear physics equipment (for example, LISE line at Ganil–Caen). The conclusion is that deceleration follows a law to the fourth power with a = 7.32×10^{42} for aluminum.

Understanding of how particles interact with matter took a major step forward with the Rutherford experiment, which demonstrated that the positively charged nucleus of the atom is concentrated in a very small volume, compared to the extension of the electron cloud surrounding it (Rutherford 1911). The first experiments to demonstrate the existence of wide-angle scattering of α particles were carried out in 1909 by Hans Geiger[3] and Ernest Marsden at Manchester University under the supervision of Ernest Rutherford. By changing the number of gold foils making up the target, they demonstrated that this reflection did not come from the surface but from the volume. Under the conditions of the device shown in Figure 3.8, one α particle in 8,000 was reflected[4]. They also established that the rate of reflected alphas increased with the atomic weight of the atoms in the reflector (Geiger and Marsden 1909). Eight months after his previous paper, E. Rutherford presented Hans Geiger's new results on the scattering of alphas by matter to the Royal Society of London (Geiger 1910b). For the first time, Geiger was able to measure the evolution of the angular distribution of alphas as a function of the thickness of gold through which they passed (Figure 3.9). He found, for example, that for a thickness of 0.4 μm, the average angle of deflection was 0.87°. At the same time, one alpha in 20,000 was scattered at 90°.

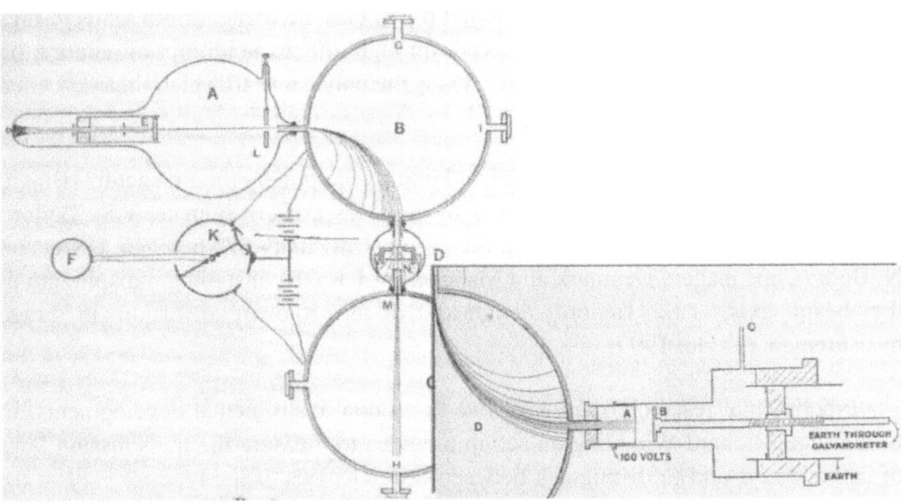

Figure 3.7. *Richard Whiddington's device for measuring the energy loss of cathode rays through metal foils. Cathode rays emitted by cathode E in chamber A are velocity selected in chamber B by the magnetic field created by a solenoid surrounding B. After passing through one of the metal foils placed on a translator, their velocity can be measured in C, a cylindrical chamber 10 cm in diameter, identical to chamber B. The insert on the right represents the detection system installed on chamber C (D in the insert figure notation) (Whiddington 1912)*

3 Hans Geiger is best known for co-inventing the Geiger–Müller counter with Walther Müller in 1928.

4 A small calculation taking into account the entire energy spectrum of α backscatter gives 1/3,000.

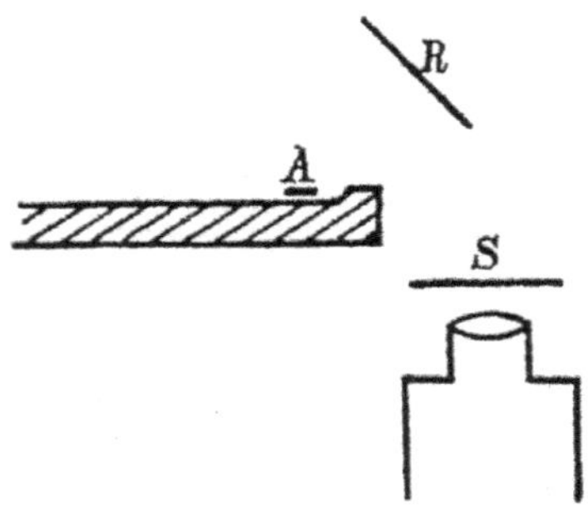

Figure 3.8. *Diagram of the Geiger and Marsden device for studying α reflected beyond 90°. A: α source, R: platinum reflector target, S: fluorescent screen with optical microscope below (Geiger and Marsden 1909)*

In his article *The scattering of α and β particles by matter and the structure of atoms*, E. Rutherford put these two results into perspective, pointing out that a simple probability calculation proves that the deflections of α particles at large angles cannot arise from a large number of small deflections. He therefore proposed considering theoretically that large deflections come from a simple collision. He posited that an atom contains +Ne charge at its center, assumed to be point-like, surrounded by an electrification sphere containing –Ne charge uniformly distributed in this sphere of radius R. He specified that the question of the atom's stability is not considered at this stage. In his calculation, he neglected the influence of the negatively charged sphere and assumed that the velocity of the particle was not significantly modified by its passage through the atom, but noted that "A discussion of these and similar cases is reserved until the question has been examined experimentally". Calculating the trajectory boils down to calculating that of a particle subjected to a repulsive force varying as the inverse square of the distance. From the conservation of angular momentum and energy, he derived the following formula for the angle of deflection: $\cot(\emptyset/2) = {}^{2p}/_{b}$ with p the impact parameter and b the minimum approach distance. Then, he calculated the angular distribution of α particles scattered by a thin metal sheet, the number of particles deflected by an angle ϕ was proportional to $\text{cosec}^4(\phi/2)$ or $1/\sin^4(\phi/2)$ and inversely proportional to the fourth power of the velocity $(mu^2)^2$, that is, the square of the energy. Theoretical and experimental results were in good agreement, as Rutherford pointed out: "Considering the difficulty of the experiments the agreement between theory and experiment is reasonably good" (Rutherford 1911).

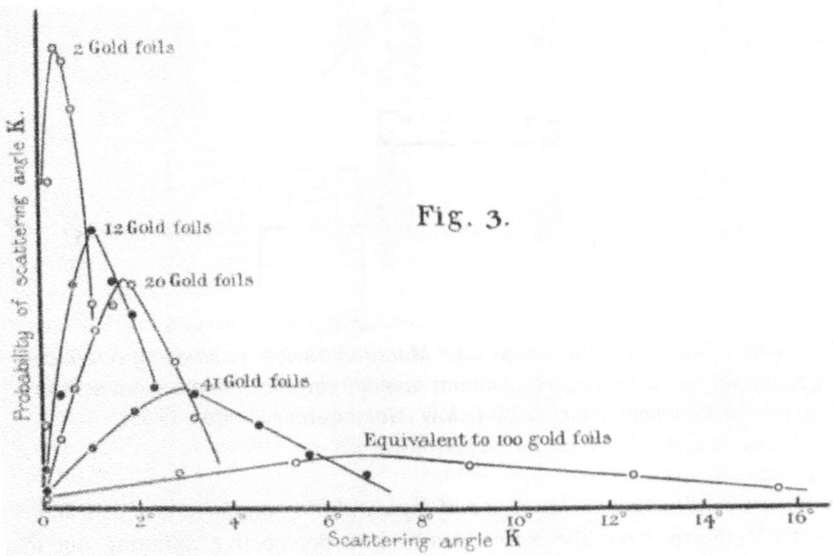

Figure 3.9. *First measurement of the evolution of the alphas scattering angle distribution for different gold foil thicknesses (Geiger 1910b)*

As for Joseph John Thomson, his 1912 paper ignored the discovery of wide-angle scattering and Rutherford's theory. He defended the idea that the atom was made up of N negatively charged corpuscles accompanied by an equal number of positively-charged quantities uniformly distributed throughout the sphere constituting the atom. In this case, wide-angle scattering was not possible by simple collision (Thomson 1912).

3.3. Particle stopping power

3.3.1. *Electronic stopping power*

Experiments with α and β particles have enabled us to gain a better understanding of their nature and properties, and opened the field to many new questions. Theorists took up these problems, particularly those concerning the slowing down by matter. One of the first to propose a theoretical approach was Niels Bohr. In 1913, he published a theory of the slowing down of charged particles based on a semi-classical approach and Rutherford's model of the atom (Bohr 1913).

He considered that the interaction with the nucleus was at the origin of angular deflection and that the decrease in velocity was due to the transfer of kinetic energy to electrons. This raised the question of the limit that needed to be introduced to prevent the stopping force from becoming infinite. As a limit, Bohr introduced the distance at which the interaction time became comparable to the electron vibration time. Developing the calculation further, he showed that the decrease in velocity was expressed as a function of the particle velocity (V) and the number (r) and frequency (n_s) of the electrons (equation 4 of Bohr (1913)):

$$\frac{dV}{dx} = -\frac{4\pi e^2 E^2 N}{mMV^3} \sum_{s=1}^{s=r} \log\left(\frac{V^3 kMm}{n_s eE(M+m)}\right)$$

Despite some questionable assumptions about the speed and frequency of certain electrons, which he considered small compared to the particle's speed and frequency, his calculation of the absorption of α rays in light elements agreed with the experiment, although this was less true for heavier elements. Neglecting interactions with the central charge of atoms, the theory corresponds to what is now called electronic stopping power.

A little over two years later, in a second article, Niels Bohr developed his theory (Bohr 1915). He introduced probability distributions for presence, collision, etc. He showed, for example, that for α rays passing through 5 cm of hydrogen gas, around half the particles suffer an energy loss that differs by less than 1% from the mean value, and less than 1% of particles suffer a loss that differs by more than 5%. He compared his calculation with experimental data, but the experiments disagreed with each other. He wrote: "Further experiments on this point would be very desirable".

In 1930, at the age of 24, Hans Bethe published his eighth paper as sole author. While his first three papers dealt with electron diffraction, this last one was a long article in which he developed a theory of inelastic collision between a fast particle and an atom (Bethe 1930). He treated these collisions in the formalism of Born wave mechanics. He justified this choice as follows:

> [...] wave mechanics immediately provide what we want, i.e. the statistics of the impact process, without having to first introduce an impact parameter beyond the physically significant quantities, as classical mechanics does, and then average all the values of this parameter (translation from German by the author).

Bethe used results from individual collisions to calculate the stopping power of fast particles in matter:

$$-\frac{dT}{dx} = NE = \frac{4\pi e^4 z^2 N}{mv^2} \sum_{nl} f_{nl} \ln \frac{(2)mv^2}{A_{nl}}$$

This formula is commonly referred to as *Bethe's formula* (for the meaning of the various symbols, see Bethe (1930, p. 328)).

Bethe ends his article with a section that makes the physical implications of his theory particularly clear. Two years later, he published a formula for the stopping of relativistic electrons. His work in Germany at the Munich Institute of Theoretical Physics was interrupted by the rise of the Nazi regime, and he emigrated first to Great Britain, then to the United States, where he helped design the first atomic bomb.

Several researchers, including F. Bloch, E. Fano, J. Lindhard and B. Firsov, developed other approaches to the slowing down of a particle in matter. In chronological order, they were as follows:

Felix Bloch, born in Switzerland, defended his doctoral thesis in 1928 at the University of Leipzig on the quantum theory of the solid state, with electrons described by waves now known as "Bloch waves". After his thesis, he remained at the Institut für Theoretische Physics der Universität, where he collaborated with Wermer Heisenberg, Wolfgang Pauli, Niels Bohr and Enrico Fermi, among others. Before leaving Germany for the United States in 1933, he published two papers on the stopping of fast particles in matter (Bloch 1933a, 1933b).

Bohr's and Bethe's theories are actually gas-phase collision theories, with the particle interacting with an isolated atom. In 1950, E. Fermi calculated the amount of energy lost by the particle at distances greater than the interatomic distances in a classical electrodynamic model. In such an approach, the field of a charged particle moving through a medium with dielectric properties is affected by the polarization of the medium. For small impact parameters, a quantum-mechanical correction must be introduced (Fermi 1940). It was Ugo Fano who made the connection between Bethe's and Fermi's theories (Fano 1956).

In his 1954 paper, Jens Lindhard of the Copenhagen School developed a quantum theory of particle transport. In this approach, a particle's motion is not determined by interaction with all other particles, but solely by a system of electromagnetic field equations (Lindhard 1954).

The field of electronic excitations in matter at low energies has been much less studied, probably because experimental data in solids are difficult to obtain. At low energies, this process is less efficient than elastic collisions and has little influence on the range. In 1959, Oleg Borisovich Firsov published a paper on low-energy atomic collisions in which he asserted that inelastic energy transfer between two atoms results from the formation of a quasi-molecule with electrons exchanged during the collision. Each electron transfers a momentum proportional to its relative velocity and mass (Firsov 1959). Two years later, Jens Lindhard and Morton Scharff[5] published an analytical formula for the electronic stopping cross-section at low speeds for $v < Z_1^{2/3} v_0$ (Lindhard and Scharff 1961).

In reality, in the field of materials under irradiation, these theoretical approaches have had little impact. Bethe's formula is an exception; it is sometimes used for high-energy experiments, where it gives the electronic stopping power with fairly good accuracy, but within a relatively restricted range. In just a few lines, we will look at the additions that have been made to extend its range of validity. The extension to high energies includes the relativistic effects $\ln\left(\frac{1}{1-\beta^2}\right) - \beta^2$, introduced by H. Bethe, and the density correction $-\delta$. At low energies, it is necessary to take into account the effective charge of ions Z_{eff} or Z_1^* and deep-layer correction $-\frac{C}{Z_2}$. Bethe's complete formula is therefore as follows :

$$S = \frac{4\pi e^2 Z_{eff}^2 Z_2}{mv^2}\left[ln\left(\frac{2mv^2}{I}\right) + ln\left(\frac{1}{1-\beta^2}\right) - \beta^2 - \frac{C}{Z_2} - \delta\right]:$$

In a calculation that takes into account polarization effects in the case of large impact parameters, Enrico Fermi showed that, in liquids and solids, energy loss depends on the dielectric properties of the medium. This effect becomes non-negligible when the particle velocity exceeds c/n, where n is the square root of the static dielectric constant (Fermi 1940). This calculation was taken up a few years later by Otto Halpern and Harvey Hall (Halpern and Hall 1948). The main difference lies in the assumed frequency dependence of the dielectric constant. In contrast to E. Fermi, O. Halpern and H. Hall considered that the reduction in losses caused by the dielectric properties of condensed materials depended very critically on the shape of the dielectric constant as a function of frequency. Both articles include an example of energy loss of mesotron in matter, an intermediate particle between the electron and the proton, currently known as a muon.

5 At the time of publication, Morton Scharff had just passed away.

The notion of effective charge was proposed by Niels Bohr in a discussion of the range of fission fragments (Bohr 1941). A few years later, he gave it a particularly simple expression: $Z_1^* \sim Z_1^{1/3} \frac{v}{v_0}$ for $v_0 < v < Z_1^{1/3} v_0$ (Bohr 1948). Also for detection purposes, Walter H. Barkas[6] proposed an empirical formula for protons: $\left(\frac{Z}{Z_{eff}}\right)^2 \approx 1 + 1.3 \times 10^{-5} \frac{Z^{5/3}}{\beta^{7/3}}$ (Barkas and Berger 1964). The aim was to improve detection of ^8Li and ^8B in proton-on-carbon nuclear reaction experiments at Berkeley's 184-inch cyclotron. The distinction between these two atoms was made by measuring the radius of curvature and the length of the traces they left in the emulsions. For this, a good knowledge of the range–energy relationship was required. Considering that at low speeds a positive particle tends to be neutralized by electrons, W.H. Barkas introduced a speed $\beta_0 c$ below which ions were no longer completely stripped (Barkas 1953). Below this speed, he gave an empirical formula for the variation in charge for fast heavy ions: $Z_{eff} = Z\left(1 - \exp\left(-125\beta Z^{2/3}\right)\right)$.

An inner-layer correction, sometimes referred to as "tight-binding", is necessary when the speed of the particle is not greater than that of the electrons in the medium through which it passes. This correction is of the form $-{}^{C}/_{Z}$ with $C = C_K + C_L + \cdots$. A detailed description is given by Ugo Fano of the National Bureau of Standards in Washington, D.C. (Fano 1963).

3.3.2. *Nuclear stopping power*

Nuclear stopping power corresponds to the energy transferred to the target nuclei in the form of kinetic energy, and it is often referred to as elastic stopping power.

The story of nuclear stopping power began, of course, with E. Rutherford's discovery of the nucleus. Then came the work of N. Bohr. This work was taken up again some 15 years later, first by Gregor Wentzel in Leipzig (Wentzel 1926) and Julius Robert Oppenheimer in Göttingen (Oppenheimer 1927).

In 1928, Nevill Francis Mott published a solution for particle diffusion by a central Coulombian force (Mott 1928). The solution to the equation $\nabla^2\phi + \frac{2m}{h^2}(E - V)\phi = 0$ makes it possible to determine the probability that a particle can be scattered in a given solid angle. The following year, he demonstrated near-perfect

6 Lawrence Radiation Laboratory, University of California, Berkeley.

agreement between experimental results and his wave-mechanical calculation of elastic electron scattering by helium (Figure 3.10) (Mott 1929).

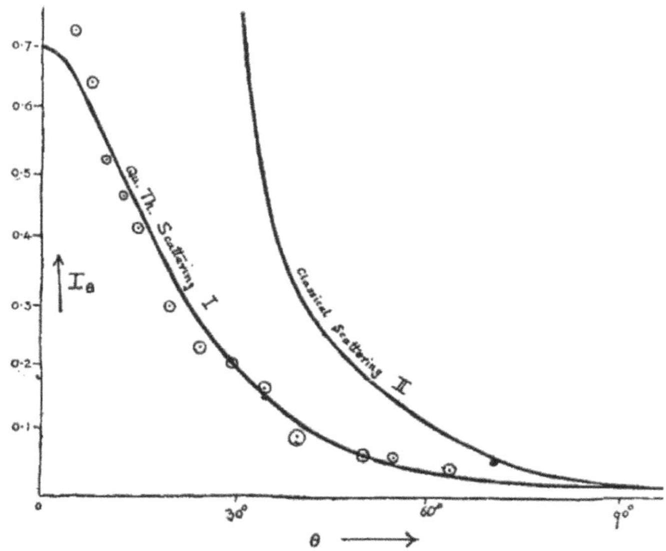

Figure 3.10. *Elastic scattering of 210 eV electrons by helium (Mott 1929). Curve I: quantum calculation; curve II: classical calculation, circles: experimental results by E.G. Dymond and E.E. Watson (Dymond et al. 1929)*

The disadvantage of a coulombic potential is its divergence as the distance tends toward zero; so, many researchers have proposed truncations or other forms of potential to get around this problem. It was Llewellyn Thomas from Trinity College, Dublin and Enrico Fermi from Rome University who first independently proposed a calculation of the electric field inside an atom. The potential is determined by the charge of the nucleus and the distribution of electrons. For L. Thomas, electrons are uniformly distributed in a phase space for six-dimensional motion, each point of which can be occupied by two electrons of different spin. The effective field at each point depends on its occupation by electrons (Thomas 1927). As for E. Fermi, he calculated the electron distribution based on a statistical approach to an electron gas obeying Pauli's exclusion principle (Fermi 1927). The Thomas–Fermi model thus consists of determining an electron density $\rho(r)$ around the nucleus that minimizes the total energy of the system.

In 1946, Von Gert Moliere of the Kaiser–Wilhelm Institut für Physik in Hechingen set out to solve the problem of alphas scattering as analytically as possible (Moliere 1947).

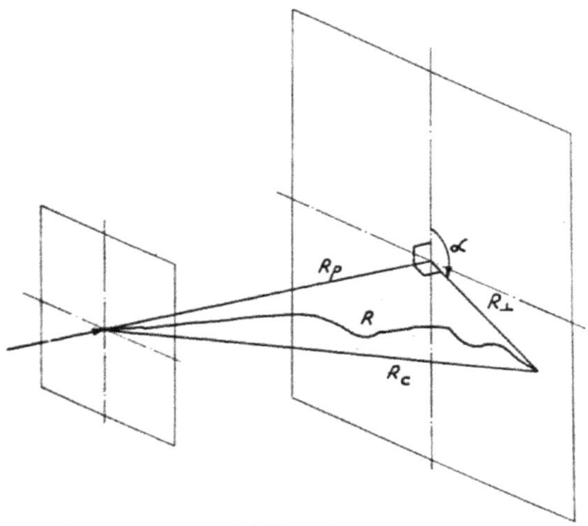

Figure 3.11. *Schematic representation of the different range concepts (Lindhard et al. 1963)*

In 1948, Niels Bohr published a very long article entitled "The penetration of atomic particles through matter" (Bohr 1948). The first chapter deals with the problem of the collision of two point charges. Starting from Rutherford's classical approach, he developed a quantum-mechanical approach, proposing the use of a coulombic interaction field:

$P_a(r) = \frac{e_1 e_2}{r} e^{-\frac{r}{a}}$ where a is the screening parameter.

The following sections deal with particle transport in matter, that is, phenomena that depend on nuclear collisions, ionizations and atomic excitations, electron capture and loss during collisions, and finally, the relationship between range and velocity.

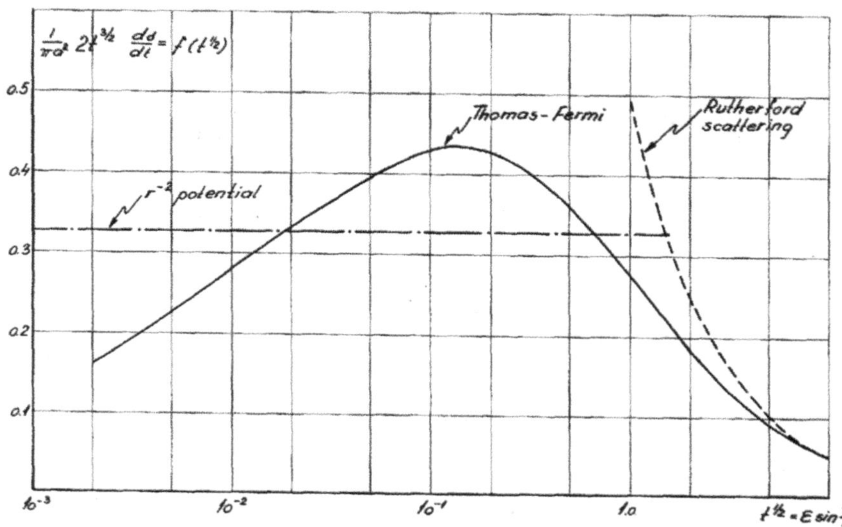

Figure 3.12. *Universal differential scattering cross-section for elastic collisions, based on a Thomas–Fermi-type potential (Lindhard et al. 1963)*

3.4. Particle range

In 1963, J. Lindhard and his collaborators, M. Scharff and H.E. Schiott, published a landmark paper on the concept of range and the ranges of heavy ions (Lindhard et al. 1963). In this article, they discuss the trajectory of heavy ions of moderate velocity, that is, in a region where electronic and nuclear stopping powers are in competition. To this end, they use a theoretical approach to quasi-elastic collisions: a simple Thomas–Fermi formula proportional to velocity for electronic stopping power and a universal formula describing the differential cross-section for angular scattering (Figure 3.12). Comparisons with range measurements show very good agreement for the projected range at low energies and in the energy range of fission fragments.

Readers interested in the state of knowledge in 1964 on the transport of charged particles in matter can refer to the report *Studies in Penetration of Charged Particles in Matter* from the National Research Council (1964). This report brings together contributions from the leading specialists of the time, such as W.H. Barkas, H. Bichsel, U. Fano, J. Lindhard and L.C. Northcliffe.

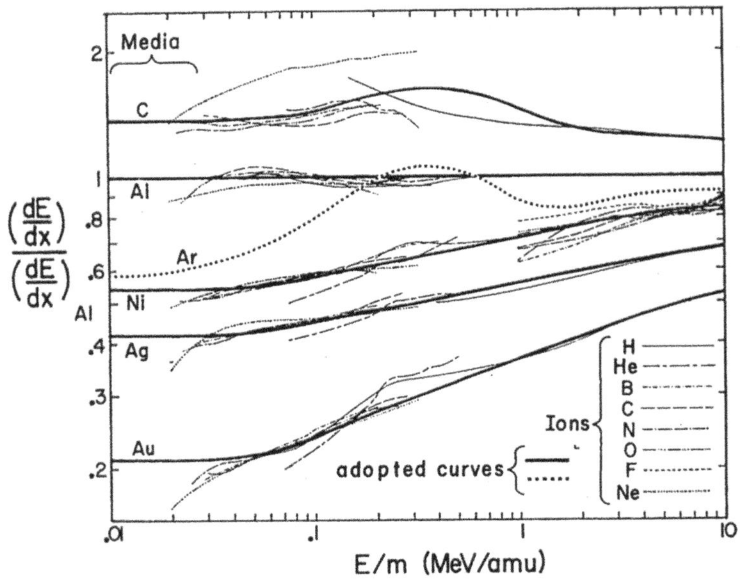

Figure 3.13. *Stopping power of some targets normalized to that of aluminum for different ions. Experimental results are represented by thin lines, and thick lines correspond to the adopted curves (Northcliffe and Schilling 1970)*

3.5. Transport simulation

The search for scaling laws enabling the stopping power of an ion in a material to be obtained from that of another ion was active in the 1940s. The most widely used empirical relationship is that between the electronic stopping power of a heavy ion $S_{HI}(v, Z_2)$ to that of the proton $S_p(v, Z_2)$ at the same speed (Ziegler and Biersack 1985):

$\frac{S_{HI}(v,Z_2)}{[Z_{HI}^*(v)]^2} = \frac{S_p(v,Z_2)}{[Z_p^*(v)]^2}$ with Z_{HI}^* the speed-dependent effective charge.

Before the widespread use of computers, it was common practice to consult data tables, but who remembers the logarithm tables of Bouvard and Ratinet in France, Briggs in the UK or Glogauer in Germany? So, it was handy to have a compilation of data on energy losses and the range of the most common particles. Most of these tables are based on empirical equations fitted to experimental data. This is the case for the first of these tables, that of Ward Whaling, professor at *Caltech's Kellogg Radiation Laboratory*, whose data are essentially experimental and extrapolated to high energies with Bethe's relation $S \sim (\ln E)/E$ (Whaling 1958). Then came the more precise and comprehensive table by L.C. Northcliffe and R.F. Schilling, both working at the Cyclotron Institute of Texas A&M University (Northcliffe and Schilling 1970). Their electronic stopping power tables are based on subtle mixtures of experimental results, interpolation and theory. The stopping power of many ions in aluminum is used as a reference (Figure 3.13). The choice of aluminum stems from the fact that it is a metal that can easily be rolled into a homogeneous thin sheet without pinholes. At low energies, the LSS theory[7] is used. In this publication, the electronic stopping power and range of all ions from hydrogen to lawrencium in 24 targets, including two polymers and water, are tabulated for 38 energies between 0.0125 and 12 MeV/amu.

The 1970s saw a sharp increase in the number of implantation experiments. The number of papers with the keyword "ion implantation" in their title increased by a factor of 7.8 between 1965 and 1980, and the number of implanters in industry rose from 10 to 1000 over the same period. At the same time, ion-beam analysis was gaining in importance. These ion beam applications require reliable data on stopping powers and ranges, but at the same time, they contribute to the enrichment of databases. J.F. Zieger and W.K. Chu's table is aimed at this ion beam analysis community. Using a semi-empirical approach, it gives stopping cross-sections for ^{4}He of energy between 0.4 and 4 MeV in all elements ($1 \leq Z \leq 92$) (Ziegler and Chu 1974).

In 1977, H.H. Andersen and J.F. Ziegler published a book on the stopping power of the proton in all elements (Andersen and Ziegler 1977). Electronic stopping power is based on two theoretical approaches: at low energies, a linear behavior with projectile velocity (Lindhard and Scharff 1961) and at high energy, a variation described by Bethe's formula, including relativistic and deep-layer effects (Fano 1963). The behavior between these two regimes is described by an interpolation formula of variable complexity. Curves are fitted to experimental points, where available.

Then, in 1978, J.F. Ziegler published the book *Helium: Stopping and Range in All Elemental Matter* (Ziegler 1977), followed by *The Stopping and Range of Ions in Matter* (Ziegler and Biersack 1985).

For electrons, O.S. Oen's table gives the displacement cross-sections in solids (Oen 1965). Less well known are the tables by L. Pages and colleagues from the Centre d'Études Nucléaires de Saclay (Pages et al. 1972).

---

7 LSS: J. Lindhard, M. Scharff, H.E. Schiott.

Figure 3.14. *Three volumes in the collection "The Stopping and Ranges of Ions in Matter" edited by J.F. Zeigler. For a color version of this figure, see www.iste.co.uk/bouffard/irradiation.zip*

3.5.1. *First simulations*

The emergence of simulation methods owes a great deal to the Manhattan Project, not for irradiation purposes, but for determining the critical mass and geometry required to build an atomic bomb. The problem was to predict the fraction of neutrons involved in fission reactions. The development of these methods was closely linked to the development of computing resources, and at the same time has been the driving force behind them. In this field, the Los Alamos Laboratory was at the forefront, starting in 1942 with the purchase of mechanical calculators to support theoretical work. The use of these mechanical calculators made it possible to solve neutron scattering numerically. Solving these kinds of complex physical problems was the task of an almost exclusively female group, many of whom were the wives of Los Alamos scientists, using desktop calculators. Each person calculated segments of a calculation by hand and passed on the value obtained to the next person to continue the calculation. However, the idea of using punch-card accounting machines (PCAMs) for scientific calculations was soon born. The first successful application was an astronomical calculation at Colombia University. Inspired by this success, Los Alamos placed an order for:

– three IBM 601 multipliers;

– an IBM 402 tabulator;

– various devices for perforating, sorting, etc., punched cards.

The IBM 601 was the first IBM machine capable of multiplication. To do this, it was a machine capable of reading two numbers from a punched card and punching their product into a blank field on the same card. The 601 was introduced in 1931. The first calculations were devoted to simulating the implosion phase of an atomic bomb. In 1944, IBM agreed to build a more powerful machine that could perform triple multiplication A × B × C and, above all, division. At the same time, at Harvard, Howard H. Aiken worked with IBM to develop the Mark I electromechanical relay calculator, which Von Neumann used for calculations he was able to have declassified. At the end of the war, the ENIAC was built at the University of Pennsylvania. These beginnings of computer science are well documented in Metropolis and Nelson (1982), Fitzpatrick (1999) and Archer (2021).

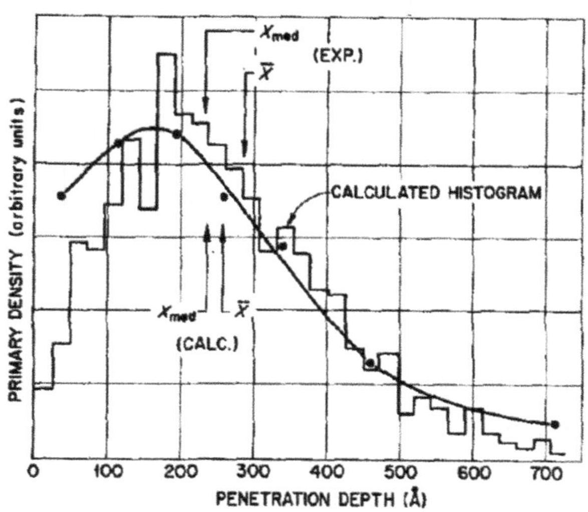

Figure 3.15. *Comparison of simulated range distribution (Oen et al. 1963) with experimental measurements (Davies and Sims 1961) for 10.5 keV Na^{24} in Al*

The idea of the Monte Carlo method is credited to Stanislaw Ulam, a mathematician born in Poland but working first at Princeton and then at Los Alamos during the Second World War[8]. In 1946, he shared his idea with John von Neumann, who together with R.D. Richtmyer described the first method for solving the neutron scattering problem (Von Neumann and Richtmyer 1947). The estimated time needed to track 100 neutrons, each undergoing 1,000 collisions, was five hours on ENIAC, the first all-electronic computer. The first neutron scattering calculation was

8 It seems that Enrico Fermi had the idea for this statistical approach years earlier, without publishing it or exploiting it.

performed in mid-1948. The MCS code written in FORTRAN (1963) was the first real code for particle transport in matter. Various improvements, including geometry processing and the extension of the cross-section base, led to a new version in 1965 (MCN). MCN and MCG, the photon code, merged in 1973 (MCNG) and became MCNP in 1977.

Transporting ions in the range from a few hundred eV to a hundred thousand eV is a difficult problem, due to the large angular deflections caused by the elastic collisions that dominate the slowdown. To overcome this problem, O.S. Oen, D.K. Holmes and M.T. Robinson of Oak Ridge National Laboratory, Solid State Division, used the Monte Carlo technique to calculate the range of ions from 1 to 100 keV (Oen et al. 1963)[9]. In this calculation, the lattice was not taken into account and binary collisions were assumed to be independent, which meant that the impact parameter had to be less than half the distance between two atoms. The interatomic potential is a shielded coulombic potential (Bohr potential), and a comparison was made with the potential of hard spheres. Although the electronic stopping power was neglected, at these low energies the calculated ranges and range distributions were comparable to those measured (Figure 3.15). The calculation was programmed on an IBM 7090 computer, one of the first transistor-based scientific computers. The authors state: "...the present calculation rests on the use of high-speed computing machines...".

Two of the authors, M.T. Robinson and O.S. Oen, are the creators of the Monte Carlo code MARLOWE, in which the crystalline organization of atoms is taken into account. In a study of the slowing down of 10 keV Cu atoms in copper, they found that around 1% of the atoms had much longer ranges. These particles undergo numerous grazing collisions that guide them along a crystallographic direction: the particles are "channeled"[10] (Robinson and Oen 1963a, 1963b).

3.5.1.1. *SRIM (Stopping and Range of Ions in Matter) software*

Designed to run on PCs using the DOS (Disk Operating System) operating system, TRIM-85 (Transport of Ions in Matter) saw the light of day in the summer of 1985. From version to version, the software evolved and improved, but its old-fashioned but highly recognizable interface remained virtually unchanged. The DOS version was replaced by a Windows version in 1989. In 2024, the SRIM-2013 version was released.

9 This article was preceded by the ORNL report no. 3213 (1961).

10 The history of the channeling is described in section 3.6.

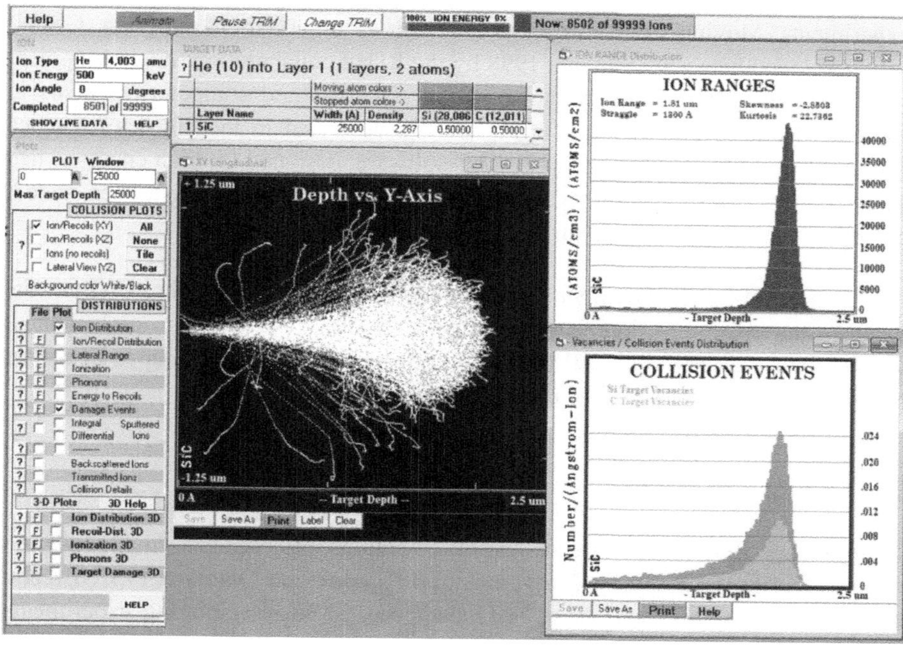

Figure 3.16. *Screenshot of SRIM-2013 software: calculation of 8,500 helium trajectories at 500 keV in SiC. For a color version of this figure, see www.iste.co.uk/bouffard/irradiation.zip*

Its operation is relatively straightforward. Between two elastic collisions, the ions have a mean free path that depends on the total cross-section of the elastic collision. At the point of collision, the energy transferred and the scattering angles are randomly selected according to the differential cross-sections. If the target atom receives an energy greater than the displacement threshold energy, it is considered displaced, and if it has sufficient kinetic energy, its trajectory is followed using the same formalism as for projectile ions. Between two collisions, the ions undergo continuous deceleration due to electronic excitations.

3.5.1.2. *Geant (Geometry And Tracking)*

Geant is a toolbox for simulating the passage of particles through matter. The first version of Geant dates back to 1974, with the aim of describing the trajectory of different types of particles in detectors (typically those installed at CERN). In 1982, Geant3 was released, and was mainly used for high-energy physics experiments. Moving from Fortran to C++ with an object-oriented design, Geant4 was developed between 1994 and 1998 by the RD44 collaboration. Its development was based on a worldwide collaboration of scientists and computer scientists in some 15 working groups. Particle transport is described in the physics reference manual (Geant4 Collaboration 2017). This software, designed for high energies, had a low particle tracking limit of 10 keV, but various processes have been added, extending the energy range down to 1 eV.

As part of an international collaboration coordinated by the CNBG, the mechanistic effects of ionizing radiation on biology have been implemented: Geant4-DNA (Incerti et al. 2018). Apart from its use in hadrontherapy, Geant4 is rarely used in the community of materials under irradiation.

3.6. Channeling effects

In the late 1950s, M.M. Bredov and his co-authors observed abnormally long ion ranges for 4 keV cesium ions in germanium (Bredov and Okuneva 1957; Bredov et al. 1958).[11] A few years later, in Gothenburg, Sweden, sputtering experiments with 45 keV krypton ions on copper single crystals showed that the sputtering rate was much lower when the beam direction coincided with a dense sample direction (Almén and Bruce 1961). They attempted, with reservations, to explain this effect by a greater penetration depth in directions of crystal transparency. At virtually the same time, a team from Atomic Energy of Canada Limited, Chalk River, Ontario, reported the detection of ions whose path was much longer than the average range (Davies et al. 1960). The technique used involved implanting a radioactive ion (^{137}Cs) in aluminum, then determining the range distribution by measuring radioactivity as a function of depth by successively dissolving very thin layers of aluminum. Figure 3.17 shows that the ion distribution is highly asymmetric. The median range (~11.4 μg/cm^2) is greater than the most probable range (~8.9 μg/cm^2). The authors point out that the existence of this tail in the range distribution is characteristic of studies in this energy range. Their interpretation is based on the assumption that penetration depth is primarily determined by the mean free path between first collisions. In reality, is it an effect of surface roughness, dissolution inhomogeneity or, as M.T. Robinson and O.S. Oen state, of ions that have undergone fewer collisions because they propagate under channeled conditions?

11 According to Davies et al. (1962). *Canadian Journal of Chemistry*, 40(1962), 1605.

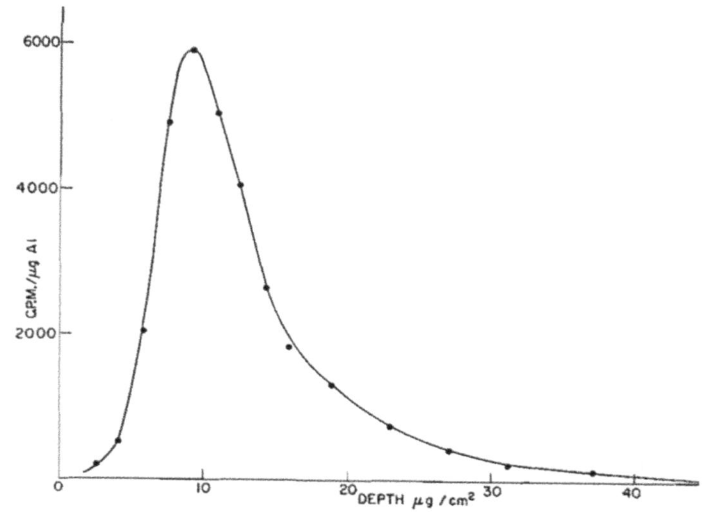

Figure 3.17. *Penetration depth of 50 keV ^{137}Cs in aluminum. The asymmetrical nature of the distribution is due to the channeling phenomenon which, by reducing the stopping power of the ions, allows for longer path (Davies et al. 1960)*

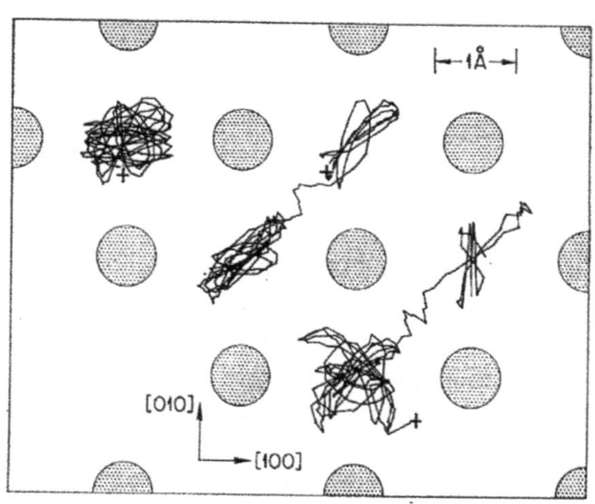

Figure 3.18. *Projection on the (001) plane of three 5 keV Cu ion trajectories channeled in the [001] direction of a copper crystal (Robinson and Oen 1963b)*

Continuing to improve their simulation of atomic transport in solids, M.T. Robinson and O.S. Oen introduced the position of atoms at lattice sites. In 1963, they published two papers on the slowing down in copper of copper atoms from

1 to 10 keV, the first in January in *Applied Physics Letters* (Robinson and Oen 1963a) and the second, more detailed, in December in *Physical Review* (Robinson and Oen 1963b). They observed that around 1% of 1 keV atoms have very long ranges. These atoms have their trajectory aligned with the [110] direction. These "channeled" particles undergo numerous grazing–incidence collisions with lattice atoms. In a face-centered cubic material, the direction most favorable to this channeling phenomenon is [011], followed by [001] and [111]. In a face-centered cubic material, on the other hand, the order is [111], [001] and [011]. It is therefore the rows with the most compact stacking of atoms that are most conducive to channeling. Figure 3.18 shows three projections on the (001) plane of channeled atom trajectories in the [001] direction.

The explanation of long trajectories by the channeling phenomenon has revived experiments in this field, notably at Chalk River Nuclear Laboratories. The role of crystalline arrangement is corroborated by the fact that the range distribution of ^{85}Kr, a β^- emitter, implanted in amorphous alumina does not exhibit the exponential tail found in crystalline metals. Furthermore, experiments on aluminum single crystals oriented along the three main crystallographic axes confirm that the [110] direction is the most favorable for channeling (Piercy et al. 1963). At the same time, H. Lutz and R. Sizmann (München, Germany) carried out the same experiment on copper and found essentially the same result (Lutz and Sizmann 1963). The surface erosion technique differed: electrolytic erosion for the Canadians and sputtering for the Germans.

At the same time, more direct observation of channeling by transmission and reflection experiments was published by R.S. Nelson and M.W. Thomson from Harwell (Nelson and Thompson 1963). Two excellent experiments are presented. The first involved measuring the transmission of a 75 keV proton beam through a gold foil with a thickness of between 3,000 and 4,000 Å. The gold foil was oriented in such a way that, as the foil rotated, the axes [011] were successively aligned with the direction of the beam. Figure 3.19(a) shows the increase in transmitted current when the beam was aligned with an axis [110]. Conversely, the number of reflected ions decreased when the beam was in a channeled condition (Figure 3.19(b)).

Disorder dechanneling experiments began a few years after the discovery of particle channeling. The experiment that was certainly the first was carried out in Canada by a team from Chalk River Nuclear Laboratories. They studied the effect of thermal agitation on the range under channeling conditions. To do so, they measured the range distribution of ^{133}Xe$^+$ 40 keV ions in the [100] direction of gold at 293 K and 20 K. The effect of temperature was very marked: the maximum penetration depth of the ions was three times greater at 20 K than at room temperature, from 1800 Å at 293 K to 5500 Å at 20 K (Channing and Whitton 1964).

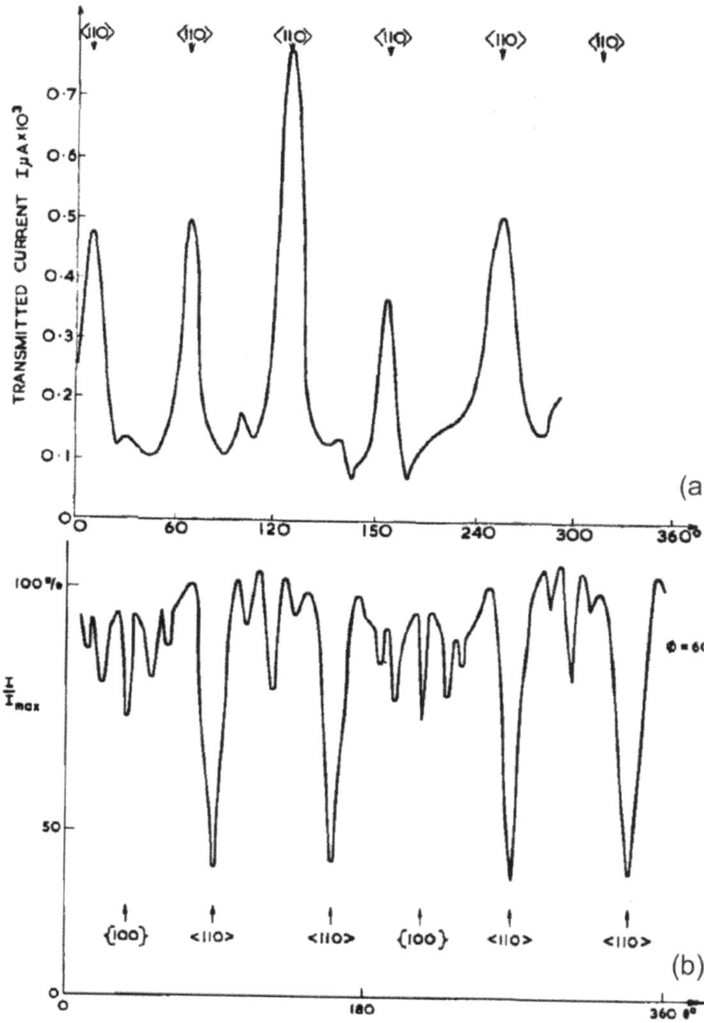

Figure 3.19. *(a) Transmission of 75 keV protons through a gold crystal as a function of the angular position of the foil. The gold foil rotates around its normal, which coincides with the [111] direction, and the beam is inclined at 35° to the normal. (b) Reflection of 50 keV Ne^+ ions on a gold single crystal oriented [110] rotating around this axis, with the ion beam inclined at 60° (Nelson and Thompson 1963)*

In 1965, B. Domeij and K. Björkqvist of Stockholm's Research Institute of Physics demonstrated the opposite of channeling: blocking. When an ion is emitted from a lattice site along a crystalline direction, there is a high probability that it will collide with the first atom in that row. As this collision deflects it by a large angle, fewer ions will be detected in dense crystalline directions. The Swedish group carried out the experiment by implanting ^{222}Rn, an α emitter, in tungsten. They observed a decrease in α counted in certain directions, indicating that radon was probably in a substitutional position in tungsten (Domeij and Björkqvist 1965). With a few months' difference, at Argonne National Laboratory, D.S. Gemmel and R.E. Holland made a similar observation, but for protons, (d,p) reactions in silicon (Gemmell and Holland 1965).

Soon after its discovery, channeling became a tool for characterizing crystalline structures. Its most frequent application was certainly the study of defects in crystal lattices. It was as early as 1967 that the Chalk River group demonstrated the advantages of 1 MeV helium ions over protons for backscattering studies under channeling conditions, based on experiments involving the implantation of antimony in silicon: improved depth resolution, excellent atomic mass separation, greater sensitivity for the detection of heavy atoms and a high counting rate. The quantities obtained were of several types. Around 3,000 silicon atoms were moved around, each 40 keV Sb ion trajectory. Antimony atoms were 95% substituted (Davies et al. 1967).

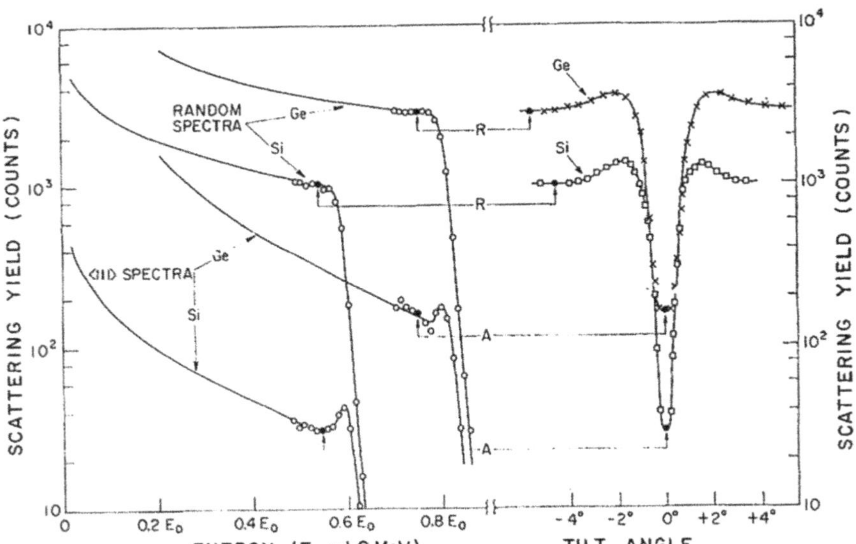

Figure 3.20. *Left: backscattering spectra in random and aligned [111] directions for Si and Ge crystals. Right: angular rotations around the [111] axis (Mayer et al. 1968)*

In fact, all deviations from crystalline order can be studied: location of impurities, measurements of elastic deformations, thermal vibrations, etc. For more recent developments in channeling experiments and applications, please refer to

André Vantomme's article entitled *50 years of ion channeling in materials science* (Vantomme 2016).

3.7. Chapter 3 references

3.7.1. *Short chronology*

Date	Person	Advances
1895	Röntgen (*1901 Nobel Prize in Physics*)	X-ray discovery
1896	Becquerel (*1903 Nobel Prize in Physics*)	Radioactivity discovery
1899	Rutherford	Becquerel radiation: alpha-beta
1905	Bragg (*1915 Nobel Prize in Physics*)	Ionization curve
1906	Thomson (*1906 Nobel Prize in Physics*)	Book "Conduction of Electricity through Gases"
1909	Geiger and Marsden	Backscatter experiment
1911	Rutherford (*1908 Nobel Prize in Chemistry*)	Nucleus atom
1912	Whiddington	Cathode ray transmission
1913	Bohr (*1922 Nobel Prize in Physics*)	Ion-matter interaction theory
1929	Mott	Elastic scattering
1930	Bethe (*1967 Nobel Prize in Physics*)	Electronic stopping power
1933	Bloch	Electronic stopping power
1963	Lindhard	The concept of a journey
1963		MCS code
1963	Robinson, Oen	Monte Carlo code MARLOWE
1965	Oen	Effective electron cross sections
1970	Northcliffe, Schilling	Stopping power of ions

Table 3.2. *A brief chronology*

3.7.2. *Biographies of some of the chapter's personalities*

Many researchers have contributed to the advancement of knowledge in particle transport. Below are brief biographies of four of them, arbitrarily chosen.

Ernest Rutherford (1871 Spring Grove, New Zealand–1937 London) was awarded an 1851 Exhibition Science Scholarship in 1894 to continue his work in England. J.J. Thomson, future discoverer of the electron, invited him to work at the Cavendish Laboratory in Cambridge. Studying the radiation emitted by uranium, he discovered that it was composed of two types of radiation, and proposed to call them α and β. In 1898, he moved to Canada to work at McGill University, Montreal. He discovered that thorium compounds continuously emit radioactive particles, which he called "emanation". Emanation became radon in 1923. More importantly, he demonstrated the existence of the nucleus of atoms containing positive charges. He was awarded the Nobel Prize for Chemistry for this.

Figure 3.21. *Ernest Rutherford, Nobel Prize in Chemistry 1908. Photo credit: Nobel Foundation archive*

William Henry Bragg (1862, Wigton, UK–1942, London, UK) graduated in 1886 as Professor of Mathematics at the University of Adelaide (Australia). However, it was only later that he began to work in physics. In 1904, his involvement in radioactivity and radiation began with the publication of two articles in *Philosophical Magazine* on the ionization curves of radium alphas. The ionization peak observed became known as the Bragg peak. In 1908, he returned to England, to Leeds University, where he worked with his son William Lawrence on the characterization of materials with X-rays. For this work, they were jointly awarded the Nobel Prize in Physics in 1915. From 1935 to 1940, he was President of the Royal Society (Jenkin 2008).

Hans Albrecht Bethe (1906, Strasbourg, German Empire–2005, Ithaca, USA) left Germany in 1933 for England, then in 1935 for the United States, where he obtained a position as assistant professor at Cornell University (New York State). In 1938, he published a paper with Charles Critchfield suggesting that the energy of stars comes from nuclear fusion. He was awarded the Nobel Prize in Physics in 1967 for his work on nuclear reactions and energy production in stars.

Figure 3.22. *Hans Albrecht Bethe, Nobel Prize in Physics 1967. Photo credit: Nobel Foundation archive*

Jens Lindhard (1922, Tystofle, Denmark–1997, Denmark). His early work focused on electrons in solids. At the age of 26, he published a paper on the theory of static electricity and magnetic susceptibility for rapidly varying fields in space. In the early 1950s, he worked closely with Niels Bohr on the penetration of particles into matter. With Morten Scharff and H.E. Schiott, he developed a theory of low-energy ion transport in solids, commonly known as LSS theory.

3.8. References

Almén, O. and Bruce, G. (1961). Collection and sputtering experiments with noble gas ions. *Nuclear Instruments & Methods in Physics Research,* 11, 257–278.

Andersen, H.H. and Ziegler, J.F. (1977). *Hydrogen Stopping Powers and Ranges in All Elements*. Pergamon Press, New York.

Archer, B.J. (2021). The Los Alamos computing facility during the Manhattan project. *Nuclear Technology*, 207(sup1), S190–S203.

Barkas, W.H. (1953). The range correction for electron pick-up. *Physical Review*, 89(5), 1019–1022.

Barkas, W.H. and Berger, M.J. (1964). Tables of energy losses and ranges of heavy charged particles. In *Studies in Penetration of Charged Particles in Matter – Nuclear Science Series*, National Research Council (ed.). The National Academic Press, Washington, D.C.

Becquerel, H. (1906). On some properties of the α rays emitted by radium and by bodies rendered active by the radium emanation. *The London, Edinburgh and Dublin Philosophical Magazine and Journal of Science (6th series)*, XI, 722–728.

Bethe, H.A. (1930). Zur Theorie des Durchgangs schneller Korpuskularstrahlen durch Materie (On the theory of the passage of fast corpuscular rays through matter). *Annalen der Physik*, 397(3), 325–400.

Bloch, F. (1933a). Zur Bremsung rasch bewegter Teilchen beim Durchgang durch Materie (For the stopping of fast-moving particles as they pass through matter). *Annalen der Physik*, 408(3), 285–320.

Bloch, F. (1933b). Bremsvermögen von Atomen mit mehreren Elektronen (Stopping power of atoms with several electrons). *Zeitschrift für Physik*, 81(5), 363–376.

Bohr, N. (1913). On the theory of the decrease of velocity of moving electrified particles on passing through matter. *The London, Edinburgh and Dublin Philosophical Magazine and Journal of Science (6th series)*, XXV, 10–31.

Bohr, N. (1915). On the decrease of velocity of swiftly moving electrified particles in passing through matter. *The London, Edinburgh and Dublin Philosophical Magazine and Journal of Science (6th series)*, XXX, 10–31.

Bohr, N. (1941). Velocity-range relation for fission fragments. *Physical Review*, 59(3), 270–275.

Bohr, N. (1948). The penetration of atomic particles through matter. *Det Kongelige Danske Videnskabernes Selskab. Matematisk-fysiske Meddelelser*, XVIII(8), 1–144.

Bragg, W.H. (1904). On the absorption of alpha rays, and on the classification of the alpha rays from radium. *The London, Edinburgh and Dublin Philosophical Magazine and Journal of Science (6th series)*, VIII(XLVIII), 719–725.

Bragg, W.H. (1907). The influence of the velocity of the a particle upon the stopping power of the substance through which it passes. *The London, Edinburgh and Dublin Philosophical Magazine and Journal of Science (6th series)*, XIII(LXXVI), 507–516.

Bragg, W.H. and Kleeman, R.D. (1904). On the ionization curves of radium. *The London, Edinburgh and Dublin Philosophical Magazine and Journal of Science (6th series)*, VIII(XLVIII), 726–738.

Bragg, W.H. and Kleeman, R.D. (1905). On the a particles of radium, and their loss of range in passing through various atoms and molecules. *The London, Edinburgh and Dublin Philosophical Magazine and Journal of Science (6th series)*, X(LVII), 318–340.

Bredov, M.M. and Okuneva, N.M. (1957). The depth of penetration of matter by moderate energy ions. *Doklady Akademii Nauk Sssr*, 113(4), 795–796.

Bredov, M.M., Lang, I.G., Okuneva, N.M. (1958). On the problem of penetration of medium-energy opns into matter. *Soviet Physics-Technical Physics,* 3(2), 228–229.

Channing, D.A. and Whitton J.L. (1964). Effect of temperature on the channeling of Xe^{133} ions in gold. *Physics Letters*, 13(1), 27–28.

Davies, J.A. and Sims, G.A. (1961). The range of Na^{24} ions of kiloelectron volt energies in aluminium. *Canadian Journal of Chemistry*, 39(3), 601–610.

Davies, J.A., McIntyre, J.D., Cushing, R.L., Lounsbury, M. (1960). The range of alkali metal ions of kiloelectron volt energies in aluminum. *Canadian Journal of Chemistry*, 38(9), 1535–1546.

Davies, J.A., Denhartog, J., Eriksson, L., Mayer, J.W. (1967). Ion implantation of silicon: I. atom location and lattice disorder by means of 1.0-MeV helium ion scattering. *Canadian Journal of Physics,* 45(12), 4053–4071.

Domeij, B. and Björkqvist, K. (1965). Anisotropic emission of α-particles from a monocrystalline source. *Physics Letters*, 14(2), 127–128.

Dymond, E.G., Watson, E.E., Rutherford, E. (1929). Electron scattering in helium. *Proceedings of the Royal Society of London. Series A,* 122(790), 571–582.

Fano, U. (1956). Atomic theory of electromagnetic interactions in dense materials. *Physical Review,* 103(5), 1202–1218.

Fano, U. (1963). Penetration of protons, alpha particles and mesons. *Annual Review of Nuclear Science*, 13, 1–66.

Fermi, E. (1927). Statistical method to determine some properties of atoms. *Rendicoti Lincei*, 6, 602–607.

Fermi, E. (1940). The ionization loss of energy in gases and in condensed materials. *Physical Review*, 57(6), 485–493.

Firsov, O.B. (1959). A qualitative interpretation of the mean electron excitation energy in atomic collisions. *Soviet Physics JETP*, 36(9), 1076.

Fitzpatrick, A. (1999). Igniting the light elements: The Los Alamos weapon project, 1942–1952. PhD Thesis, University of California, Los Alamos National Laboratories.

Geant4 Collaboration (2017). Geant4 Physics Reference Manual, Release 10.4. Report, CERN.

Geiger, H. (1909). The ionisation produced by an α-particle. Part I. *Proceedings of the Royal Society of London. Series A*, 82(557), 486–495.

Geiger, H. (1910a). The ionisation produced by an α-particle. Part II – Connection between ionisation and absorption. *Proceedings of the Royal Society of London. Series A*, 83(565), 505–515.

Geiger, H. (1910b). The scattering of α-particles by matter. *Proceedings of the Royal Society of London. Series A*, 83(565), 492–504.

Geiger, H. and Marsden, E. (1909). On a diffuse reflection of the α-particles. *Proceedings of the Royal Society of London. Series A*, 82, 495–500.

Gemmell, D.S. and Holland, R.E. (1965). Blocking effects in the emergence of charged particles from single crystals. *Physical Review Letters*, 14(23), 945–948.

Halpern, O. and Hall, H. (1948). The ionization loss of energy of fast charged particles in gases and condensed bodies. *Physical Review*, 73(5), 477–486.

Incerti, S., Kyriakou, I., Bernal, M.A., Bordage, M.C., Francis, Z., Guatelli, S., Ivanchenko, V., Karamitros, M., Lampe, N., Lee, S.B. et al. (2018). Geant4-DNA example applications for track structure simulations in liquid water: A report from the Geant4-DNA Project. *Medical Physics*, 45(8), e722–e739.

Jenkin, J. (2008). *William and Lawrence Bragg, Father and Son. The Most Extraordinary Collaboration in Science.* Oxford University Press, Oxford.

Kucera, B. and Masek, B. (1906a). Über die Strahlung des Radiotellurs II (About the radiation of the radiotellurium II). *Physikalische Zeitschrift*, 7(18), 630–640.

Kucera, B. and Masek, B. (1906b). Über die Strahlung des Radiotellurs III: Die Sekundärstrahlung der a-Strahlen (About the radiation of the radiotellurium III: The secondary radiation of the a-rays). *Physikalische Zeitschrift*, 7(19), 650–654.

Kucera, B. and Masek, B. (1906c). Über die Strahlung des Radiotellurs I (About the radiation of the radiotellurium I). *Physikalische Zeitschrift*, 7(10), 337–340.

Lindhard, J. (1954). On the properties of a gas of charged particles. *Matematisk-fysiske Meddelelser Det Kongelige Danske Videnskabernes Selskab*, 28(8), 1–57.

Lindhard, J. and Scharff, M. (1961). Energy dissipation by ions in the keV region. *Physical Review*, 124, 128–130.

Lindhard, J., Scharff, M., Schiott, H.E. (1963). Range concepts and heavy ion ranges. *Matematisk-fysiske Meddelelser Det Kongelige Danske Videnskabernes Selskab*, 33(14), 1–42.

Lutz, H. and Sizmann, R. (1963). Super ranges of fast ions in copper single crystals. *Physics Letters*, 5(2), 113–114.

Mayer, J.W., Eriksson, L., Picraux, S.T., Davies, J.A. (1968). Ion implantation of silicon and germanium at room temperature. Analysis by means of 1.0-MeV helium ion scattering. *Canadian Journal of Physics*, 46(6), 663–673.

Metropolis, N. and Nelson, E.C. (1982). Early computing at Los Alamos. *Annals of the History of Computing,* 4(4), 348–357.

Moliere, G. (1947). Theorie der Streuung schneller geladener Teilchen I. Einzelstreuung am abgeschirmten Coulomb-Feld (Theory of scattering of fast charged particles I. Single scattering at the shielded Coulomb field). *Zeitschrift für Naturforschung A*, 2(3), 133–145.

Mott, N.F. (1928). The solution of the wave equation for the scattering of particles by a coulombian centre of force. *Proceedings of the Royal Society of London. Series A*, 118(780), 542–549.

Mott, N.F. (1929). Elastic collisions of electrons with helium. *Nature*, 123(3106), 717.

National Research Council (1964). *Studies in Penetration of Charged Particles in Matter.* National Academy of Sciences, Washington, D.C.

Nelson, R.S. and Thompson, M.W. (1963). The penetration of energetic ions through the open channels in a crystal lattice. *Philosophical Magazine A*, 8(94), 1677–1690.

Northcliffe, L.C. and Schilling, R.F. (1970). Range and stopping-power tables for heavy ions. *Atomic Data and Nuclear Data Tables*, 7(3), 233–463.

Oen, O.S. (1965). Cross sections for atomic displacements in solid by fast electrons. Technical Report, Oak Ridge National Laboratory, ORNL-3813.

Oen, O.S., Holmes, D.K., Robinson, M.T. (1963). Ranges of energetic atoms in solids. *Journal of Applied Physics*, 34(2), 302–312.

Oppenheimer, J.R. (1927). Bemerkung zur Zerstreuung der α-Teilchen (Remark on the dispersion of α-particles). *Zeitschrift für Physik*, 43(5), 413–415.

Pages, L., Bertel, E., Joffre, H., Sklavenitis, L. (1972). Energy loss, range, and bremsstrahlung yield for 10-keV to 100-MeV electrons in various elements and chemical compounds. *Atomic Data and Nuclear Data Tables*, 4, 1–27.

Piercy, G.R., Brown, F., Davies, J.A., McCargo, M. (1963). Experimental evidence for the increase of heavy ion ranges by channeling in crystalline structure. *Physical Review Letters*, 10(9), 399–400.

Robinson, M.T. and Oen, O.S. (1963a). The channeling of energetic atoms in crystal lattices. *Applied Physics Letters*, 2(2), 30–32.

Robinson, M.T. and Oen, O.S. (1963b). Computer studies of the slowing down of energetic atoms in crystals. *Physical Review*, 132(6), 2385–2398.

Röntgen, W.C. (1895). Über eine neue Art von Strahlen (About a new type of rays). *Sitzungsherichte der Physicalisch-Medicinischen Gesellshaft zu Würsburg*, 137, 132–141.

Rutherford, E. (1897). The velocity and the rate of recombination of the ions of gases exposed to Röntgen radiation. *The London, Edinburgh and Dublin Philosophical Magazine and Journal of Science (5th series)*, XLIV (CCLXX), 422–440.

Rutherford, E. (1899). Uranium radiation and the electrical conduction produced by it. *The London, Edinburgh and Dublin Philosophical Magazine and Journal of Science (5th series)*, XLVII, 109–163.

Rutherford, E. (1905). Some properties of a rays from radium. *The London, Edinburgh and Dublin Philosophical Magazine and Journal of Science (6th series)*, X, 163–177.

Rutherford, E. (1906a). The retardation of the velocity of the α particles in passing through matter. *The London, Edinburgh and Dublin Philosophical Magazine and Journal of Science (6th series)*, XI, 553–554.

Rutherford, E. (1906b). Retardation of the a particles in passing through matter. *The London, Edinburgh and Dublin Philosophical Magazine and Journal of Science (6th series)*, XII, 134–146.

Rutherford, E. (1911). The scattering of α and β particles by matter and the structure of the atom. *The London, Edinburgh and Dublin Philosophical Magazine and Journal of Science (6th series)*, XXI, 669–688.

Rutherford, E. and McClung, R.K. (1901). Energy of Röntgen and Becquerel rays, and the energy to produce an ion in gases. *Philosophical Transactions of the Royal Society of London, Series A: Mathematical and Physical Sciences*, 196, 25–59.

Sklodowska-Curie, M. (1900). Sur la pénétration des rayons de Becquerel non déviables par le champ magnétique. *Comptes Rendus Hebdomadaire des Scéances de l'Académie des Sciences*, 130, 76–79.

Strutt, R.J. (1901). On the conductivity of gases under Becquerel rays. *Philosophical Transactions of the Royal Society of London, Series A: Mathematical and Physical Sciences*, 196, 507–527.

Taylot, T.S. (1909). On the retardation of alpha rays by metals and gases. *The London, Edinburgh and Dublin Philosophical Magazine and Journal of Science (6th series)*, XVIII(CVI), 604–619.

Thomas, L.H. (1927). The calculation of atomic fields. *Mathematical Proceedings of the Cambridge Philosophical Society*, 25(5), 542–548.

Thompson, S.P. (1905). The quadrant electrometer. In *Elementary Lessons in Electricity and Magnetism*. Thompson and Thomas, Chicago, IL.

Thomson, J.J. (1896). On the discharge of electricity produced by the Röntgen rays, and the effects produced by these rays on dielectrics through which they pass. *Proceedings of the Royal Society of London*, 59(353–358), 274–276.

Thomson, J.J. (1898). On the charge of electricity carried by the ions produced by Rontgen rays. *The London, Edinburgh and Dublin Philosophical Magazine and Journal of Science*, 46, 528–545.

Thomson, J.J. (1906). *Conduction of Electricity Throught Gazes.* Cambridge University Press, Cambridge.

Thomson, J.J. (1912). Ionization by moving electrified particles. *The London, Edinburgh and Dublin Philosophical Magazine and Journal of Science (6th series)*, XXIII, 449–457.

Vantomme, A. (2016). 50 years of ion channeling in materials science. *Nuclear Instruments & Methods in Physics Research Section B: Beam Interactions with Materials and Atoms*, 371, 12–26.

Von Neumann, J. and Richtmyer, R.D. (1947). Statistical methods in neutron diffusion. Report, Los Alamos National Laboratory, LAMS-551.

Wentzel, G. (1926). Zwei Bemerkungen über die Zerstreuung korpuskularer Strahlen als Beugungserscheinung (Two remarks on the dispersion of corpuscular rays as a diffraction phenomenon). *Zeitschrift für Physik,* 40(8), 590–593.

Whaling, W. (1958). The energy loss of charged particles in matter. In *Corpuscles and Radiation in Matter II / Korpuskeln und Strahlung in Materie II*, Flügge, S. (ed.). Springer, Berlin, Heidelberg.

Whiddington, R. (1912). The transmission of cathode rays through matter. *Proceedings of the Royal Society of London. Series A*, 86(588), 360–370.

Wilson, C.T.R. (1897). XI. Condensation of water vapour in the presence of dust-free air and other gases. *Philosophical Transactions of the Royal Society of London, Series A: Mathematical and Physical Sciences*, 189, 265–307.

Ziegler, J.F. (1977). *Helium: Stopping Powers and Ranges in All Elemental Matter.* Pergamon Press, New York.

Ziegler, J.F. and Biersack, J.P. (1985). The stopping and range of ions in matter. In *Treatise on Heavy-Ion Science*, Bromley, D.A. (ed.). Springer, Boston, MA.

Ziegler, J.F. and Chu, W.K. (1974). Stopping cross sections and backscattering factors for ^{4}He ions in matter Z = 1–92, E(^{4}He) = 400–4000 keV. *Atomic Data and Nuclear Data Tables*, 13(5), 463–489.

Ziegler, J.F., Biersack, J.P., Littmark, U. (1985). *SRIM – The Stopping and Range of Ions in Solids.* Pergamon Press, New York.

4

First Notions of Defects

Having studied energetic particles and their interactions with matter, we now turn our attention to disorder in the structural organization of materials. The aim of this chapter is to show how the notion of irradiation-controlled disorder gradually emerged. The focus of this book is on materials under irradiation, and in this context we will only consider irradiation that permanently modifies the structure of materials.

4.1. First observations of defects

The notion of defects is closely linked to the scale at which a material is observed. Sapphire, for example, goes from being a precious stone to an aluminum oxide containing impurities when the resolution goes from a fraction of a millimeter to a nanometer. The first observations of defects were guided by the most widely-used instrument: the eye. So it was materials whose color changes under irradiation that were at the heart of the studies. This change in color can be traced back to photographic processes, which were discovered prior to the study of material irradiation.

4.1.1. *Photographic processes*

Attempts to capture images in camera obscuras were the first irradiation experiments, although they are rarely classified in this category. The effect of light on certain materials has been known since antiquity. In the 13th century, Albertus Magnus is said to have observed the blackening of silver salts by light (Amar 2020). In 1727, German-born Johann Heinrich Schulze discovered by chance that sunlight could darken matter and give it a dark color. In an interesting short text[1], he described his sometimes trial and error approach to understanding this phenomenon. While studying phosphorescence, he observed that chalk soaked in a solution of silver in nitric acid turned dark red when exposed to bright sunlight. He came up with the idea of writing sentences on a sheet of paper and hollowing out the letters with a scalpel. Illumination by the sun's rays through this stencil reproduced the text on the chalk, making this the first example of contact printing. Seeking the origin of the color change through a truly scientific approach, he showed that the phenomenon was linked to the presence of silver (Schulze 1727). In 1782, the Swiss Jean Senebier published a physical chemistry memoir on the influence of sunlight on the three kingdoms of nature: plant, animal and mineral (Senebier 1782). After describing phosphorescent materials, he gave a series of examples of materials that change color: "The mineral Turbith[2], well washed and exposed to the action of light in well-stoppered glass flasks, blackens on the surface, but retains its beautiful yellow color under this black layer" (p. 189, author's translation). Then, referring to the work of Beccari, Meyer, Schulze and Scheele, he undertook a systematic study of the effect of light on lunar caustic (silver nitrate). He showed that only light induced this coloration, and studied the effects of interposing various objects – paper, porcelain, etc. – and of the color of the light. He ended his chapter with a comparison of the effects of light and phlogiston[3]. In 1802, two British scientists, Thomas Wedgwood and Humphrey Davy, made public their photochemical work in the form of photograms of objects obtained on paper coated with silver nitrate, then exposed to light (Wedgwood and Davy 1802). These images remained ephemeral, however, as the silver salts continued to darken in the light. The problem was to make permanent the ephemeral images created by exposure to light. In 1819, English astronomer John Herschel demonstrated the possibility of fixing the contrast with sodium hyposulfite[4]. But the first real successes came with a different approach in 1824. The process called heliography, developed by Nicéphore Niepce, was based on the cross-linking by light of bitumen of Judea (a heavy hydrocarbon of the tar family). Insolated areas of a bitumen layer deposited on a silver plate became insoluble, while areas with little or no insolation were removed by soaking them in diluted lavender oil. The resulting image was a negative that could be reversed by exposure to iodine vapor. The silver iodide formed in this manner turned black under the action of light. This process required very long exposure times of several days and was soon abandoned. In association with Louis Jacques Mandé Daguerre,

1 The original Latin text was translated by Danielle Bouffard.

2 Mineral turbich: mercuric sulfate.

3 Phlogiston is a fluid present in materials that causes combustion when it leaves the body. This theory was invalidated by Lavoisier in the 1780s.

4 Now called sodium thiosulfate.

Niepce explored the use of resins, which led, in 1832, to a new process based on the residue of lavender oil distillation, the physautotype[5]. The improved process was presented on August 19, 1839 by François Arago under the name of Daguerreotype at the Paris Academy of Sciences (Arago 1839). All the details of how these first photographs were taken are described in publications by Jean-Louis Marignier, from the Laboratoire de Chimie Physique d'Orsay. He reproduced the historic experiments of Niepce and Daguerre (Marignier 1997, 1999). Early photographic work therefore relied on irradiation mechanisms: photo-crosslinking and photo-reduction. These two mechanisms are not specific to light, and photography was involved in two major discoveries at the end of the 19th century, that of X-rays and natural radioactivity (Figure 4.1).

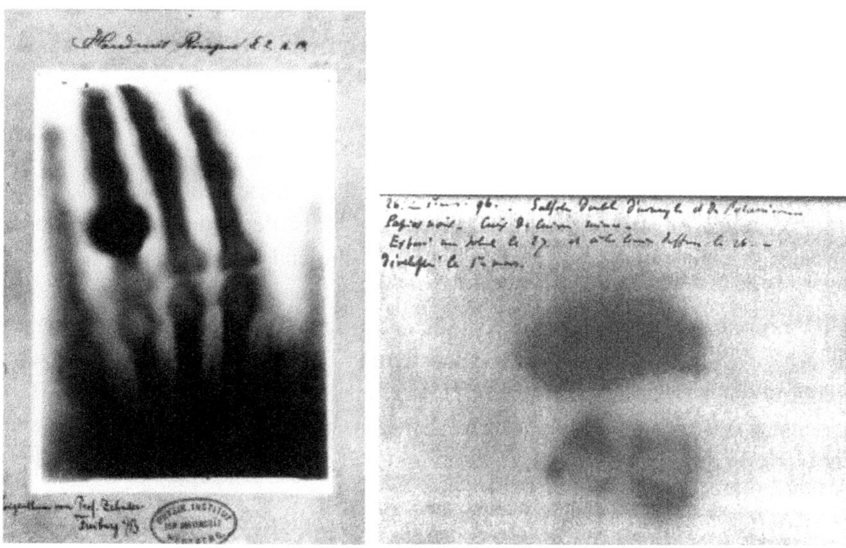

Figure 4.1. *Two photographs linked to two major discoveries at the end of the 19th century. On the left, an X-ray image of Anna Bertha Röntgen's hand taken on December 22, 1895, 44 days after the discovery of X-rays. On the right, an X-ray image developed by Henri Becquerel on March 1, 1896, after having remained in a drawer in contact with a slice of uranium salt*

The development and democratization of photography was only made possible by Georges Eastman's development of a simple, portable camera (Kodak) with a much more sensitive film. A layer of emulsion, or several for color film, containing silver bromide crystals formed the active part of this film. When exposed to light,

5 "Physautotype image" means an image of nature produced by a natural process..

photons can release an electron which, when trapped on an Ag^+ ion, reduces it to Ag^0. Development consists of completely reducing the grains containing a small silver aggregate.

In the rest of this book, we will no longer address photonic irradiation, unless it helps us to understand the mechanisms of defect creation. This is the case, for example, with pulsed laser radiolysis experiments.

4.1.2. *First experiments: an approach guided by sight*

The most widely used tool for characterizing the effects of irradiation at the end of the 19th century was certainly the eye. Changes in color and transparency were immediately visible after irradiation, despite the difficulty of describing changes in color or nuances in words.

4.1.2.1. *Color of defects*

During the 19th century, the interaction of cathode rays with matter was mainly used to excite the phosphorescence of certain materials. Permanent modifications of luminous substances were not truly studied until the end of the 19th century, when Eugen Goldstein in particular devoted part of his time to exploring the effects of cathode rays on salts, after having studied the operation of discharge tubes and discovering what he called Kathodenstrahlen or cathode rays and which J.J. Thomson characterized and named electron. Recruited in 1888 at the Berlin Observatory to study the nature of electricity in space, Goldstein spent his entire career working on gas discharges.

In particular, he showed that cathode rays stain sodium chloride buff-yellow, but if the salt is brought closer to the cathode the color darkens to brownish yellow. This coloring only occurs on a thin layer on the surface of the salt on the cathode side. In addition, if the cathode rays are condensed over a small area, or if the irradiated salt is heated, it takes on a blue color. To understand the origin of these colorations and their evolution, he studied their stability over time and their thermal stability as well as their dissolution in water, and found that the color was not transferred to the water. These experiments were carried out on a whole series of salts, including virtually all combinations of halogens and alkali metals[6]. He deduced:

> I am not yet in a position to give a positive explanation of the phenomena described here; for the time being, I must limit myself to

6 With the exception of astatine and francium, which were discovered only half a century later.

ruling out a number of initially conceivable causes. The probability that the impurities initially present in the salts are the cause of the coloration phenomena is very low [...] It is more difficult to answer the question of whether, under the influence of cathode rays, new chemical compounds are formed by reaction from the salts, for example following the partial release of a component. There is no doubt that cathode rays have the ability to break down chemical compounds after various phenomena, which I won't list here [...].

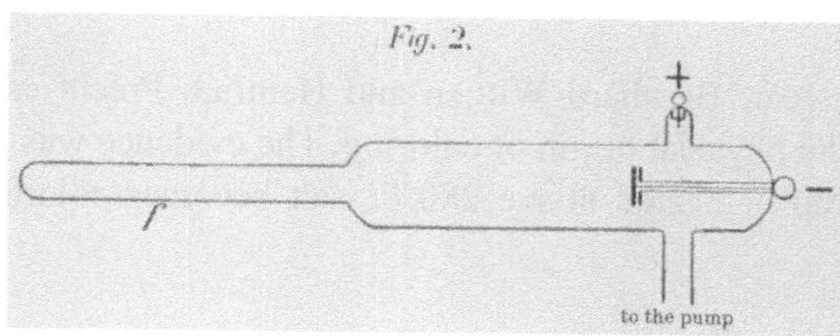

Figure 4.2. *Schematic representation of the discharge tube used by Goldstein. The salts are placed in the cylinder marked f, 1.5 cm in diameter and 15–20 cm long (Goldstein 1894)*

But the reversibility of the coloring made him abandon the chemical hypothesis:

> [...] The hypothesis of chemical modifications is made improbable by the fact that some of the colored substances return of their own accord to what appears to be the normal nature of the salt used [...] If decomposition had taken place, for example a release of chlorine in the chlorides, the chlorine would be sucked out after pumping again, i.e. restoration of the previous modification would no longer be possible.

Establishing a link between phosphorescence and coloration, he then introduced the hypothesis of atomic displacements:

> All substances that have shown coloration by irradiation phosphoresce under the influence of cathode rays; their particles are therefore brought out of their normal position or movement by irradiation and are inevitably subjected to violent shocks. As post-coloration occurs or becomes more intense, the phosphorescence of the salts becomes duller and, in some cases, eventually completely imperceptible. It is

> therefore conceivable that particles, without undergoing chemical changes, could eventually modify their phosphorescence; for most of them, the phosphorescence would be quite unstable, and over time the particles would return to their initial or less unstable disposition (gray sodium chloride). This suggestion is only a possibility, not an explicit hypothesis. Before an explanatory hypothesis can be constructed, the experiment must first answer a number of questions. Work will continue in this direction. Such was the situation in 1894. (Goldstein 1894; translated from German by the author)

The following year, Bernhard Wittjen and Heinrich Precht came to the same conclusion about the physical origin of coloring. The evidence was based on the fact that when powdered or heated above 280°C, salt becomes colorless (Wittjen and Precht 1883).

In 1899, a short text by Pierre and Marie Curie reported on some of the "chemical" effects of Becquerel's rays: glass darkened on contact with radium salt, barium platinocyanide turned yellow then brown.... They compared these observations to those made with Röntgen's rays (Curie and Curie 1899).

The origin of rock salt's blue color had long been the subject of debate. In 1908, Wilhelm Prinz, a professor at the University of Brussels, published a paper in which he attributed the blue color to the presence of an organic dye in small cavities (Prinz 1908).

In 1920, Samuel Colville Lind wrote a critical analysis of published explanations of the violet or brown color of glasses (Lind 1920). He disputed Meyer and Pzribram's observation that a flask partially filled with a solution of a radium salt could turn purple above the liquid and brown below. Lind and Dwight Bardwell of the Rare and Precious Metals Station of the U.S. Bureau of Mines in Reno (Nevada) described the coloration, discoloration and thermo-phosphorescence of transparent minerals and precious stones subjected to irradiation by radium rays. No fewer than 18 materials and their variants[7] were described (Lind and Bardwell 1923). For one material (rock salt), X-ray powder diffraction did not reveal any differences between two crystals, one uncolored and the other colored by radium. In their discussion, Lind and Bardwell proposed "a simple and plausible general theory". A group of electrons is displaced by radiation from their normal

7 For fluorite (CaF_2), four varieties with initial colors of mauve, green, pink and colorless were studied.

position to a metastable position in which they vibrate at a frequency that may be in the visible range, hence the colorization under irradiation. No atomic displacement is involved. When there is more than one group, thermoluminescence may be due to the return of one group to its original position, and decolorization by the return of another group. Heat or excitation can induce these returns.

In the introduction to *Das Radium und die Farben*[8] (Doelter 1910), Cornelio Doelter reported on the effects of radium:

> However, radium has other properties that are closely related to color changes: the effect on colloids, the precipitation of colloids in the form of gels, the transformation of amorphous materials into crystalline materials [...] The color change is not always a purely chemical effect, but is based in part on molecular rearrangements, on changes in particle size, on the degree of dispersion of a solution. [...] But the changes caused by radiation also have some connection with the changes caused by heat, whether in air or in various gases. In particular, it is very useful to examine the stability of radium-induced changes in relation to heat, especially in the case of inorganic materials. As the effects of radium are partly chemical, it was also interesting to know how the effect of these rays behaves when the materials in question are irradiated in air, in an oxygen atmosphere or in a neutral or reducing gas (translated from German by the author).

For those who want a complete bibliography on salts up to 1928, see the titanic work of J.F.T. Berliner, who has compiled 3,987 references and their abstracts in "Potash bibliography to 1928" (Berliner 1930).

From all these observations, Alex Hamberg concluded in 1914 that the metamict state originates from bombardment by α particles over geological timescales (Hamberg 1914). The first simulation of the creation of the metamict state by external irradiation was unsuccessfully attempted in 1939 (Stackelberg and Rottenbach 1939). The first studies of defects in materials were therefore carried out within the framework of geology.

These last few pages give an idea of the scientific approach of the time. The scarcity of atomic-scale characterization tools was offset by a highly developed practical sense and a great deal of experimental freedom. The stories of these defects will continue in the chapters on different types of materials.

---

8 Translation: 'Radium and the colors'.

4.1.3. *Defects, a useful concept for diffusion*

In reality, the notion of defects or disorder is not consubstantial with the effects of radiation. Indeed, solid-phase diffusion was a few years ahead of its time in using the notions of defects and disorder to explain diffusion processes. Since diffusion plays an important role in the evolution over time of irradiation-induced disorder, studies on irradiation defects were able to draw on the expertise on defects acquired by the "diffusion in materials" community. Cold working (or cold forming) is another field whose disorder has been compared to that induced by irradiation (Eggleston 1952). This process, which only concerns metals, will be explored in Chapter 6.

4.1.3.1. *Defects and diffusion in materials*

Diffusion is a phenomenon that has long been used in processes such as the separation of silver, gold and copper, and in the carburization of iron. As far as the scientific approach is concerned, it was in 1855 that Adolf Fick published an article on the laws defining diffusion phenomena (Fick 1855) (see the beginning of his paper in Figure 4.3). However, the first studies of solid-phase diffusion only began at the end of the 19th century. These early experimental investigations used standard analytical chemistry techniques, which were cumbersome to implement. The use of radioactive atoms as tracers was a major step forward. Measurements were simplified, their precision greatly improved, and self-diffusion experiments became accessible. The pioneer of radioactive tracers was György von Hevesy in Budapest. In 1920, he and Gróh published a self-diffusion experiment in liquid lead. They used ^{212}Pb, then known as thorium B (Gróh and Hevesy 1920). Hevesy was awarded the Nobel Prize in Chemistry in 1943 "for his work on the use of isotopes as tracers in the study of chemical processes". Together with Dick Coster, he also discovered hafnium, element 72. Details of the history of the first diffusion experiments would take us too far away from the irradiation of materials. I invite those who want to know more about the beginnings of diffusion to read the article by William Chandler Roberts-Austen, Professor of Metallurgy at the Royal College of Science, who in 1896 gave an overview of diffusion in molten and solid metals (Roberts-Austen 1896).

In 1922, Hevesy, then working at the *Institut für theoretische Physik* in Copenhagen, became interested in the electrical conductivity of crystals. He was referring, of course, to ionic or iono-covalent crystals, the two materials he used as examples being sodium nitrate and calcium carbonate. He noted that the electrical conductivity of a perfect crystal is almost zero, but that any perturbation of the crystalline state results in an increase in conductivity. This disturbance can have many origins: heating close to the melting temperature, irradiation by "absorbed" light, crystal clusters made from molten salt, etc. He concluded his article by pointing out that the transport of matter in a crystal is linked to the appearance of scattered places in the lattice, within which a change in the position of ions can take place (Hevesy 1922).

IV. ***Ueber Diffusion; von Dr. Adolf Fick,***

Prosector in Zürich.

Die Hydrodiffusion durch Membranen dürfte billig nicht blofs als einer der Elementarfactoren des organischen Lebens sondern auch als ein an sich höchst interessanter physika-lischer Vorgang weit mehr Aufmerksamkeit der Physiker in Anspruch nehmen als ihr bisher zu Theil geworden ist. Wir besitzen nämlich eigentlich erst vier Untersuchungen, von Brücke [1]), Jolly [2]), Ludwig [3]) und Cloetta [4]) über diesen Gegenstand, die seine Erkenntnifs um einen Schritt weiter gefördert haben.

Figure 4.3. *The beginning of Adolf Fick's article on diffusion*[9] *(Fick 1855)*

In this type of approach, we find a significant number of publications describing electrical conduction through ionic diffusion associated with lattice imperfections. For example, at the general meeting of the *Deutsche Bunsen-Gesellschaft für angewandte physikalische Chemie* in Munich in May 1928, Adolf Smekal of Vienna concluded his speech by stating that the mechanism of electrical conductivity and diffusion processes in crystallized compounds are essentially related to deviations of the crystal structure from the ideal crystal lattice (Smekal 1928). He did not specify what this deviation consisted of, whereas 2 years earlier Jacov Frenkel had published a microscopic description of diffusion. In this model, migration is associated with the presence of dissociated atoms (spontaneous formation of the

9 Author's translation of the beginning of Fick's 1855 article (Figure 4.3): "About Diffusion. Hydrodiffusion through membranes is not only one of the essential factors of organic life, but also a highly interesting physical process in itself, and should therefore attract much more attention from physicists than has been paid to it so far. We actually only have four studies on this subject, by Brücke, Jolly, Ludwig and Cloetta, which have advanced knowledge in this field".

vacancy–interstitial atom pair) and the diffusion of these dissociated atoms (Frenkel 1926).

In 1937, Robert F. Mehl, Director of the Metals Research Laboratory (Carnegie Institute of Technology, Pittsburgh), published a review on diffusion in solid alloys (Mehl 1937). Structural factors influencing diffusion, such as the presence of grain boundaries and impurities, were discussed. He also described "dissociation", the Frenkel process in which it is the interstitial that diffuses. On the other hand, H.B. Huntington and Frederick Seitz, after testing different mechanisms, showed unambiguously that energetically, the vacancy mechanism is the most favorable (Huntington 1942; Huntington and Seitz 1942).

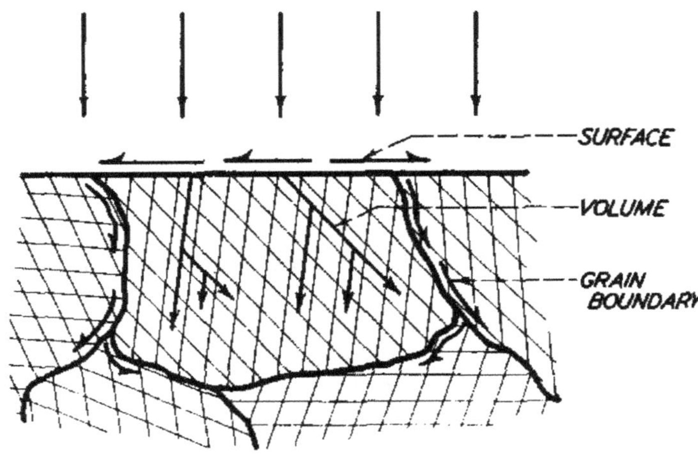

Figure 4.4. *The different types of diffusion in a polycrystalline material (Mehl 1937)*

For a more in-depth description of the history of diffusion models, please refer to C. Tuijn's *On the history of models for solid-state diffusion* (Tuijn 1997) and H. Mehrer and G.E. Murch's *History and people of solid-state diffusion – An overview* (Mehrer and Murch 2021).

4.2. Notions of defects

The first to suggest the existence of atoms outside crystalline sites was Abram Joffé (or Abram Fiodorovitch Ioffé, 1880–1960). To explain electrical conduction in dielectrics, he hypothesized that only the ions he called dissociated, or interstitial, were involved in electrical conduction. To this end, he introduced the concept of "lattice interstitial space" ("*Zwischengitterraum*"), in which the dissociated ions are located (Joffé 1923). As future director of the Physical-Technical Institute in Leningrad, he noticed a student who was to become his collaborator, Jacov Frenkel (or Yakov Illich Frenkel, 1894–1952). Based on the work of Joffé, Frenkel took up the concept of dissociated ions to explain migration in solids (Frenkel 1924, 1926). By analogy with sublimation, he assumed that within solids there was an internal sublimation to dissociated states. Thus, he wrote in his 1924 article:

> The fact that solid materials sublimate indicates that surface atoms can, due to the irregularities of thermal motion, acquire such great energy that they completely leave their equilibrium position and pass into the gas phase, tearing themselves away from neighbouing atoms. The same applies not only to surface atoms, but also to the inner atoms that form the crystal lattice of a solid material. In this case, we are dealing with a process of "internal sublimation" or dissociation of the solid material. Atoms that, by chance, stray too far from their equilibrium position, can remain in a "dissociated" state between other atoms, and then occupy another place that has become free (author's translation).

These dissociated ions must also move in the absence of electric fields. He assumed that dissociated atoms can jump to certain equilibrium positions in the "space between lattice sites" following a temperature-dependent probability. The dissociated atom jumps from interstitial site to interstitial site after a large number of vibrations at each site. This process continues until the dissociated atom encounters a normal unoccupied site. He concluded:

> [...] in solid materials, each atom must move progressively throughout the space occupied by that material. This movement or 'change of place' of atoms can be considered as a kind of diffusion.

In his 1926 article, he pointed to a second mode of migration: lattice atoms can jump to a site left vacant by the formation of the dissociated atom, and this can be understood as an elementary jump to the vacant site. If we permit ourselves an anachronism, we find here the interstitial and the vacancy of Frenkel's famous pair. He called their recombination an association. In addition to the fact that this article is cited as a reference for Frenkel's pair, it gives a coherent vision of thermal displacement in a crystalline solid, which, according to Frenkel, depended on oscillations of atoms around their normal or dissociated position, dissociation of bonded atoms, migration of dissociated atoms, association, and migration of empty sites. Spontaneous generation of the interstitial was the main error; in a pure metal, the energy of formation of an interstitial was estimated to be of the order of 3–5 eV.

In the years 1931–1935, Walter Schottky and Carl Wagner refuted Frenkel's proposal and argued for another approach (Wagner 1933; Schottky 1935)[10]. Schottky began his 1935 article with:

> The theory of electrolytic conduction in heteropolar structures should not be developed on the assumption that there are as many interstitial ions as there are ionic holes of each ion type. On the contrary, at thermal equilibrium, such structures may contain almost exclusively holes of both ion types, or only interstitial ions of both ion types. In dense structures, e.g. of the rock salt type, due to the repulsion energy between the ions increasing roughly inversely to the power 9 of the distance, the energy state of an interstitial ion is further away from the normal state of the structure than the energy state of an ionic hole, the equilibrium concentration of interstitial ions must be negligible compared to the concentration of holes of both types of ions at thermal equilibrium (author's translation, the original text is in German).

He then went on to demonstrate that a thermodynamic calculation eliminates the hypothesis of migration by the Frenkel mechanism. In fact, he referred more to the work of Jost than to that of Frenkel. Jost of the Massachusetts Institute of Technology took Frenkel's qualitative arguments and brought a quantitative approach to them (Jost 1933).

4.3. Chapter 4 references

4.3.1. *Biographies of some of the chapter's personalities*

Numerous researchers have contributed to the advancement of knowledge on the notion of defects. Below are brief biographies of four of them, arbitrarily chosen.

Samuel Colville Lind (1879, McMinville, Tennessee–1965[11]) is renowned for his work on the kinetics of various chemical reactions, particularly those induced by ionizing radiation. After 2 years at the *Institüt für Physikalische Chemie* in Leipzig,

10 There is an article that is unanimously cited: Wagner, C. and Schottky, W. (1931). *Z. Phys. Chem.*, 11B, 163, but I have not been able to find it.

11 While fishing, he was swept away by a flood caused by the emptying of a dam on the Clinch River near Oak Ridge.

he joined the University of Michigan from 1905 to 1913, where he collaborated with Marie Curie (Paris) and Stefan Meyer (Vienna). In particular, he studied the formation of ozone through the action of alphas on oxygen. From 1913 to 1925, he was employed by the U.S. Bureau of Mines, where he first worked on the extraction and refinement of radium from carnotite[12]. He then studied the chemical effects of radiation, notably on water and on color changes induced in diamonds. In 1926, he became Director of the School of Chemistry at the University of Minnesota.

Figure 4.5. *Samuel Colville Lind. Photo credit: AIP Emilio Segrè Visual Archives*

Figure 4.6. *György von Hevesy, winner of the 1943 Nobel Prize in Chemistry. Photo credit: Nobel Foundation archive*

12 Carnotite is a hydrated vanadate of uranyl and potassium: $K_2(UO_2)_2(VO_4)_2 \cdot 3H_2O$.

After graduating from university, *György von Hevesy* (1885, Budapest, Austro-Hungarian Empire–1966, Freiburg im Breisgau, Germany) moved to Manchester in 1910 to study under E. Rutherford. In 1913, he took part in the first experiment using radioactive tracers. Like many researchers of the time, he frequently changed research environments (Manchester, Vienna, Copenhagen, Fribourg-en-Brislau and New York). He is best known for his work on radioactive isotopes and their use as radioactive tracers. In 1943, he was awarded the Nobel Prize in Chemistry for his work on the use of isotopes as tracers in the study of chemical processes.

Figure 4.7. *Walter Hans Schottky (1886–1976). Photo credit: AIP Emilio Segrè Visual Archive*

Walter Hans Schottky (1886, Zurich, Switzerland–1976, Forchheim, West Germany) studied at university in Berlin. He defended his thesis under the supervision of Max Planck in 1912. Until 1927, he changed research locations several times: Jena University, Humboldt University Berlin, Würzburg University, Rostok University and Siemens. He then became director of the Siemens & Halske research center in Berlin. His name is associated with numerous discoveries: Schottky effect, Schottky diode, Schottky defects and Schottky equation. He was one of the founders of semiconductor physics and electronics.

Yakov Illich Frenkel (1894, Rostov-on-Don, Russia–1952, Leningrad, USSR) was a mathematical genius. At the age of 18, he wrote a 250-page paper on the mathematical physics of the Earth's magnetism. He obtained his doctorate from the University of St. Petersburg in 1918. He then left to teach for 3 years at the new Tavrichisky University in Simferpol (Crimea) and returned to the St. Petersburg Polytechnic Institute. At the age of 32, he had already published six books: *The Structure of Matter I* (1922), *The Theory of Relativity* (1923), *The Structure of Matter II* (1924), *Vector and Tensor Analysis* (1925), *Electricity and Matter* (1925), and *Electrodynamics* (1926). He became head of theoretical physics at Kazan University, in the town where his institute had been evacuated during the Second

World War. When he returned to Leningrad, his work was criticized for not having contributed to the building of socialism. His work on quantum mechanics was seen as opposed to socialist determinism.

Figure 4.8. *Yakov Illich Frenkel (1894–1952) Physico Technical Inst. St Petersburg. Photo credit: AIP Emilio Segrè Visual Archives, Frenkel collection*

4.4. References

Amar, P.-J. (2020). Histoire des techniques. In *Histoire de la photographie*. Presses Universitaires de France, Paris.

Arago, F. (1839). Le Daguerréotype. *Comptes Rendus des Séances de l'Académie des Sciences*, 9, 250–267.

Berliner, J.F.T. (1930). *Potash Bibliography to 1928*. U.S. Department of Commerce, Washington, D.C.

Curie, P. and Curie, M. (1899). Effets chimiques produits par les rayons de Becquerel. *Comptes Rendus Hebdomadaires des Séances de l'Academie des Sciences*, 129, 823–825.

Doelter, C. (1910). *Das Radium und die Farben (Radium and the colours)*. Verlag von Theodor Steinkopff, Dresden.

Eggleston, R.R. (1952). Cold work studies on Copper at low temperature. *Journal of Applied Physics*, 23(12), 1400–1401.

Fick, A. (1855). Ueber diffusion (About diffusion). *Annalen der Physik*, 170(1), 59–86.

Frenkel, J. (1924). Theorie der Adsorption und verwandter Erscheinungen (Theory of adsorption and related phenomena). *Zeitschrift für Physik*, 26(1), 117–138.

Frenkel, J. (1926). Über die Wärmebewegung in festen und flüssigen Körpern (About the heat transport in solids and liquids). *Zeitschrift für Physik*, 35(8–9), 652–669.

Goldstein, E. (1894). Über die Einwirkung von Kathodenstrahlen auf einigen Salze (On the action of cathode rays on some salts). *Sitzungsberichte der Königlich Preussischen Akademie der Wissenschaften zu Berlin*, 937–945.

Gróh, J. and Hevesy, G. (1920). Die Selbstdiffusionsgeschwindigkeit des geschmolzenen Bleis (The self-diffusion rate of molten lead). *Annalen der Physik*, 368(17), 85–92.

Hamberg, A. (1914). Die radioaktiven Substanzen und die geologische Forschung (Radioactive substances and geological research). *Geologiska Föreningen i Stockholm Förhandlingar*, 36, 31–96.

Hevesy, G. (1922). Über Materietransport im Kristall und Kristallit (About matter transport in crystal and crystallite). *Zeitschrift für Physik*, 10(1), 80–83.

Huntington, H.B. (1942). Self-consistent treatment of the vacancy mechanism for metallic diffusion. *Physical Review*, 61(5–6), 325.

Huntington, H.B. and Seitz, F. (1942). Mechanism for self-diffusion in metallic copper. *Physical Review*, 61(5–6), 315.

Joffé, A. (1923). Elektrizitätsdurchgang durch Kristalle (Electricity conduction through crystals). *Annalen der Physik*, 72(22), 461–500.

Jost, W. (1933). Diffusion and electrolytic conduction in crystals (ionic semiconductors). *The Journal of Chemical Physics*, 1(7), 466–475.

Lind, S.C. (1920). Notes on the coloring and thermoluminescence of glass produced by radium radiation. *Journal of Physical Chemistry*, 24(6), 437–443.

Lind, S.C. and Bardwell, D.C. (1923). The coloring and thermophosphorescence produced in transparent minerals and gems by radium radiation. *Journal of the Franklin Institute*, 196(3), 375–390.

Marignier, J.-L. (1997). L'invention de la photographie. *Comptes Rendus Hebdomadaire des Scéances de l'Académie des Sciences*, 325(7), 415–420.

Marignier, J.-L. (1999). *Nicéphore Niépce 1765-1833 : l'invention de la photographie*. Belin, Paris.

Mehl, R.F. (1937). Rates of diffusion in solid alloys. *Journal of Applied Physics*, 8(3), 174–185.

Mehrer, H. and Murch, G. (2021). History and people of solid-state diffusion – An overview. *Diffusion Foundations*, 29, 1–30.

Prinz, W. (1908). Observations sur le sel gemme blanc et bleu. *Bulletin de la Société Belge de Géologie - Mémoires*, 22, 63–82.

Roberts-Austen, W.C. (1896). X. Bakerian lecture – On the diffusion of metals. *Philosophical Transactions of the Royal Society of London, Series A: Mathematical and Physical Sciences*, 187, 383–415.

Schottky, W. (1935). Über den Mechanismus der Ionenbewegung in festen Elektrolyten. *Zeitschrift für Physikalische Chemie*, 29B(1), 335–355.

Schulze, J.H. (1727). Scotophorus pro Phosphoro inventvs seu experimentum curiosum de effectu radiorum solarium. *Acta Physico-Medica Academiæ Cæsareæ Leopoldino-Carolina Natura Curiosorum exhibentia Ephemerides, Nürnberg*, 528–533.

Senebier, J. (1782). *Mémoires physico-chimiques sur l'influence de la lumière solaire pour modifier les ëtres des trois règnes de la Nature, et surtout ceux du règne végétal.* Barthelemi Chirol, Geneva.

Smekal, A. (1928). Zeitschrift Fur Elektrochemie Und Angewandte Physikalische Chemie (Electric conductor and diffusion in crystallised compounds). *Zeitschrift für Elektrochemie Und Angewandte Physikalische Chemie*, 34, 472–480.

Stackelberg, M. and Rottenbach, E. (1939). Dichte und Struktur des Zirkons IV. Die Ursache des Isotropisierung des Zirkons (Density and structure of the zircon IV. The cause of the isotropisation of the zircon). *Zeitschrift für Kristallographic, Kristallgeometrie, Kristallphysiks, Kristallchemie*, 102, 207–208.

Tuijn, C. (1997). On the history of models for solid-state diffusion. *Defect and Diffusion Forum*, 143–147, 11–20.

Wagner, C. (1933). Theorie der geordneten Mischphasen. III Fehlordnungserscheinungen in polaren Verbindungen als Grundlage für Ionen- und Elektronenleitung. *Zeitschrift für Physikalische Chemie*, 22B(1), 181–194.

Wedgwood, T. and Davy, H. (1802). An account of a method of copying paintings upon glass, and of making profiles by the agency of light upon nitrate of silvea. *Journals of the Royal Institution of Great Britain*, 1, 170–174.

Wittjen, B. and Precht, H. (1883). Zur Kenntnis des blaugefärbten Steinsalzes (On the knowledge of the blue colored rock salt). *Berichte der Deutschen Chemischen Gesellschaft*, 16(1), 1454–1457.

5

Defect Creation Mechanisms

In *Radiation Effects in Solids*, published in 1957, G.J. Dienes and G.H. Vineyard of Brookhaven National Laboratory traced interest in radiation-induced effects in solids back to the construction of the first nuclear reactor:

> While some aspects of the interaction of various nuclear radiation effects in solids have been studied for many years, the very active interest in radiation effects in solids at the present time is largely the result of wartime research on nuclear reactor development (Dienes and Vineyard 1957).

It is more precise in 1942. In that year, Eugene Paul Wigner suggested that neutrons could affect the graphite moderator of nuclear reactors, all the more so as defects increase graphite's internal energy. During a temperature excursion, this stored energy can be suddenly released, inducing a sudden rise in temperature (this effect is now known as the Wigner effect) (Wigner 1942). The Radiation Chemistry section of the Manhattan Project was responsible for studying this phenomenon. In a 1956 Scientific American article, F. Seitz and E.P. Wigner wrote that this group was very pessimistic about the reactor's ability to survive radiation. They are quoted as saying, "It would be unscientific to claim a useful life longer than about 100 days" (Seitz and Wigner 1956). Milton Burton[1] shows in Figure 5.1 the diversity of possible effects produced by very high-energy radiation (Burton 1947). The complexity of this diagram is intended to illustrate the difficulty of fully understanding the phenomena induced by this energy input. Furthermore, the "chemical reaction" box hides all the specific chemistry of the irradiated medium, as in the case of water, where solvation must be taken into account: $H_2O^+ + aq \rightarrow H_3O^+ + OH$.

---

1 At the time, Milton Burton was working for Monsanto Chemical Company in Oak Ridge, Tennessee.

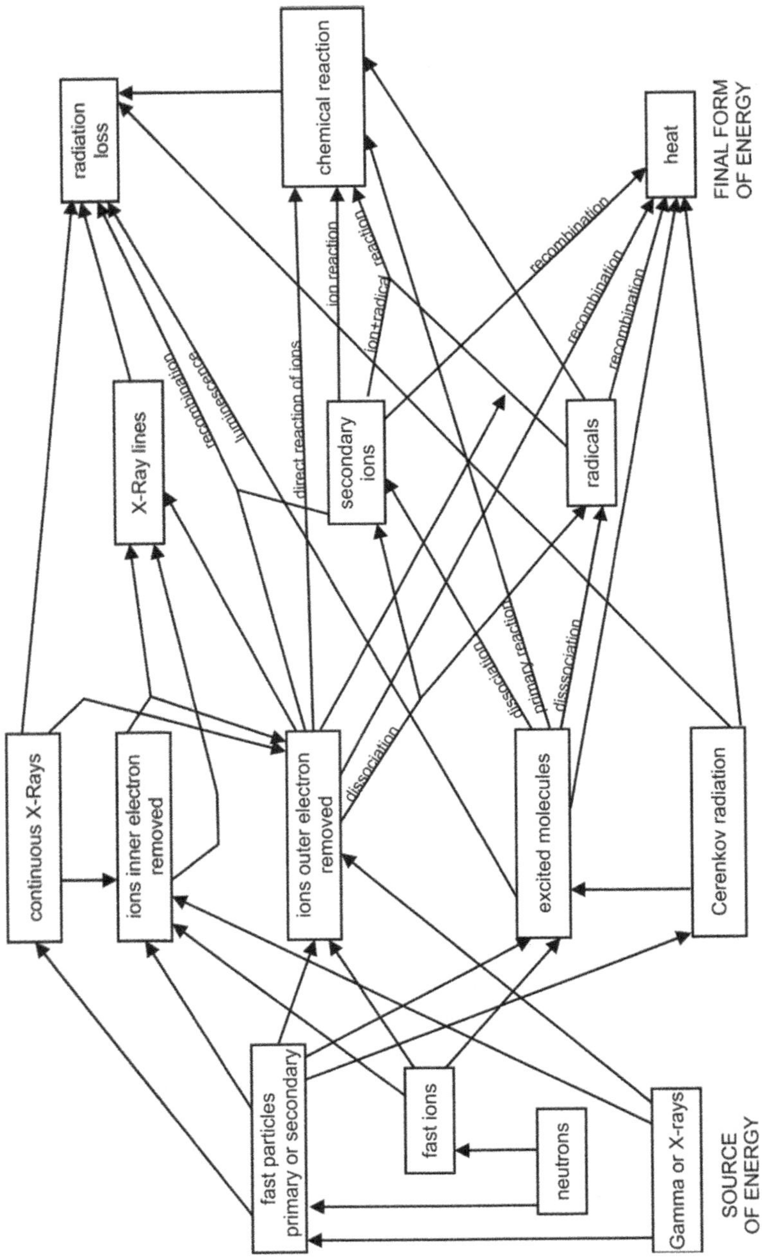

Figure 5.1. *Possible effects of high-energy radiation (after Burton (1947))*

Thus, given the scale of the research effort surrounding the Manhattan Project, those involved tended to focus on the 1940s. This is a bit hasty, and overlooks all the work carried out between 1900 and 1939, mainly on crystal coloring under irradiation. But there is no doubt that during World War 2, under technological pressure, work on the irradiation of materials took off. However, this work was classified as "defence confidential" and very little was subsequently published in the open literature, but even then, how can we be sure that it has not been updated? It is worth noting that, since 1952, the journal *Annual Review of Nuclear Science* has been publishing journal articles on the various subjects of nuclear science. In the journal's first 10 years, three articles reviewed the effects of radiation on solids (Dienes 1953; Brooks 1956; Goland 1962).

5.1. Production of defects by irradiation

We know that two processes can lead to the creation of defects. One is the direct transfer of kinetic energy to the nucleus of the target atom; the other is the conversion of ionization or excitation energy into atomic displacement. In this case, defects are created by an indirect mechanism whose effectiveness depends on the material's electronic properties, whereas the first mode of defect production is universal[2]. The effects of ionization were first observed on ionic crystals. As the effect is a change in color, it is easy to observe.

5.1.1. *Creation of defects by electronic excitations and ionizations*

The first observation of a change in a material due to electronic excitations was made by Eugen Goldstein at the end of the 19th century, while working at the *Physikalisch-Technichen Reichsanstaft (Berlin).* He observed that a salt such as lithium chloride changed color when exposed to cathode rays (Goldstein 1894). However, the mechanism behind this phenomenon was totally unknown to him. Robert Wichard Pohl and Rudolf Hilsch of Göttingen provided the beginnings of an explanation in 1930, when they presented a study of colored centers using optical spectroscopy[3] (Hilsch and Pohl 1930). Thanks to the absorption constant calculated

2 Inelastic collisions with the nucleus can induce nuclear reactions, followed by particle emission and nucleus recoil, leading to defects and the introduction of foreign atoms; this can be seen as a third process.

3 These authors quote an article by Kasimir Fajans in J.M. Eder's "Handbuch der Photographie" (1926), in which he writes: "Es kann keinem Zweifel unterliegen, daß diese Verfärbung des Steinsalzes im kurzwelligen Licht als ein vollkommenes Analogon zu der photographischen Zersetzung der Silberhalogenide angesehen werden kann". (There is no doubt that this coloration of rock salt in short-wave light can be considered a perfect analogue to the photographic decomposition of silver halides.)

according to the formula proposed by Smakula[4] (1930), they showed, for example, that the photochemical yield is between 0.33 and 0.4 colored centers per quanta of absorbed light, and the saturation concentration is around 10^6–10^7 salt molecules per colored center. As for the mechanism, the clear separation between the energy levels speaks in favor of simple elementary processes: most likely, the passage of an electron from the halogen anion to the alkali ion. They therefore indicated that the colored centers in alkali halides should be regarded as neutral alkali atoms in any lattice bond. It should be noted that as early as 1923, Karl Przibram hypothesized that coloration was associated with the presence of neutral atoms (Przibram 1923).

5.1.1.1. *Models based on ionization processes*

In 1946, Frederik Seitz put forward a first, admittedly rather improbable, vision of the creation of F centers, which can be summarized as follows. There are small clusters of four vacancies, which he calls “quartets”. If a quartet captures two electrons, it ejects two positive ion vacancies, so the cluster is reduced to a pair of linked F centers (F_2 center). Then, in a second act, the capture of two holes by this F_2 center produces two halogen vacancies that repel each other. In a third act, one of the halogen vacancies captures an electron and recombines with a positive ion vacancy. This leaves one isolated halogen vacancy, the F center! (Seitz 1946).

A few years later, F. Seitz modified his model, writing: “The writer proposes the following speculative solutions”. X-rays produce photoelectrons, which lose their energy by generating electrons, holes and excitons. Initially, the electrons and holes are trapped on the vacancies of positive or negative charge compensation associated with the edges of the dislocations. The existence of these vacancies is postulated to ensure charge balance at the dislocation edges. Exciton de-excitation induces a local rise in temperature that is sufficient to “evaporate” the charge–compensation vacancies, which become vacancies in the vicinity of the dislocations. These vacancies can aggregate. The disappearance of charge–compensation vacancies induces a dislocation climb. F. Seitz describes this process in Figure 5.2. Based on this event, we find the previous description of the creation of F centers (Seitz 1953, 1954).

At the same time, J.H.O. Varley (AERE, Harwell, UK) published two papers on an ion displacement mechanism in ionic crystals; first, a letter in *Nature* and then a long article in the *Journal of Nuclear Energy* (Varley 1954a, 1954b). He first noted that certain experiments were difficult to explain within the framework of a “billiard

---

4 In 1930, Alexander Smakula, 30, was assistant to R.W. Pohl (46), while R. Hilsch was only 27.

ball"-type collision mechanism, such as the dependence of the thickness of the colored zone on the electrons' energy, or the coloration induced by X-rays whose photoelectrons have too low an energy to directly displace atoms. He then proposed a new mechanism for producing atomic displacements in ionic crystals, which can be summarized as follows: incident particles, including X-rays and gammas, can ionize negative ions. In the case of multi-ionization, the halogen becomes positive. It then undergoes significant repulsion from nearby cations. If it does not recombine with electrons too quickly, it can be ejected into the interstitial position with the help of lattice vibrations that dissymmetrize the site. Its initial site becomes an anionic vacancy that can easily trap an electron to become an F center. He refined his model a few years later, notably by calculating the probabilities of multi-ionizations by direct, successive processes or by Auger effect (Varley 1962). Noriaki Itoh (Nagoya University, Japan) showed that, for low-energy electrons, direct and Auger processes have equivalent probabilities (Itoh 1968).

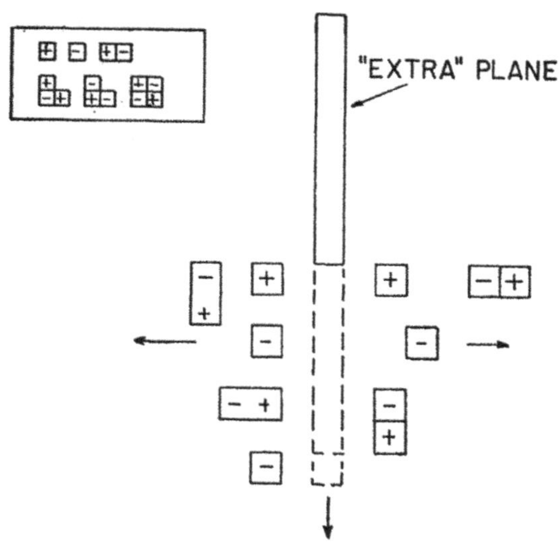

Figure 5.2. *Schematic representation of vacancy production on a Taylor–Orowan dislocation. The inset at top left shows the vacancy aggregates that can be expected from this process (Seitz 1954)*

However, Varley's mechanism, which has received sustained attention, is not universally accepted. D.L. Dexter (University of Rochester, USA) writes:

> Though no direct evidence has been adduced for this mechanism, it has increasingly been invoked because of its apparent consistency

> with low-temperature experiments for which most other mechanisms seem inappropriate (Dexter 1960).

The first objection concerns the lifetime of a doubly ionized halogen. This is measured by the time taken for the two positive holes to separate due to their coulombic repulsion: they cover an interatomic distance in less than 10^{-15} s. Dexter concludes that there is no reason for the ion to be ejected interstitially: "the Varley mechanism does not apply".

Not all calculations agree, however. Indeed, R.E. Howard and R. Smoluchowski estimate the lifetime of the doubly ionized state by calculating the return time of an electron. They found that, in NaCl at 200 K, this time is greater than 10^{-12} s, giving the doubly ionized halogen time to escape to an interstitial position (Howard and Smoluchowski 1959).

Clifford C. Klick took into account the criticisms of Varley's model and proposed a mechanism in which the transfer of an electron from a halogen to the doubly ionized anion is the basis for the creation of the F + H centers. The sequences of this model are summarized in Figure 5.3 (Klick 1960).

Another possible approach, still assuming double ionization, is based on potential curves between chlorine ions. This model, known as "de-excitation collision", assumes the formation of a quasi-molecule between a Cl^+ and a Cl^- (curve A in Figure 5.4). The equilibrium distance of Cl^+–Cl^- lies at the distance where curve A crosses the dissociative curve of Cl–Cl (curve B). Mutual neutralization of the two chlorines leads to their separation with a kinetic energy of around 3 eV. The production of a sufficiently separated interstitial and chlorine vacancy can be envisaged if the two chlores move in the [110] direction through a series of focused collisions of the dynamic crowdion type (Howard et al. 1961; Smoluchowski and Wiegand 1961).

In 1965, freeing himself from the problem of multiple ionization, Derek Pooley proposed a mechanism based on electron–hole recombination in a configuration where one electron is bound to a V_k center. The transition between the excited state represented by the electron–hole pair and the ground state can be radiative or non-radiative. In the latter case, energy is transmitted to the two halogens in the form of kinetic energy, with the momentum vectors aligned along the direction [110].

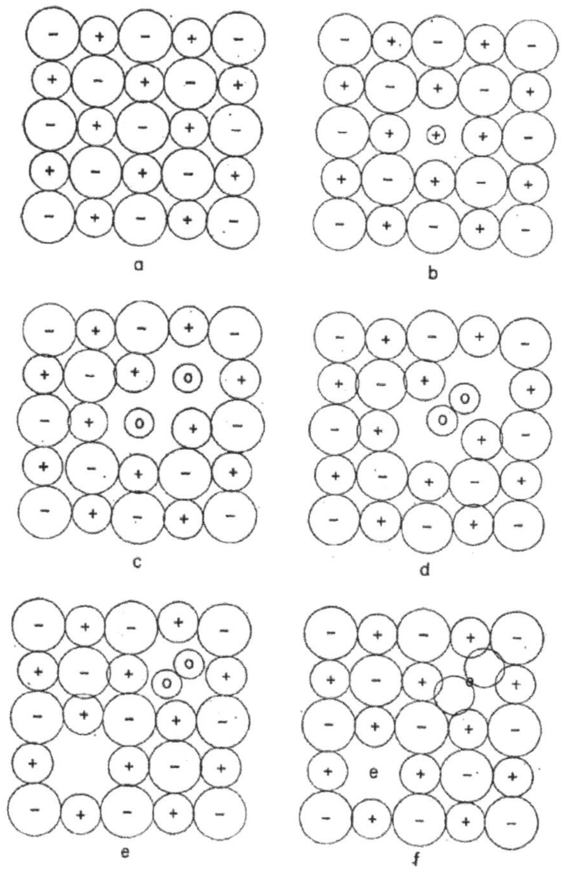

Figure 5.3. *Schematic representation of the Klick model for the creation of F and H centers at 4 K. (a) Perfect lattice, (b) double ionization of a halogen ion, (c) the doubly ionized halogen captures an electron from a neighboring halogen ion, (d) formation of a neutral halogen molecule, (e) formation of a vacancy and an interstitial by the jump of a halogen ion and (f) capture of an electron by the vacancy and by the halogen molecule to form the F and H centers, respectively (Klick 1960)*

The sequence of replacement collisions in this direction can lead to the formation of a vacancy and an interstitial (Pooley 1965, 1966a, 1966b). The low-temperature polarization of the luminescence confirms the hypothesis of an excited state consisting of an electron trapped on a V_k center (Kabler 1964). A few years later, D. Pooley and W.A. Runciman carried out a study of the luminescence of almost all alkali halides, 20 different ionic crystals. They concluded that the extinction of luminescence is due to the non-radiative process that leads to the production of F centers (Pooley and Runciman 1970).

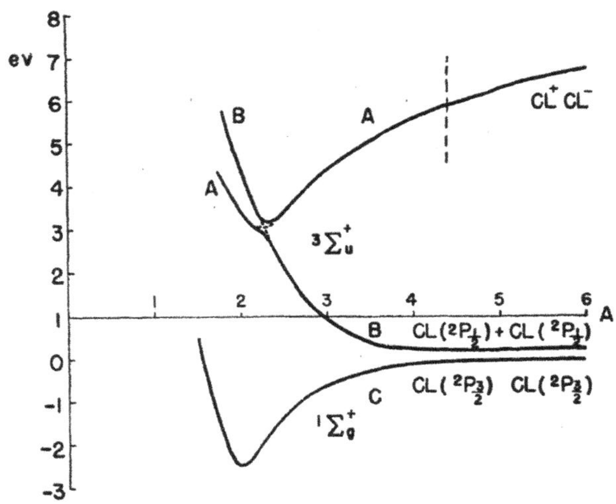

Figure 5.4. *Adiabatic potential curves of a* Cl_2 *molecule. Curve C represents the ground state* $^1\Sigma_g^+$ *curve B the repulsive state* $^3\Sigma_g^+$ *and A the attractive state* Cl^+ Cl^-*(Howard et al. 1961)*

The theory of defect creation in alkali halides owes much to research carried out at the Atomic Energy Research Establishment in Harwell, where J.H.O. Varley and D. Pooley were members. The reason why these defect-creating mechanisms are so well known is that the structure of alkali halides is simple and held together by strong coulombic forces. In the case of other materials sensitive to electronic excitations, such as polymers, the creation mechanism is more complex and less precisely described.

5.1.2. *Models for the creation of defects by elastic collisions*

The displacement of atoms by elastic collision has been described since the 1940s as a succession of collisions between the incident particle and the atoms of the solid. These first displaced atoms, if they have sufficient energy, in turn become projectiles that transfer all or part of their energy to other atoms, the secondary recoil atoms. The number of displacements can be calculated theoretically by following the history of each particle and taking into account the energy lost in the form of ionization. Knowing the interaction potential between the incident particles (ions, neutrons and electrons) is the first step in understanding collision and energy transfer probabilities (total and differential cross sections). The first of the potentials used is of course the Coulomb or Rutherford potential, which has the disadvantage of having an infinite range and diverging when the distance between particles tends toward zero. The second is the hard sphere potential, often used to describe neutron–atom collisions or collisions between atoms in early molecular dynamics simulations.

To avoid divergence of the coulombic potential at low atomic separations, Niels Bohr proposed a screening term $P_a(r) = \frac{e_1 e_2}{r} e^{-\frac{r}{a}}$, where a is the screening parameter (Bohr 1948). Numerous proposals have been put forward to describe this screening term: Firsov potentials (Firsov 1957), Lindhard potentials (Lindhard and Scharff 1961), etc. Analytical forms have also been found for empirical potentials (Moliere 1947; Ziegler et al. 1985). However, the most frequently used case remains the Coulomb potential, but considering that below a certain transmitted energy, the atom cannot leave its site.

In a journal article *On the disordering of solids by action of fast massive particles* (Seitz 1949), based on declassified notes from a series of lectures given at Oak-Ridge National Laboratory in the winter of 1946–1947, Frederick Seitz[5] describes the behavior of knocked-on atoms in solids. A few years later, he repeated this description in a long journal article (Seitz and Koehler 1956). Considering that a charged particle loses its kinetic energy through elastic interactions with solid atoms and through electron excitation and ionization, he evaluates both energy losses with a Fermi–Thomas atom model. Electron excitations lose their efficiency at low energies. He indicated that atomic displacements by elastic collision are only possible if the energy transmitted is greater than a certain energy E_d. He estimated that this minimum energy must be of the order of magnitude, but greater, than the energy required to adiabatically transport an atom from its normal site to an interstitial position. He therefore indicated that E_d can be expected to be of the order of 25 eV for atoms bound to their position with an energy of around 10 eV, without further justifying this 25 eV value. This value of 25 eV will become the reference value. In the absence of measurement or calculation, the threshold displacement energy will be 25 eV.

5 During this period, Frederic Seitz was working at the Carnegie Institute of Technology, Pennsylvania, USA.

F. Seitz demonstrated that, according to his assumptions, the number of displacements induced by a primary is given by the following formula:

$$N = \frac{R_d E_c}{\bar{E}} n_s = \frac{R_d E_c}{\sqrt{\bar{E} E_d}}$$

where $\bar{E}$ is the average energy transferred to an atom displaced by the primary (the number of secondary n_s is approximately equal to $\sqrt{\bar{E}/E_d}$), R_d is the fraction of energy lost through elastic collisions and E_c is the initial energy of the primary). He found that a 20 MeV proton displaces around 87 atoms in germanium (a calculation with SRIM-2013 gives 133 displacements per proton).

J. Ozeroff of the Knolls Atomic Power Laboratory of the General Electric Company[6] in a confidential memo dated June 29, 1949, now declassified, developed the calculation of the number of atomic displacements for fission fragments and fission neutrons (Ozeroff 1949). The title "Atomic displacements produced by fission fragments and fission neutrons in matter" is highly simplistic, as it also calculates elastic and electronic stopping powers. To calculate the number of displacements due to elastic collisions of fission fragments, he used three different approaches. At high energies, he considered two cases depending on whether the distance r between the two nuclei was greater or smaller than an impact parameter b. If $r < b$, he considered the case to be one of Rutherford scattering, and if $r > b$, the potential decreased exponentially with distance. In the case of a collision, the first case was explored only if the moving atom had an energy greater than $\tilde{E} = zZe^2/b$. When E was greater than $\tilde{E}$, he made the additional assumption that in all cases, the recoil atom received a minimum energy which, for example, was equal to 390 eV for an 80 MeV Ag ion in a uranium target. The primary recoil atoms all received sufficient energy to be displaced. The number of primary recoil atoms was equal to the fission fragment energy lost through elastic collisions, divided by the average energy transferred to the primary $\langle E \rangle$. He estimated this elastic energy at 5.5% of the fission fragment energy. The number of primaries in the chosen example was equal to $N_1 = E \times 5.5\%/\langle E \rangle$ = 2,600 primaries of average energy 1,690 eV. When $E < \tilde{E}$, he considered that the collision was of the hard sphere type and that the loss of electronic energy was zero, the calculation for secondary recoils gave an average energy of the secondaries of $\langle E_2 \rangle = 163$ eV and $\langle E_3 \rangle = 20$ eV. For tertiaries whose average energy was close to the chosen threshold displacement energy (25 eV), the number of atoms displaced was given by a law of the form $\left(1 - e^{-E/E_d}\right)$. This gave

6 General Electric built the Vallecitos power plant in California, the first nuclear power plant to be connected to the U.S. power grid (August 3, 1957). Worldwide, the first 5 MW power plant was connected to the power grid in Obninsk, Soviet Union, on June 27, 1954.

2,600 primaries per fission fragment, 7.5 secondaries per primary and 1.6 tertiaries per secondary, that is, 31,000 displacements per fission fragment. A Monte Carlo calculation with SRIM "full cascades" under the same conditions gives 134,000 displacements, that is, 4.3 times more. This discrepancy is reduced if we consider that J. Ozeroff found that elastic collisions account for 5.5% of the energy of fission fragments; the SRIM software indicates a ratio of 8.9%. With this value, J. Ozeroff would have found 50,000 displacements; the difference is only a factor of 2.7.

5.1.2.1. *Kinchin and Pease model*

In 1955, with the publication of the review article *The displacement of atoms in solids by radiation*, George H. Kinchin and Rendel Sebastian Pease of the Atomic Energy Research Establishment in Harwell (United Kingdom) disseminated a simple model for calculating the number of displacements created by a primary. This model became universally known as the Kinchin and Pease model. It should be noted that the article not only describes the production of displacements, but also includes defect annealing, irradiation techniques and experimental results for different types of material (Kinchin and Pease 1955b). After pointing out the limitations of previous calculations, those by F. Seitz, J. Ozeroff and J.A. Brinkman, G.H. Kinchin and R.S. Pease justified the use of hard sphere collisions by electron screening, which introduces a cutoff in the interaction at a distance $a_0Z_0^{1/3}$ (a_0 Bohr radius). This approximation is valid when the kinetic energy of the incident particle is less than the interaction potential of the two nuclei separated by the screening radius. For the majority of primaries, this approximation is possible: carbon primary <5 keV, aluminum <30 keV, copper <200 keV. Two further hypotheses were necessary for their calculation: (1) there is no loss of energy by ionization and (2) the atoms all have the same mass. With this in mind, they calculated the number of atoms displaced, assuming that at the *p*th collision stage, the energy is shared between 2^P atoms. The number of atoms in the *dE* energy domain is given by:

$$N_p(E)dE = 2^p\left(\ln(\bar{E}/E)\right)^{p-1} dE/\bar{E}(p-1)!$$

This gives the number of atoms displaced for $\bar{E} > 2E_d$:

$$N_d = \sum_{p=1}^{p=\infty} E_d N_p(2E_d) = \frac{\bar{E}}{2E_d}$$

E_d being the displacement threshold energy and $\bar{E}$ average primary energy, for $E_d < \bar{E} < 2E_d, N_d = 1$. At high energies, energy loss through excitation–ionization becomes the dominant process. They then considered that above an energy E_1, the number of displacements was equal to $E_1/2E_d$.

This article was preceded the same year by a publication on irradiation-induced disorder in alloys in which they put forward the hypothesis that all atoms have the same mass (!) (Kinchin and Pease 1955a). Assuming hard-sphere collisions, the number of displacements was equal to $D = E_{max}/4E_d$ (the average energy in the case of hard spheres being $E_{max}/2$, both papers are consistent).

For the value of the displacement threshold, they followed F. Seitz's proposal:

> Since the displacement energy varies with direction of recoil, must vary from one material to another, and cannot be at all easily calculated, the easiest practical procedure is to follow Seitz (1949) in taking an isotropic value of 25 eV, and to modify this value only when directly indicated by experiment. This has generally been done in work so far published (Kinchin and Pease 1955b).

The major shortcoming of the Kinchin and Pease model was that it neglected electronic energy loss, whereas Seitz and Koehler took it into account. It was not until the mid-1970s that a new model, by Norgett, Robinson and Torrens, became the standard. These were then referred to as NRT displacements.

5.1.2.2. *NRT model*

The international collaboration of Michael J. Norgett (Atomic Energy Research Establishement, Harwell, UK), Mark T. Robinson (Oak Ridge National Laboratory, Tennessee, USA) and Ian M. Torrens (Centre d'Études Nucléaires de Saclay, France) proposed, in 1975, an analytical model for calculating the number of displacements, taking into account energy losses in the electronic system (Norgett et al. 1975).

Two of the authors, Robinson and Torrens, first studied atomic displacement cascades using computer simulations[7]. Unlike early numerical simulations where the energy of the primary was limited by the small box size used for molecular dynamics calculations, the use of a Monte Carlo approach enabled the simulation of high-energy primary cascades. With the Marlowe program, the calculation of a 50 keV cascade takes 23 s. The ingredients of this program are based on an approximation of binary collisions with a Moliere screen potential, including an inelastic energy loss described by Firsov's theory (Robinson and Torrens 1974). These simulations give them access to numerous quantities characteristic of atomic

7 Within a decade, computing capacity had increased enormously: the computer used, an IBM system/360 model 91, was capable of executing up to 16.6 million instructions per second. Until 1968, it was the most powerful computer on the market.

displacement cascades (number of Frenkel pairs and their spatial distribution, amount of energy lost through inelastic processes, vacancy-interstitial recombinations, etc.).

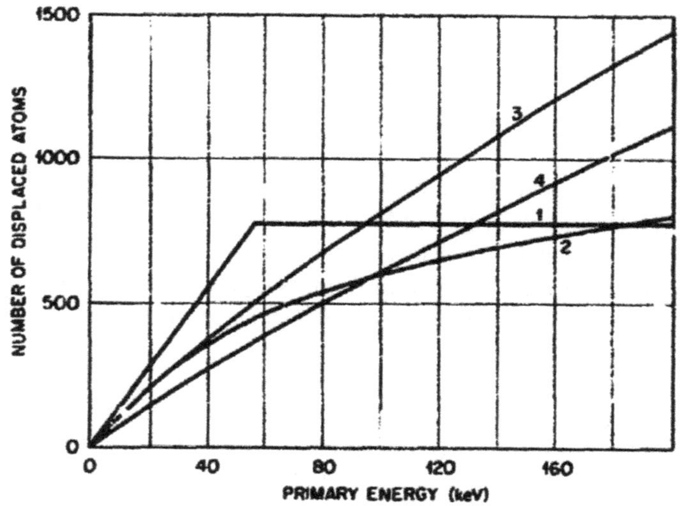

Figure 5.5. *Comparison of the number of displacements created in iron by a primary. (1) Kinchin and Pease model with E_d = 40 eV, (2) half-Nelson formula, (3) first Norgett calculation using a Torrens–Robinson computer simulation program and (4) Norgett, Robinson and Torrens model. Figure taken from Norgett et al. (1975)*

After comparing five methods for calculating the number of displacements, Norgett, Robinson and Torrens proposed a simple procedure for estimating the number of atoms displaced. First, they used a modified formula from Kinchin and Pease: $N_d = \kappa\hat{E}/E_d$ where $\hat{E}$ is the energy available to create defects by elastic collisions and κ is the displacement efficiency (0.8). The inelastic energy loss was calculated using the results of numerical simulations and following Lindhard's formalism in which energy is measured by a dimensionless variable (ϵ) (Lindhard et al. 1963):

$$\epsilon = [A_2E/(A_1 + A_2)][a/Z_1Z_2e^2] \text{ with } a = 0.8853a_0\left[Z_1^{2/3} + Z_2^{2/3}\right]^{-1/2}$$

$$\hat{E} = \frac{E}{[1+kg(\epsilon)]}$$

with $g(\epsilon) = 3.4008\epsilon^{1/6} + 0.40244\epsilon^{3/4} + \epsilon$ and $k = 0.1337Z_1^{1/6}\sqrt{Z_1/A_1}5$

Despite some limitations, this formalism quickly became a standard for calculating the number of atomic displacements by elastic collisions. However, with the advent of desktop computing and the free availability in 1985 of the SRIM software (Ziegler et al. 1985), NRT displacements were supplanted by the use of the latter software.

5.2. Determination of threshold displacement energy

In the preceding sections, we have used the notion of threshold displacement energy without defining its origin, so let us go back a few years to look at the work on this quantity. The number of defects depends directly on the minimum energy E_d that an atom must receive to be ejected from its crystallographic site: the threshold displacement energy. This quantity has been estimated theoretically and measured experimentally.

At the American Physical Society meetings in March 1951 and 1952, E.E. Klontz and K. Lark-Horowitz probably presented one of the first measurements of a displacement threshold. In germanium, to displace an atom, electrons must have an energy greater than 630 keV (Klontz and Lark-Horowitz 1951, 1952). This threshold energy corresponds to a displacement threshold of 31 eV. Using the same method based on conductivity variation, a group from the USSR found the threshold to be 22 eV (Vavilov et al. 1956). Then J.J. Loferski and P. Rappaport of Princeton, deploying a more precise measurement based on the lifetime of minority carriers, obtained a lower value: 14.5 ± 0.4 eV (12.9 ± 0.6 for silicon) (Loferski and Rappaport 1958). More detailed measurements by DLTS (Deep Level Transient Spectroscopy) show that above 20 ± 5 eV (T_d), the E_1 and E_2 traps attributed to the association of a vacancy (or interstitial) with an impurity are detected, while above 40 ± 10 eV, that is, around $2T_d$, the E_4 and E_5 traps associated with divacancies appear (Poulin and Bourgoin 1980). Similar results have been obtained for silicon (Corbett and Watkins 1965).

It was not until a few years later that the first threshold displacement energy calculations were published. H.B. Huntington of Brookhaven National Laboratory calculated the final position of the ejected atom as a function of the amplitude and direction of its initial impulse (Huntington 1954). The calculation was made for a metal with compact structure, copper, using a semi-empirical Born–Mayer potential. This potential is actually only used to determine the size of the spheres in the hard sphere model used. It shows that the minimum energy required to extract an atom from its site (the threshold energy) depends on the direction in which the atom is ejected from the crystallographic axes: 18.5 eV for the [111] direction and 17.5 eV for [100]. However, the steepness of the potential has a much greater effect than the direction of ejection, from 18.5 to 43 eV. Huntington was aware of the difficulty of deciding on the equilibrium configuration of the interstitial atom. This is probably the case for the [111] direction, where subsequent results show that for an energy of around 20 eV, the atom certainly remains in the spontaneous recombination volume of the vacancy.

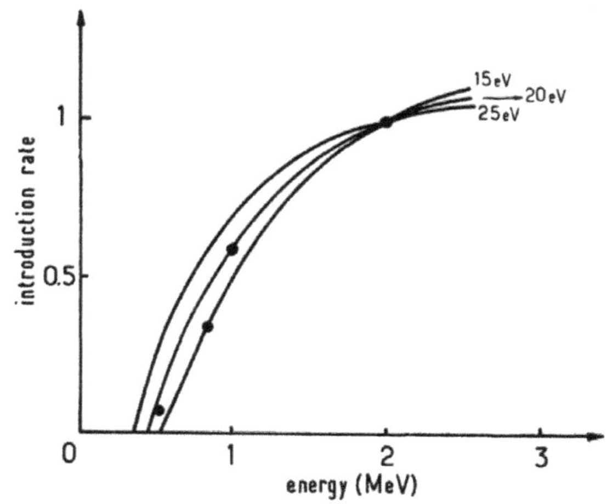

Figure 5.6. *E_1 trap production rate in germanium, normalized to 1 MeV as a function of electron energy. The curves correspond to theoretical calculations for three values of the displacement threshold (Poulin and Bourgoin 1980)*

The threshold displacement energy has been determined in many materials. Such experiments can only be carried out with an electron beam. In fact, protons in the 300–1,000 eV range would have to be used to measure the threshold energy, and therefore samples would only be a few nanometers thick. Figure 5.6 illustrates how the threshold value is determined. In general, data are normalized to an electron energy, often 1 MeV, to avoid the physical property–defect concentration relationship. The defect creation rate measured experimentally is compared with that calculated theoretically for different values of the threshold displacement energy. This requires a variable-energy electron accelerator and a parameter sensitive to the presence of point defects (electrical resistivity, optical absorption, etc.). It is preferable to operate at very low temperatures, in a temperature where defects are stable. These operating conditions have been met in many laboratories since the 1960s. Table 5.1 gives some examples of experimental determinations of threshold displacement energy.

Materials	Observables	Tp (K)	Threshold (eV)	References
Germanium	Electrical resistivity		31	(Klontz and Lark-Horowitz 1952)
Copper	Electrical resistivity	10	22	(Corbett et al. 1957)
Silicon	Carrier lifetimes	290	12.9 ± 0.6	(Loferski and Rappaport 1958)
α-Al_2O_3	Optical absorption	77	50 ± 5 (O) 90 ± 5 (Al)	(Arnold and Compton 1960)
Iron	Electrical resistivity	20	24	(Lucasson and Walker 1961)
Gold	Electrical resistivity	10	33–36	(Bauer and Sosin 1964)
Graphite	Electron microscopy	~473	33	(Montet 1967)
Silica	Compaction rate	293	~25	(Primak 1972)
Bismuth	Electrical resistivity	20	13 ± 12	(Quelard et al. 1978)
a-$Fe_{75}B_{25}$	Electrical resistivity	21	22 ± 3	(Lesueur et al. 1983)

Table 5.1. *Some experiments to determine threshold displacement energy*

5.2.1. *Threshold displacement energy mapping*

In the above, the authors considered a single threshold energy value, which implicitly means that all ejection directions are equivalent. However, if we consider a crystalline structure, it seems obvious that the energy required to produce a permanent displacement must depend on the direction of ejection with respect to the crystallographic axes. We saw at the start of this chapter that H.B. Huntington's numerical approach found a small difference between the threshold displacement energies in the [100] and [111] directions of a face-centered cubic. A few years later, a team from Brookhaven National Laboratory repeated the calculation and found a threshold energy of 85 eV in the [111] direction and only 25 eV for the [100] and [110] directions (Gibson et al. 1960). Similar work has been carried out in the case of iron, a body-centered cubic lattice (Erginsoy et al. 1964).

While it is easy in a numerical approach to impose an initial direction on the first atom hit, it is impossible experimentally. Indeed, it would be necessary to control the impact parameter. Experimental mapping of threshold displacement energies is difficult, and it has only been carried out for a few materials.

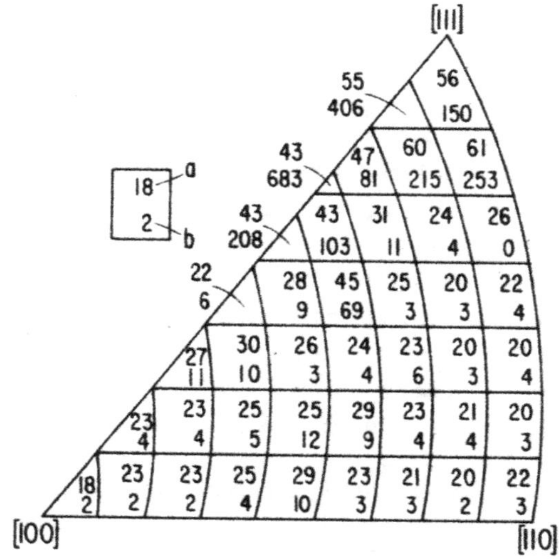

Figure 5.7. *Threshold energy as a function of primary ejection direction relative to copper crystallographic axes (a) threshold energy values; (b) sensitivity to fit giving values (a) (King et al. 1981)*

However, the experiment was carried out in 1970 by two researchers from Northwestern University (Illinois) in association with a researcher from Argonne National Laboratory (King et al. 1981). Defect creation was measured at 10 K in situ by electrical resistivity for six energies in 41 incident directions of the electron beam from Argonne National Laboratory's Kratos high-voltage (1.2 MeV) electron microscope. From the differential scattering cross sections and an arbitrary choice of threshold energy surface[8], the probabilities of moving an atom in the different directions were calculated. The best fit of the threshold energy surface was obtained when the difference between experimental and theoretical values was minimized over all incident directions and all five electron beam energies. Figure 5.7 shows, for example, that the minimum threshold energy was 18 ± 2 eV in the <100> direction.

8 Distribution of threshold energies as a function of initial direction in a [100][110][111] frame of reference.

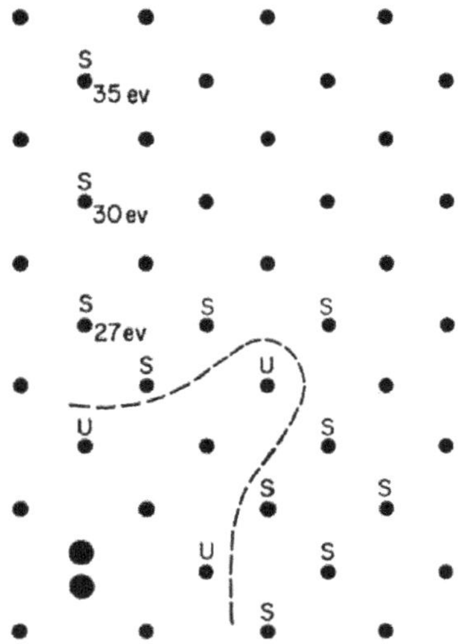

Figure 5.8. *Frenkel pair stability domain in the (100) copper plane. Positions: (S) stable; (U) unstable (Gibson et al. 1960)*

Two years later, the same authors re-analyzed their data, leaving the resistivity of the Frenkel pair as a free parameter (King et al. 1983). With $\rho_F = (2.75^{+0.6}_{-0.2})\Omega.\text{cm}$, the threshold energy surface was characterized by two domains of low values around axes [100] and [110]. On the other hand, areas of high threshold energy had lower values. From this mapping, they extracted the number of atoms displaced as a function of the energy transmitted to the first atom ejected: the damage function $\nu(T)$ of copper.

5.3. Numerical simulations

5.3.1. *Creation and stability of point defects*

The development of the first computers opened up new possibilities for the study of solid-phase collisions. In 1960, a study was carried out on the effect of atomic potential on defect creation. This stochastic calculation was carried out on an IBM 650[9]. The authors found only a slight decrease in the number of displaced atoms

9 The IBM 650 was the second computer and the first to be built in series. It ran on vacuum tubes. It performed around 17 multiplications per second.

when moving from a hard sphere potential to a screened potential (Brown and Goedecke 1960).

The same year, a group composed of J.B. Gilbon, A.N. Goland, M. Milgram and G.H. Vineyard from Brookhaven National Laboratory published a long paper entitled "Dynamics of radiation damage", in which they studied various irradiation effects in copper using a computing machine (Gibson et al. 1960). This was one of the first studies of defect creation in molecular dynamics. The computing power limited the study to crystallites of 500–1,000 atoms (a cube of 8–10 atoms on a side). Indeed, with 500 atoms, each time step required around one minute of calculation on an IBM 704[10]. In this article, the operating conditions and working hypotheses are described in detail: model representing copper, calculation method, equations of motion, units used, energy conservation tests, calculation time steps, etc. Then there are numerous results on the static properties of defects and on the dynamics of their creation. They found, for example, that the split interstitial was the only stable interstitial configuration in copper. Figure 5.8 shows a result on the unstability domain of the Frenkel pair (the spontaneous recombination volume). On the dynamic side, many of their results deal with collision chains. They take the analytical calculation of R.H. Silsbee of Oak Ridge National Laboratory a step further with their numerical approach (Silsbee 1957). The directions favorable to the propagation of collision chains are [110] and [100]. The focusing conditions are studied as a function of the angle and energy of the first atom set in motion. The consequences of these focused collisions, which transport the interstitial away from the vacancy, are then studied. They thus describe one of the first representations of a displacement cascade: a vacancy-rich core surrounded by an interstitial-rich periphery (Figure 5.9).

The same group developed an identical calculation for the case of centered cubic metals and presented some very nice images of the dynamics of defect creation (see Figure 5.10) (Erginsoy et al. 1964).

By neglecting crystal structure, the Monte Carlo approach makes it possible to describe higher energy events. Thus, in his paper, Mazayuki Yoshida of the Japan Atomic Energy Research Institute, Tokaimura, describes the spatial distribution of interstitials and vacancies for a 10 keV primary (Yoshida 1961).

10 The IBM 704 was the first widely marketed computer with floating-point computing capability. It performed around 4,000 operations per second.

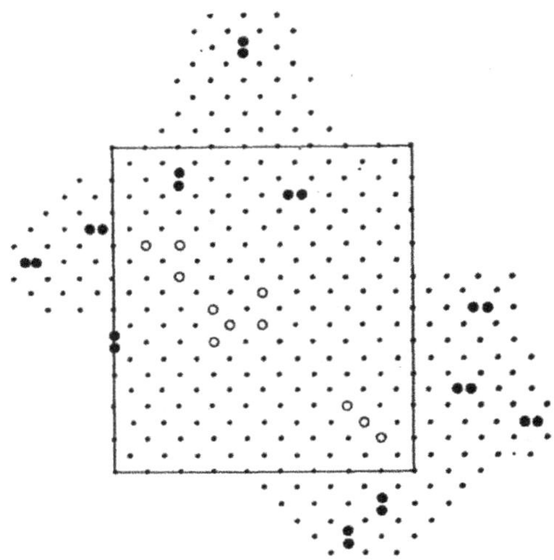

Figure 5.9. *Distribution of vacancies and interstitials created by a 400 eV primary. At the time in question, the arrangement of vacancies was not yet stable (Gibson et al. 1960)*

5.3.2. *Thermal spike*

The hypothesis of a thermodynamic approach has also been considered to explain the disorder induced by irradiation. The primary particle suddenly deposits a large amount of energy, inducing local heating followed by rapid cooling. This process is known as a thermal spike.

As early as the early 1920s, Von Friedrich Dessauer (Institut für Physik der Universität Frankfurt a. Main) was considering the effects of local heating on biological molecules. He envisaged that X-ray energy could be converted into heat at the molecular level.

A rough calculation showed that an energy of 100–1,000 eV deposited on a protein molecule with a molecular mass of 4,000–5,000 increases its temperature by 1,000–10,000°C. It also indicated that energy diffusion must remain locally limited for a sufficient time for a destructive effect, biological or otherwise, to occur (Dessauer 1923). This was not quite a thermal spike as currently defined, but the idea was there.

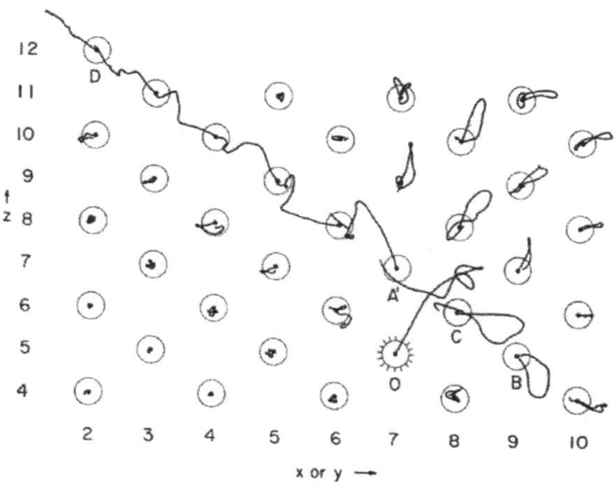

Figure 5.10. *Atom trajectories initiated by an atom emitted at 65 eV (O atom) (Erginsoy et al. 1964)*

Sidney Siegel of Oak Ridge National Laboratory referred to the notion of a thermal spike to explain the discrepancy between experiments and calculations of neutron-induced disorder in the ordered Cu_3Au alloy. Indeed, at the end of irradiation, the material is almost totally disordered, whereas Seitz's calculation indicates that the fraction of displaced atoms represents only a few percent. S. Siegel estimated that heating due to elastic collisions can, along the path of recoil atoms, reach temperatures of the order of 1,500 K for around 10^{-10} s. This temperature spike, followed by rapid quenching, may be the cause of the disorder (Siegel 1949). Similarly, W. Primak defended the idea that the convergence of quartz and silica under irradiation toward the same state of disorder could only be explained by a thermal spike model (Primak 1955).

It seems, if we are to believe the numerous articles quoting an unpublished report, that it was Harvey Brooks who developed the first thermal spike model *(I have not found this report, which would have been written in 1950 and could be number 360 in the Knolls Atomic Power Laboratory Report series)*. However, this primacy is also attributed to Frederick Seitz. In a long review article entitled *Displacement of atoms during irradiation*, Frederick Seitz and his colleague James Stark Koehler devote a chapter to the theory of thermal spike (Seitz and Koehler 1956). They considered two types of thermal spike: those with spherical symmetry and those with cylindrical symmetry. The former was thought to be created by end-of-path collisions, while the latter was associated with electron excitations along the trajectory of a charged particle. The basic assumption was that at t = 0, the energy was deposited on a point or line, creating a singularity.

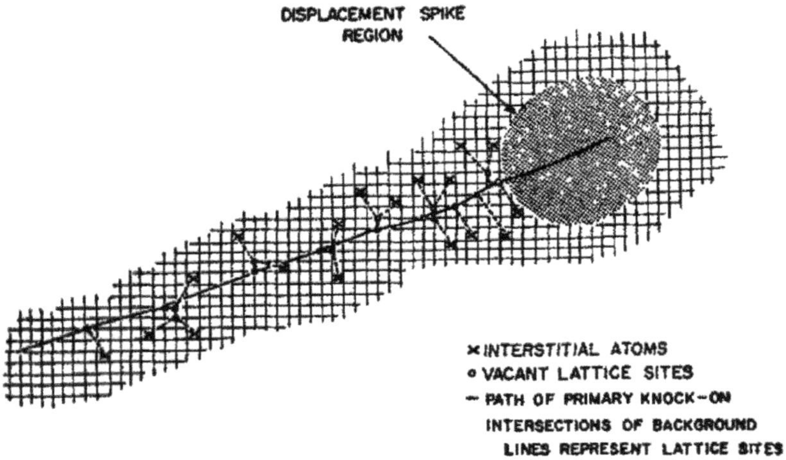

Figure 5.11. *Schematic representation of the damage created by a primary knock-out atom (pka) of very high energy (Brinkman 1955)*

Irradiation damage in metal therefore has two origins, the creation of vacancy–interstitial pairs and the consequences of the thermal spike. Authors used to consider one or the other mechanism, rarely the coexistence of both. This is what John A. Brinkman of North American Aviation, California, proposed in his article *On the nature of radiation damage in metals*. In his view, the two processes were mutually exclusive. They dominate in distinct zones: in the high-energy range of the primary, the creation of Frenkel pairs dominates, while at low energies, the thermal spike brings the material locally into a "liquid" phase that lasts long enough to annihilate vacancies and interstitials and induce numerous exchanges between atoms. J.A. Brinkman called this mechanism a displacement spike (Brinkman 1954, 1955). This displacement spike explains the rapid disordering of ordered alloys such as Cu_3Au.

5.4. Irradiation-induced sputtering

When particle interaction takes place close to the surface, the movement of atoms can lead to their ejection from the material. The rate of sputtering depends on a number of parameters, such as the projectile-target atom mass ratio, energy, angle of incidence, etc. The term sputtering appeared at the beginning of the 20th century in the expression "cathode sputtering" ("*Kathodenzerstäubung*"). However, the process behind this sputtering is more in the realm of plasma-surface interaction than that of irradiation. Nevertheless, we can cite the experiment carried out by

G. Sensel at the University of Vienna, who unsuccessfully attempted to separate uranium X from uranium[11] by sputtering (Sensel 1910).

The study of the disintegration of a tungsten cathode struck by positive ions from various gases at different speeds can be considered a sputtering study. This study, entitled *Cathode disintegration*, was published by a collective author (The Research Staff of the General Electric Company 1923).

Sputtering phenomena are exploited in a number of materials analysis techniques for in-depth analysis using secondary ion mass spectrometry (SIMS), or for atomic-scale cutting of a material using a scanning electron microscope equipped with a focused ion beam (SEM-FIB). The release of uranium into the cladding-fuel space has been the subject of numerous studies for safety reasons.

5.4.1. *Metal sputtering*

The first sputtering experiments on metals involved actinides (U, Pu). The results are reported in the following section.

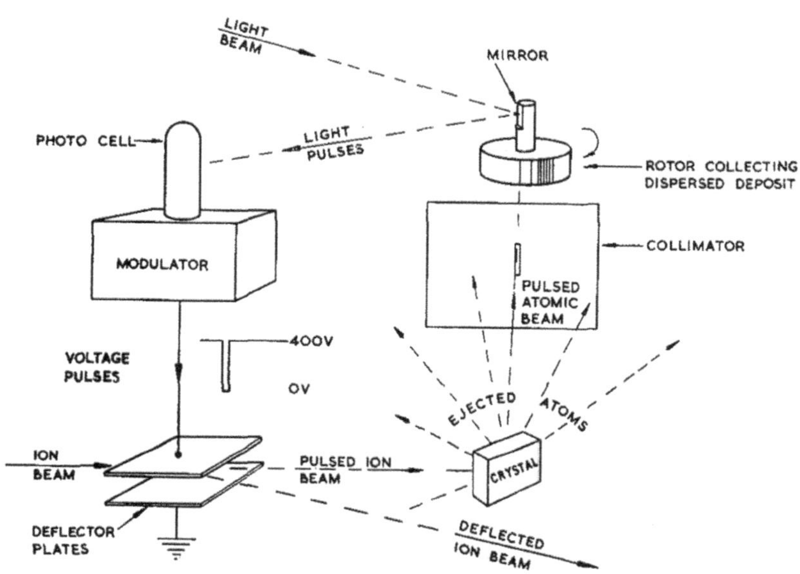

Figure 5.12. *Schematic representation of an experiment to measure the velocity of atoms ejected from a metal surface (Thompson et al. 1968)*

11 Uranium X is the radioactive element that accompanies uranium, probably the ^{234}Th direct descendant of ^{238}U.

M.W. Thompson's group at the University of Sussex measured the velocity distribution of atoms ejected from a copper and gold surface using a mechanical time-of-flight spectrometer (see Figure 5.12). The secret of the experiment was in the targets, containing radioactive isotopes ^{64}Cu or ^{198}Au, both of which emit β, enabling detection by autoradiography (Thompson et al. 1968). Analysis of the energy spectrum of ejected atoms shows that they come mainly from cascades of random displacements and sequences of focused collisions (Farmery and Thompson 1968; Thompson 1968).

5.4.2. *Uranium sputtering*

Uranium sputtering is most often associated with the action of fission fragments. In the Soviet Union in the 1950s, B.V. Ershler and F.S. Lapteva were interested in the evaporation of the atoms of two metals by fission fragments and alpha particles. These metals are somewhat special in that they are thin layers of uranium-233 and plutonium-239. The thin layer of fissile metal was placed opposite a foil that collected the atoms emitted by the thin layer. The whole, evacuated in a quartz container, was exposed to a flux of neutrons. The number of atoms evaporated was determined from the alpha activity of the collecting foil. The average number of atoms evaporated per fission fragment was 1,200 for^{233}U and 3,500 for ^{239}Pu. They explained the results within the framework of a thermal spike model, and their dispersion by the presence of a more or less thick oxide layer (Ershler and Lapteva 1957). M.D. Rogers set up a similar experiment at the Metallurgy Division of the Atomic Energy Research Establishment in Harwell. The target consisted of uranium enriched to 93% in ^{235}U and containing an "adequate" quantity of ^{234}U. The result, 2,000 atoms ejected per fission fragment, was in complete agreement with those of Ershler and Lapteva. On the other hand, they observed saturation at high fluence (Rogers and Adam 1962). In another experiment, evaporated atoms were collected on a thin sheet of carbon and observed by transmission electron microscopy. Figure 5.13 shows the appearance of spots whose size changed little with the fluence of the fission fragments. From analysis of these images, M.D. Rogers estimated that each spot was created by a single fission fragment. The average spot contained around 10,000 atoms. A cluster of uranium ejected for every 16 fission fragments crossing the surface suggested to him that displacement spikes were at the origin, but to explain the change as a function of fluence, he took into account the vaporization of spots by fission fragments (Rogers 1965a).

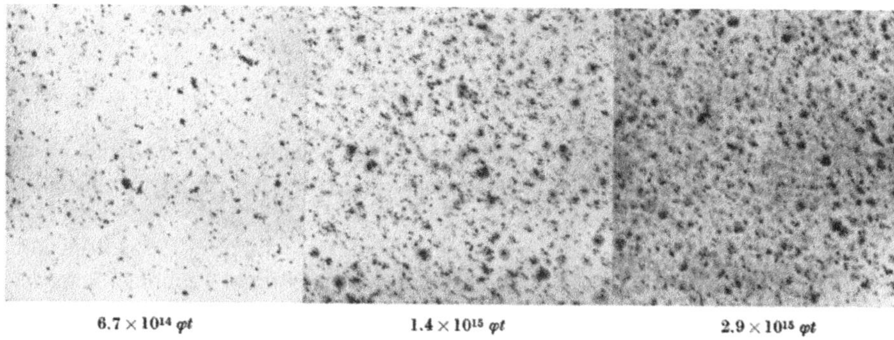

Figure 5.13. *TEM observation of uranium atoms ejected by fission fragments and collected on a carbon film (Rogers 1965a)*

Figure 5.14. *Fission traces obtained by post-irradiation of the collector and mica, indicating the presence of chunks of UO_2 (Biersack et al. 1974)*

The same experiment on UO_2 gave significantly different results. It highlighted a significant dependence of the number of ejected atoms on grain size. Atoms were ejected as vapors rather than as clusters (Rogers 1965b). In 1974, J.P. Biersack, D. Fink and P. Mertens from the Hahn-Meitner-Institute in Berlin carried out the same experiment using a different detection technique. At the end of the experiment, the collecting foil, placed against a sheet of mica, was again subjected to neutron irradiation to fission the collected uranium-235. The fission fragments created tracks (latent tracks) in the mica that could be revealed chemically (etched tracks). The presence of tracks, organized in a star shape, indicated that real "pieces" of UO_2 had been expelled (Figure 5.14). These pieces of UO_2, which they called chunks, were said to contain more than 10^9 atoms, that is, a diameter greater than 0.43 μm (Biersack et al. 1974).

5.5. Chapter 5 references

5.5.1. *Biographies of some of the chapter's personalities*

Numerous researchers have contributed to the advancement of knowledge on the notion of defects. Below are brief biographies of three of them, arbitrarily chosen.

Jenö Pál Wigner (1902, Budapest, Austro-Hungarian Empire–1995, Princeton, USA) began his university studies at the Budapest University of Technical Sciences, continuing at the Technische Hochschule Berlin. He also attended the colloquia of the German Physical Society, which brought together researchers such as Max Planck, Max von Laue, Werner Heisenberg, Walther Nernst, Wolfgang Pauli and Albert Einstein. Wigner met Leó Szilárd, who became his friend. In 1921, he took up the post of assistant to Davis Hilbert, where he studied the theory of symmetries in quantum mechanics. In 1930, he was recruited by Princeton University. It was at this time that he anglicized his first name to Eugene Paul. During the Manhattan Project, he led a group designing a plutonium nuclear reactor. In the irradiated materials community, J.P. Wigner is best known for predicting that the energy stored in materials when defects are created can be suddenly released during a thermal excursion. He was awarded the Nobel Prize in Physics in 1963 "for his contributions to the theory of the atomic nucleus and elementary particles", notably through the discovery and application of fundamental principles of symmetry.

Figure 5.15. *Eugene-Paul Wigner. Photo credit: AIP Emilio Segrè Visual Archives, Wigner Collection*

Frederick Seitz (1911, San Francisco, California–2008, New York) began his career as a solid-state physics researcher at Princeton University under the direction of Eugene Wigner. This work led to the crystallographic definition of the Wigner–Seitz cell. In 1940, he published *The Modern Theory of Solid.* On defects, he and H. Huntington made the first calculations of the energy of vacancy formation and interstitial formation. He published numerous journal articles on the mechanisms of defect creation, colored centers, irradiation effects in solids, etc. Among his administrative duties, he was President of Rockefeller University, and President of the United States National Academy of Sciences. F. Seitz was also a consultant to numerous industries (e.g. Dupont and Texas Instruments). R.J. Reynold Tobacco Company put him in charge of the company's biomedical research grants program. In 1981, the tobacco industry paid $6.3 million to researchers who demonstrated the absence of proof that tobacco was harmful. Over the years, F. Seitz's opinions became increasingly conservative: he supported the war in Vietnam, the anti-missile shield and downplayed the importance of the hole in the ozone layer and global warming.

Figure 5.16. *Frédéric Seitz. Photo credit: AIP Emilio Segrè Visual Archives, Physics Today Collection*

Rendel Sebastian Pease (1922, Cambridgeshire, UK–2004, Oxford, UK) had to interrupt his studies at Trinity College Cambridge because of the war, during which, as a member of the Royal Air Force's Bomber Command, he was tasked with evaluating the effectiveness of radar guidance. In 1947, R.S. Pease joined the Physics Division of the United Kingdom Atomic Energy Authority (UKAEA) at Harwell. His research, in collaboration with George Kinchin, focused on the structure of materials for fission reactors. In 1955, he joined the Compact Toroidal Reactor (CTR) program, still at Harwell. He was appointed Director of Fusion

Research at the Culham laboratory in 1968, Director of Culham from 1968 to 1981, and then Director of the fusion program until his retirement in 1987.

J.H.O. Varley would have merited a short biography, but I have found strictly no information on his scientific background. His published address is Atomic Energy Research Establishment, Harwell. I welcome any information on his biography.

5.6. References

Arnold, G.W. and Compton, W.D. (1960). Threshold energy for lattice displacement in a-Al_2O_3. *Physical Review Letters*, 4(2), 66–68.

Bauer, W. and Sosin, A. (1964). Threshold displacement energies and subthreshold displacements in copper and gold near 10°K. *Journal of Applied Physics*, 35(3), 703–709.

Biersack, J.P., Fink, D., Mertens, P. (1974). Sputtering and chunk ejection from UO_2 and metallic layers deposited on UO_2. *Journal of Nuclear Materials*, 53, 194–200.

Bohr, N. (1948). The penetration of atomic particles through matter. *Det Kongelige Danske Videnskabernes Selskab. Matematisk-fysiske Meddelelser*, XVIII(8), 1–144.

Brinkman, J.A. (1954). On the nature of radiation damage in metals. *Journal of Applied Physics*, 25(8), 961–970.

Brinkman, J.A. (1955). Production of atomic displacements by high-energy particles. *American Journal of Physics*, 24(4), 246–267.

Brooks, H. (1956). Nuclear radiation effects in solids. *Annual Review of Nuclear Science*, 6(1), 215–276.

Brown, E. and Goedecke, G.H. (1960). Effect of atomic interaction potential on defect production in metals. *Journal of Applied Physics*, 31(5), 932–934.

Burton, M. (1947). Radiation chemistry. *Journal of Physical and Colloid Chemistry*, 51(2), 611–625.

Corbett, J.W. and Watkins, G.D. (1965). Production of divacancies and vacancies by electron irradiation of silicon. *Physical Review*, 138(2A), A555–A560.

Corbett, J.W., Denney, J.M., Fiske, M.D., Walker, R.M. (1957). Electron irradiation of Copper near 10K. *Physical Review*, 108(4), 954.

Dessauer, F. (1923). Über einige Wirkungen von Strahlen. IV (About some effects of radiation. IV). *Zeitschrift für Physik*, 20(1), 288–298.

Dexter, D.L. (1960). Varley mechanism for defect formation in alkali halides. *Physical Review*, 118(4), 934–935.

Dienes, G.J. (1953). Radiation effects in solids. *Annual Review of Nuclear Science*, 2, 187–220.

Dienes, G.J. and Vineyard, G.H. (1957). *Radiation Effects in Solids*, Interscience Publishers Ltd., New York.

Erginsoy, C., Vineyard, G.H., Englert, A. (1964). Dynamics of radiation damage in a body-centered cubic lattice. *Physical Review*, 133(2), A595–A606.

Ershler, B.V. and Lapteva, F.S. (1957). The evaporation of metals by fission fragments. *Journal of Nuclear Energy (translated from Atomnaya Energiya)*, 4(4), 471–474.

Farmery, B.W. and Thompson, M.W. (1968). III. Energy spectra for copper. *Philosophical Magazine A,* 18(152), 415–424.

Firsov, O.B. (1957). Interaction energy of atoms for small nuclear separations. *Soviet Physics JETP*, 5(6), 1192–1196.

Gibson, J.B., Goland, A.N., Milgram, M., Vineyard, G.H. (1960). Dynamics of radiation damage. *Physical Review*, 120(4), 1229–1253.

Goland, A.N. (1962). Atomic displacement in solids by nuclear radiation. *Annual Review of Nuclear Science*, 12, 243–284.

Goldstein, E. (1894). Über die Einwirkung von Kathodenstrahlen auf einigen Salze (On the action of cathode rays on some salts). *Sitzungsberichte der Königlich Preussischen Akademie der Wissenschaften zu Berlin*, 937–945.

Hilsch, R. and Pohl, R.W. (1930). Zur Photochemie der Alkali- und Silberhalogenidkristalle (On the photochemistry of the alkali and silver halide crystals). *Zeitschrift für Physik*, 64(9–10), 606–622.

Howard, R.E. and Smoluchowski, R. (1959). Formation of interstitials in alkali halides by ionizing radiation. *Physical Review*, 116(2), 314–315.

Howard, R.E., Vosko, S., Smoluchowski, R. (1961). Mechanism for production of interstitiels in KCl by X-rays at low temperature. *Physical Review*, 122(5), 1406–1408.

Huntington, H.B. (1954). Creation of displacements in radiation damage. *Physical Review*, 93(6), 1414.

Itoh, N. (1968). Degradation of low-energy electrons in alkali halides. *Physica Status Solidi B*, 30(1), 199–207.

Kabler, M.N. (1964). Low-temperature recombination luminescence in alkali halide crystals. *Physical Review*, 136(5A), A1296–A1302.

Kinchin, G.H. and Pease, R.S. (1955a). The mechanism of the irradiation disordering of alloys. *Journal of Nuclear Energy*, 1, 200–202.

Kinchin, G.H. and Pease, R.S. (1955b). The displacement of atoms in solids by radiation. *Reports on Progress in Physics*, 18, 1–51.

King, W.E., Merkle, K.L., Meshii, M. (1981). Determination of the treshold-energy surface for copper using in-situ electrical-resistivity measurements in the high-voltage electron microscope. *Physical Review B (Condensed Matter and Materials Physics)*, 23(12), 6319–6334.

King, W.E., Merkle, K.L., Meshii, M. (1983). Threshold energy surface and Frenkel pair resistivity for Cu. *Journal of Nuclear Materials*, 117, 12–25.

Klick, C.C. (1960). Mechanism for coloration of Alkali Halides at low temperature. *Physical Review*, 120(3), 760–762.

Klontz, E.E. and Lark-Horowitz, K. (1951). Electron bombardment to Ge. *Physical Review*, 82(5), 763.

Klontz, E.E. and Lark-Horowitz, K. (1952). Displacements produced by electron bombardment of germanium (abstract). *Physical Review*, 86(4), 643.

Lesueur, D., Morillo, J., Mutka, H., Audouard, A., Jousset, J.C. (1983). Displacement threshold energies in binary compounds – amorphous metallic alloy $Fe_{75}B_{25}$, low dimensional conductors TaS_2 and TaS_3, homogeneous intermediate valence compound SmS. *Radiation Effects and Defects in Solids*, 77(1–2), 125–144.

Lindhard, J. and Scharff, M. (1961). Energy dissipation by ions in the keV region. *Physical Review*, 124, 128–130.

Lindhard, J., Nielsen, V., Scharff, M., Thomsen, P.V. (1963). Integral equations governing radiation effects. *Matematisk-fysiske Meddelelser Det Kongelige Danske Videnskabernes Selskab*, 33(10), 1–42.

Loferski, J.J. and Rappaport, P. (1958). Radiation damage in Ge and Si detected by carrier lifetime changes: Damage thresholds. *Physical Review*, 111(2), 432.

Lucasson, P. and Walker, R.M. (1961). Energy dependence of electron-induced atomic displacements in Al, Ag, Cu, Fe and Ni. *Discussions of the Faraday Society*, 31, 57–66.

Moliere, G. (1947). Theorie der Streuung schneller geladener Teilchen I. Einzelstreuung am abgeschirmten Coulomb-Feld (Theory of scattering of fast charged particles I. Single scattering at the shielded Coulomb field). *Zeitschrift für Naturforschung A*, 2(3), 133–145.

Montet, G.L. (1967). Threshold energy for displacement of atoms in graphite. *Carbon*, 5(1), 19.

Norgett, M.J., Robinson, M.T., Torrens, I.M. (1975). A proposed method of calculating displacement dose rates. *Nuclear Engineering and Design*, 33, 50–54.

Ozeroff, J. (1949). Atomic displacements produced by fission fragments and fission neutrons in matter. Report, General Electric Company, contract no. W-31-109 eng 52.

Pooley, D. (1965). F centre production in alkali halides by radiationless electron hole recombination. *Solid State Communications*, 3, 241–243.

Pooley, D. (1966a). [110] anion replacement sequences in alkali halides and their relation to F-centre production by electron-hole recombination. *Proceedings of the Physical Society*, 87(1), 257.

Pooley, D. (1966b). F-centre production in alkali halides by electron–hole recombination and a subsequent [110] replacement sequence: A discussion of the electron–hole recombination. *Proceedings of the Physical Society*, 87, 245–256.

Pooley, D. and Runciman, W.A. (1970). Recombination luminescence in alkali halides. *Journal of Physics C: Solid State Physics*, 3(8), 1815–1824.

Poulin, F. and Bourgoin, J.C. (1980). Threshold energy for atomic displacement in electron-irradiated germanium. *Revue de Physique Appliquée*, 15(1), 15–19.

Primak, W. (1955). Experimental evidence for thermal spikes in radiation damage. *Physical Review*, 98(6), 1854.

Primak, W. (1972). Threshold for radiation effects in silica. *Physical Review B (Condensed Matter and Materials Physics)*, 6(12), 4846–4851.

Przibram, K. (1923). Verfärbung und Lumineszenz durch Becquerelstrahlen (Discolouration and luminescence due to Becquerel rays). *Zeitschrift für Physik*, 20, 196–208.

Quelard, G., Dural, J., Ardonceau, J., Lesueur, D. (1978). Threshold energy of displacement in bismuth. *Radiation Effects and Defects in Solids*, 39(1), 45–46.

Robinson, M.T. and Torrens, I.M. (1974). Computer simulation of atomic-displacement cascades in solids in the binary-collision approximation. *Physical Review B (Condensed Matter and Materials Physics)*, 9, 5008–5024.

Rogers, M.D. (1965a). Mass transport of uranium by fission fragments. *Journal of Nuclear Materials*, 15(1), 65–72.

Rogers, M.D. (1965b). Mass transport and grain growth induced by fission fragments in thin films of uranium dioxide. *Journal of Nuclear Materials*, 16(3), 298–305.

Rogers, M.D. and Adam, J. (1962). Ejection of atoms from uranium by fission fragments. *Journal of Nuclear Materials*, 6(2), 182–189.

Seitz, F. (1946). Color centers in alkali halide crystals. *Reviews of Modern Physics*, 18(3), 384–408.

Seitz, F. (1949). On the disordering of solids by action of fast massive particles. *Discussions of the Faraday Society*, 5, 271–282.

Seitz, F. (1953). Generation of F centers at low temperature. *Physical Review*, 89(6), 1299.

Seitz, F. (1954). Color centers in alkali halide crystals. II. *Reviews of Modern Physics*, 26(1), 7–94.

Seitz, F. and Koehler, J.S. (1956). Displacement of atoms during irradiation. *Solid State Physics*, 2, 305–448.

Seitz, F. and Wigner, E.P. (1956). The effects of radiation on solids. *Scientific American*, 185(2), 76–84.

Sensel, G. (1910). Versuche zur Trennung des Uran und Uran X auf elektrolytischem Wege und durch Kathodenzerstäubung (Experiments on the separation of uranium and uranium X by electrolytic means and by cathode sputtering). *Zeitschrift für physikalische Chemie-Stochiometrie und verwandtschaftslehre*, 71, 563–570.

Siegel, S. (1949). Effect of neutrons bombardment on order in the alloy Cu_3Au. *Physical Review*, 75(12), 1823–1824.

Silsbee, R.H. (1957). Focusing in collision problems in solids. *Journal of Applied Physics*, 28(11), 1246–1250.

Smakula, A. (1930). Über die Verfärbung der Alkalihalogenidkristalle durch ultraviolettes Licht (Coloration of crystals of alkali halides by ultra-violet light). *Zeitschrift für Physik*, 63(11–12), 762–770.

Smoluchowski, R. and Wiegand, D.A. (1961). Formation of interstitials and vacancies in alkali halides. *Discussions of the Faraday Society*, 31, 151–155.

The Research Staff of the General Electric Company (1923). VII. Cathode disintegration. *The London, Edinburgh, and Dublin Philosophical Magazine and Journal of Science*, 45(265), 98–112.

Thompson, M.W. (1968). II. The energy spectrum of ejected atoms during the high energy sputtering of gold. *Philosophical Magazine A*, 18, 377–414.

Thompson, M.W., Farmery, B.W., Newson, P.A. (1968). I. A mechanical spectrometer for analysing the energy distribution of sputtered atoms of copper or gold. *Philosophical Magazine A*, 18, 361–376.

Varley, J.H.O. (1954a). A new interpretation of irradiation-induced phenomena in alkali halides. *Journal of Nuclear Energy*, 1, 130–143.

Varley, J.H.O. (1954b). A mechanism for the displacement of ions in an ionic lattice. *Nature*, 174, 886.

Varley, J.H.O. (1962). Discussion of some mechanisms of F-centre formation in alkali halides. *Journal of Physics and Chemistry of Solids*, 23, 985–1005.

Vavilov, V.S., Smirnov, L.S., Galkin, G.N., Spitsyn, A.V., Patskevich, V.M. (1956). Formation of crystal lattice defects in germanium by fast-electron bombardment. *Soviet Physics-Technical Physics*, 1(9), 1805–1808.

Wigner, E.P. (1942). U.S. Atomic Energy Commission Report, CP-387.

Yoshida, M. (1961). Distribution of interstitials and vacancies produced by an incident and fast neutron. *Journal of the Physical Society of Japan*, 16(1), 44–50.

Ziegler, J.F., Biersack, J.P., Littmark, U. (1985). *SRIM: The Stopping and Range of Ions in Solids*. Pergamon Press, New York.

6

Metals Under Irradiation

Studies of metals under irradiation began after those on ionic crystals for several reasons linked to their properties: (1) their strong resistance to irradiation requires high fluences; (2) X-rays, gammas and low-energy electrons are inoperative; (3) in most metals, defects are mobile at ordinary temperature. Many defects disappear by recombination, elimination on wells and formation of defect clusters; (4) defect detection, initially mainly by electrical conductivity, is more delicate than by optical spectroscopy. As we shall see, research into metals under irradiation only really began with the advent of particle accelerators and nuclear reactors, and the ability to carry out irradiation and measurements at low temperatures.

The first mention of an irradiation effect in a metal was published in the *Journal of the Russian Society of Physics and Chemistry (Zhurnal Russkogo Fisiko-Khimicheskogo Obshchestva)* by Orloff, about whom I have found no information. Having sealed a capsule containing radium bromide with aluminum foil, four months later he observed protuberances on the surface of the foil in contact with the radioactive substance. Orloff (1904) believed that an alloy had been formed by the accumulation of particles released by the radium. An origin he did not consider was the appearance of blistering due to the formation of helium bubbles. Apart from this anecdotal article, there are no publications on metal irradiation prior to the early 1940s. But everything (almost everything) was in place for the rapid development of studies on the effects of irradiation on the properties of metallic materials. The notions of defects, diffusion and dislocation were well known, particularly from studies on the cold workings of metals.

A brief bibliographical study using the keywords "irradiation–neutron–metal" and "irradiation–ion–metal" showed the quasi-exponential increase in the number of publications from the 1950s onwards (Figure 6.1). A large number of works existed before 1949, but these were marked with the seal "restrained diffusion" or "confidential defense". These reports have often been declassified, but access to them remains difficult.

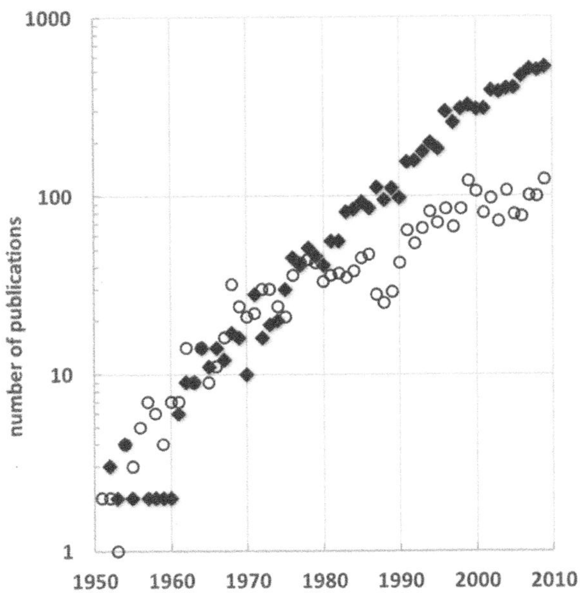

Figure 6.1. *Evolution of the annual number of publications with the keywords "irradiation–ion–metal" (solid symbols) and "irradiation–neutron–metal" (hollow circles)*

Metals play a major role in the reliability and safety of nuclear reactors. Therefore, it is hardly surprising that research into the irradiation of metallic materials has developed rapidly, in all its aspects, from the most fundamental to developments in usage properties. The methods and techniques used range from density-functional theory to mechanical property testing, including simulations and experiments at different resolutions, with a view to describing the multi-scale evolution of metal properties under irradiation.

In view of the limited number of pages in this book, the history of these subjects will not be developed, and only that of point defects will be covered.

6.1. Notions shared with other disciplines

When research into metal irradiation began in earnest, a great deal was already known about defects thanks to work on diffusion and the cold working of metals. The first notions of diffusion phenomena in metals owe much to the work of Williams Chandler Roberts-Austen, Professor of Metallurgy at the Royal School of Mine. Roberts-Austen gained his reputation through his knowledge of metal properties and his interest in industrial applications. Austenite was named in his honor. He studied the diffusion of gold, platinum and rhodium in liquid lead by chemical analyses of sections of solidified samples. He also applied this procedure to the study of the diffusion of gold in solid lead (Roberts-Austen 1896). As the subject of diffusion has already been dealt with in section 4.1.3.1, only a brief review of self-diffusion in metals is given here.

6.1.1. *Self-diffusion in metals*

Experimental measurements of self-diffusion use radioactive isotopes to avoid the chemical effects that are always possible with simulants. The first of these experiments was a measurement of self-diffusion in liquid lead. The marker of diffusion was thorium B, the name of ^{212}Pb at the beginning of the 20th century (Gróh and Hevesy 1920). The following year, the same team virtually published the same experiment, but in solid lead (Gróh and Hevesy 1921). The diffusion of lead in lead at 280°C over a period of one year remained low. It gave an upper limit for the diffusion coefficient: $D < 0.0001$ cm²/day ($< 1.1 \times 10^{-13}$ m²/s). In this paper, the authors make no mention of diffusion mechanisms. Hevesy and Gróh worked in Budapest at the *Chemisches Institut der Tierärztkichen Hochschule*[1]. In 1926, Jacov Frenkel published a microscopic description of diffusion based on the presence of dissociated atoms (spontaneous formation of the vacancy-interstitial atom pair) (Frenkel 1926).

In 1939, Rollin (working at the Radiation Laboratory, University of California, Berkeley) published the self-diffusion coefficient of copper at 940°C and 1030°C, that is, at 87% and 95% of the melting temperature. The experiment involved producing ^{64}Cu, a radioactive isotope, in a layer about a tenth of a millimeter thick, by interaction of an 8 MeV deuteron beam with copper. Diffusion was obtained by heating at high temperature for around 40 h, and the diffusion profile by measuring the activity of successive sections of the sample. The period of ^{64}Cu (12.7 h) only

1 Veterinary School Chemistry Institute.

allowed short diffusion times, so only measurements of diffusion coefficients at temperatures close to the melting temperature were accessible (Rollin 1939).

In the same year, Steigman (Columbia University, New York), Shockley and Nix (Bell Telephone Laboratories, New York), in a paper entitled "The self-diffusion in copper", examined the mechanism of diffusion. At the start of the article, they announced that three mechanisms could be considered possible: (1) interatomic exchanges between neighboring atoms; (2) atoms in interstitial positions diffusing from interstitial site to interstitial site; and (3) vacant sites, with diffusion resulting from atoms jumping to vacant sites (Steigman et al. 1939). These three possibilities were the starting point for an article by Huntington and Seitz in 1942. Calculation of the equilibrium energies and saddle-point configuration for each scenario showed that the vacancy mechanism is greatly preferable (Huntington and Seitz 1942). The following article in *Physical Review* presents a more detailed calculation of the energy required to displace a vacancy in copper. The value found was comparable to the activation energy of self-diffusion in copper (Huntington 1942). It should be remembered that these studies were carried out at thermodynamic equilibrium, that is, close to the melting temperature, only the vacancies were in non-negligible concentration, interstitials only playing a role under irradiation.

6.1.2. *Cold metalworking*

My grandfather, a blacksmith, had ancestral knowledge of the role of temperature on the ease with which iron can be deformed and of quenching speed on hardness. The steels he worked with were mild steels, with a carbon content of less than 0.5%. But he had no idea of the microstructure he was creating within the material. Plumbers have acquired the same knowledge about copper. They speak of annealed copper or work-hardened copper.

The earliest mention I have found of a scientific approach to annealing and work-hardening is a paper by Brandsma (1928) of the Philips Natuurkunding Laboratorium in Eindhoven. The object of study was a copper wire that was annealed and then cold-worked along half its length. The effects of thermal annealing were monitored by measuring the thermoelectric power created by the junction of the two copper states. No empirical law was derived, but he concluded that electrical power was a measure of internal strains.

The 1950s saw a wealth of experiments on the effect of cold working on electrical resistivity at low temperatures (Molenaar and Aarts 1950; Eggleston 1952).

A comparison between irradiation and the cold working of metals was made, particularly with regard to the concepts of defects and their investigation techniques. In 1951, an article by John Clarke Slater of the Massachusetts Institute of Technology described the effects of radiation on materials and compared the effects of radiation with those of cold working:

> Radiation damage will produce displacements which in some ways are similar, in other ways undoubtedly different, from those resulting from cold work. It should be clear by now, however, that the methods of investigating it, both experimentally and theoretically, will be identical to those used for cold work. It thus becomes evident that any well-thought-out program of research on radiation damage should be combined with a program for studying cold work by the identical methods (Slater 1951).

The following year, a team from the Atomic Energy Research Department, North American Aviation[2] published an article on the annealing of cold-worked copper (Bowen et al. 1952). The degree of cold working[3] is characterized by the residual electrical resistance, that is, the resistance value in the temperature range 4–16K. Their hypothesis was that cold working creates network distortions, about which they wrote:

> In the present fashion these distortions would be called dislocations, but since no detailed description of the distortion is required for this study the more general terms will be retained.

These distortions would change the electron diffusion cross-section. So, if Mattiessen's rule applies, it can be used to determine the concentration of distortion in the lattice. This rule has been known since the 1860s (Matthiessen et al. 1860).

Numerous experiments have been carried out combining low-temperature plastic deformation with dimensional and electrical measurements. For example, a paper by Molenaar and Aarts (1950, South Africa) shows that after plastic deformation at liquid air temperature, a return to room temperature restores resistivity without being

2 North American Aviation had a nuclear power program for its space business. The SNAP 2, 8, 10A, 10A/2 reactors are optimized for space applications. Their fuel is a zirconium-uranium hydride. SNAP 10A/2 had a thermal output of 100 kW, and went into operation in 1967 (Cockeram D.J. (1966). *Progress in Astronautics and Rocketry*, 16, 393–415).

3 In this example, the copper wire has been cold drawn, reducing its cross-section by 96.9%.

accompanied by a restoration of mechanical properties. The annealing curve for gold shows two domains in which resistivity decreases, attributed to the disappearance of vacant sites (at room temperature) and to conglomerates of two or more vacant sites (between –100 and –50°C). The authors also state that: "the absence of any change in the mechanical properties suggests that the dislocations, which are responsible for the strain-hardening, do not change at these temperatures" (Druyvesteyn and Manintveld 1951). These initial experiments by Delft's *Laboratory for Technical Physics* were followed by a more quantitative study by a group from North American Aviation, California (Bowen et al. 1952). Experiments on the softening kinetics of cold-worked copper, carried out at numerous temperatures (from 100 to 250°C) and over a wide range of annealing times (from 1 to 300 min), can be described within the framework of a fourth-order reaction: $dn/dt = -\nu n^4 e^{-E/RT}$. What remains is to find the processes that work together to give fourth-order kinetics.

In 1952, Frederick Seitz published a long article entitled "Generation of vacancies by moving dislocations", in which he explains the creation of vacancies by Franck-Read sources when they generate dislocations under increasing stress (Seitz 1952).

The analogy between irradiation and cold working has often been evoked; for example, Dienes writes:

> The analogy to cold work is based on the notion that the properties of crystals are a unique function of atomic positions regardless of how the atoms got there. In many cases this has been a fruitful notion and many of the radiation effects are similar to those produced by cold work. However, the analogy must break down if the nature of the disturbances introduced by cold work and by radiation is quite different. First of all, plastic deformation involves the motion and multiplication of dislocations while irradiation is not expected to introduce such an organized lattice disturbance. Some significant differences in behavior are, therefore, to be expected on this basis. Further, differences are also expected in the vacancy interstitial type of disturbances left behind in the lattice as indicated by the following considerations. It is known from recent theoretical work by Seitz that cold work very probably introduces vacancies. Interstitials may also be introduced by plastic deformation but it requires much higher energies to produce interstitials than vacancies, at least in metals, so that one may expect the number of interstitials created by cold work to be rather small (Dienes 1953).

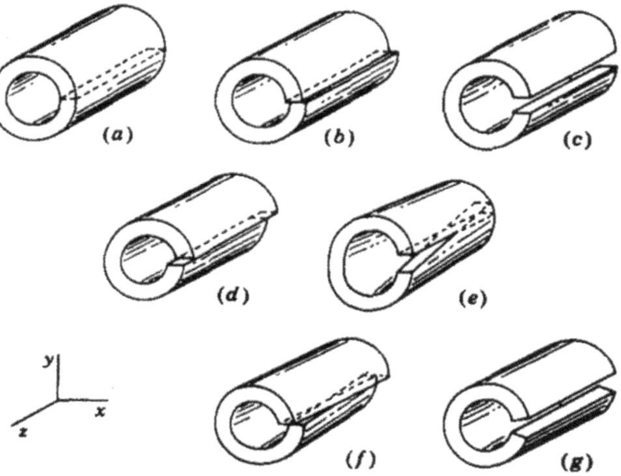

Figure 6.2. *The different types of dislocation according to Voltera, after (Hirth 1985). In reality, only figures (b), (c) and (d) are dislocations, while figures (e), (f) and (g) represent disclinations*

6.1.3. *Dislocation theory*

The emergence of dislocation theory alone deserves a book of its own. So as not to stray too far from radiation, I will just give a few important dates, based on the excellent article "A brief history of dislocation theory" by John Price Hirth[4] (1985).

The forerunners of dislocation theories can be found in ether[5], which contains localized defects that allow it to deform. The stress field around these objects has been calculated by several researchers, including Vito Volterra. The latter, a professor at Rome's La Sapienza University, published a very long article in French on a study of the deformation field, the distortions, in the cases presented (Voltera 1907)[6].

4 Hirth: Professor at Ohio State University, Columbus in 1985, then at Washington State University.

5 Ether: a mysterious substance that was supposed to fill the universe and whose role was to transmit effects between bodies such as gravitational force, light, electric and magnetic fields... Michelson and Morley's experiments at the end of the 19th century put an end to this hypothesis.

6 I confess I have only skimmed the 117-page article, packed with equations.

Understanding the plasticity of metals has progressed considerably since their crystalline structure was first described. For example, Polanyi[7] (1923) studied the planes and directions of sliding in single-crystal metal wires by observing the steps created on the surface during a tensile test. Yet, he did not address the mechanism of sliding. The same year, Taylor and Elam (1923, Cambridge) published a similar study on aluminum.

Between 1934–1935, Orowan (Budapest) published five articles on crystal plasticity. The subjects were low-temperature plasticity, a dynamic view of plasticity, slip mechanisms, justification of the plasticity dynamics model and a complement to the shear stress formula (Orowan 1934a, 1934b, 1934c, 1935a, 1935b).

This very brief overview of knowledge on defects in metallic materials was intended to remind us that work on irradiation is part of a wider context.

6.2. Creation of defects in metals by irradiation

It was in the 1950s that irradiation with neutrons or charged particles made a real contribution to understanding the fundamental processes involved in metal damage. This basic research was accompanied by equally active technological research. In March 1950, Billington and Siegel (1950, Oak Ridge National Laboratory) reported for the U.S. Atomic Energy Commission on the state of knowledge concerning the effects of nuclear reactor radiation on metals. The materials used in this review were of three types: two simple metals (Al and Cu), two alloys Cu_3Au and Co-Be, and three stainless steels. They showed that irradiations have major effects on the properties of these materials, but that the available data were insufficient to understand the origin of the changes observed. This work, which mobilized major research centers, evolved rapidly.

As early as 1955, relatively extensive reviews were published by researchers at the Royal Society Mond Laboratory, Cambridge (Glen 1955), the University of Illinois, Urbana (Seitz and Koehler 1956) and an extensive bibliographic database containing 779 references (about half of them on metals) was compiled by Friedemann (1956, 1957). The 1957 edition, the fourth, was specially compiled for the *Colloquium on the effects of irradiation on materials* (John Hopkins University, Baltimore, March 27–29, 1957).

---

7 Polanyi: Kaiser Wilhelm Institut für Faserstoffchemie, Berlin-Dahlen.

6.2.1. *Irradiation of pure metals*

The first irradiation combined with in situ resistivity measurement was probably carried out by a team from North American Aviation. The experiment involved the irradiation at −150°C of aluminum, copper and two alloys AuCu and $AuCu_3$ by alphas accelerated by Berkeley's 60-inch cyclotron. They observed a significant change in electrical resistivity and noted that the annealing of the induced effects begins as early as −80°C.

In contrast, induced effects in ordered alloys are stable at room temperature (Martin et al. 1951). In 1952, Marx et al. (1952a, University of Illinois) published the evolution of the electrical resistivity of several metals subjected to irradiation with 12 MeV deuterons[8] at two temperatures (−140°C and −150°C). The resistivity–fluence curves show a negative curvature that becomes more pronounced the higher the irradiation temperature. They attributed the origin of the curvature to the migration of vacancy pairs. As for the difference in resistivity evolution rates, they considered that copper, silver and gold have similar electronic structures, and that this difference is due to an atomic number effect in the processes involved. But this did not work for nickel and tantalum. A new analysis of these results was presented in Columbus at the *1952 March meeting* of the *American Physical Society* (Henderson et al. 1952; Marx et al. 1952b). They proposed a completely different approach based on the increase in the probability of defect annihilation as their concentration increases: $\Delta\rho = KN$ with N the number of vacancy-interstitial pairs, $dN = Adn - BNdn - C_{(N,T)}dt$ where n is the number of deuterons per unit area, A is the pair creation term, B is an annihilation term due to the action of an incident particle and C represents thermal annealing (here considered to be absent). This gives $\Delta\rho = KA/B \times (1 - \exp(-Bn))$. The agreement with the experiment is remarkable (Figure 6.3).

Monitoring damage by measuring resistivity is a particularly powerful method, enabling in situ measurements but giving no information on the type of defect causing the resistivity increment. Other experimental techniques need to be deployed in parallel. Strain measurement, one of the simplest methods to implement, has been widely used in cold deformation studies. In his review article, Broom (1954, Department of Industrial Metallurgy at the University of Birmingham) states that over 150 publications have appeared on this subject since Becquerel's work in 1846, but he adds "but much of the data is of little present value owing to the neglect of experimental precautions now recognized as essential".

---

8 The cyclotron at the University of Illinois delivers deuterons from 10 to 30 MeV.

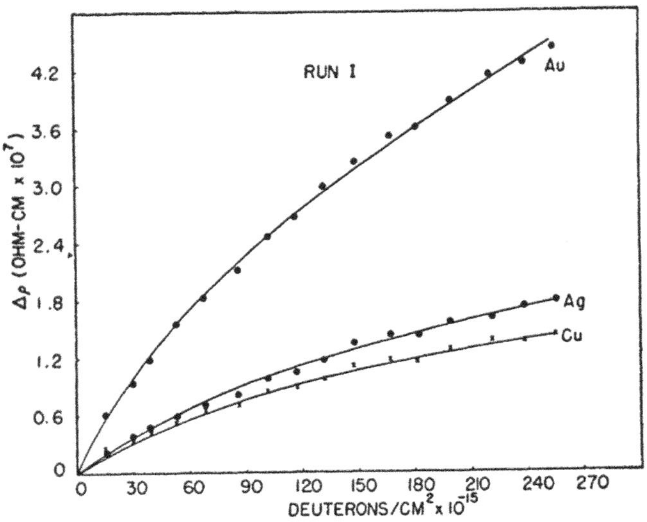

Figure 6.3. *Evolution of the electrical resistivity of three metals irradiated at –140°C with 12 MeV deuterons. The curves are taken from a model based on the annihilation of defects as the concentration increases (Marx et al. 1952a)*

In 1945, Andrade, a researcher at University College, London, had the idea of verifying the role of surface imperfections on glide. To this end, he modified the surface by bombardment with polonium alphas[9]. The result was that the creep rate[10] was increased under α bombardment with an efficiency that depended on the initial strain-hardening rate of the copper wire. Before concluding, the author pointed out that the data were obtained before the war and published afterwards: "owing to the destruction of this laboratory during the air attacks, facilities are at present lacking for any such measurements" (Andrade 1945). Andrade's experiment was unsuccessfully reproduced on a cadmium single crystal and a source producing a higher flux (5×10^8 α/cm^2/s) (Makin 1954).

Thus, ten years after the first experiment, an understanding was far from reach, since a North American Aviation publication concluded: "since the effect is small, apparently independent of beam density, and is comparable to the usual deviations in the absence of radiation it can be considered a null result" (Jeppson et al. 1955).

9 This source had to be particularly intense to have an effect on creep. The author only says "a strong polonium source".

10 The author uses the term "rate of flow".

It is remarkable that at this time it was possible to publish an article describing a method, only to conclude that it was unsuccessful.

For a 1950s–1960s view of the effect of irradiation on creep rate, we recommend consulting the review articles by Broom (1954) and Glen (1955).

Indentation hardness measurements are another easy-to-use method. Billington and Siegel (1950) studied the effects of nuclear reactor radiation on the Rockwell F hardness of annealed or cold-worked copper. For a fluence of 5×10^{19} neutrons/cm^2, the hardness of annealed or cold-worked copper increases from R_F 43 to R_F 90 for annealed copper, and from R_F 93 to R_F 97 for cold-worked copper. From their results on several materials, they extracted seven behaviors that seem universal: for example, the hardness of all samples increases provided that the exposure is long enough, and the effects of irradiation on hardness and electrical resistance are all the greater, the softer the material is initially.

The first studies of the effects of irradiation on stress–strain curves date from the same period, that is, the early 1950s (Blewitt and Coltman 1951; Blewitt et al. 1960). Blewitt and his colleagues worked at the Solid State Division, Oak Ridge National Laboratory. Given the technological interest of this research, the effects of irradiation on all mechanical properties were studied. As a result, the number of publications rapidly grew.

6.2.2. *Irradiation of ordered alloys*

The notion of order in substitutional metal alloys has been known at least since the early 1920s, particularly in the gold–copper alloy. The effect of temperature on the order–disorder transformation was studied very early by X-ray diffraction (Nix and Shockley 1938). So it is hardly surprising that ordered alloys were the focus of the first irradiations of metallic materials.

Siegel (1949, Oak Ridge National Laboratory, Tennessee) published a paper on the effect of neutron bombardment on order in Cu_3Au. He justified his studies of ordered alloys on the grounds that, in a mono-atomic metal, interstitial migration and their recombination mean that, at equilibrium, the number of atoms displaced (in the sense of atoms out of normally occupied positions) depend on neutron flux and diffusion rate, and was probably too small to produce a significant change in macroscopic properties. In an ordered alloy, on the other hand, the change in degree of order was permanent as long as the temperature did not exceed 200°C for Cu_3Au. Under these experimental conditions, the fraction of displaced atoms did not exceed a few percent (calculated according to Seitz's formula). This concentration seemed too low to account for almost complete disorder. He therefore suggested that elastic collisions produce local heating, followed by rapid quenching of the disordered high-temperature state.

If the material was initially disordered, irradiation induced little change. However, this hypothesis contradicted irradiation results for Cu–Be alloys (Billington and Siegel 1950). In addition, experiments on Cu_3Au by a team from Westinghouse Electric Corporation, Pittsburgh, showed that irradiation in a reactor at 80°C tended to disorder initially ordered samples and to order initially disordered materials (see Figure 6.4) (Glick et al. 1952).

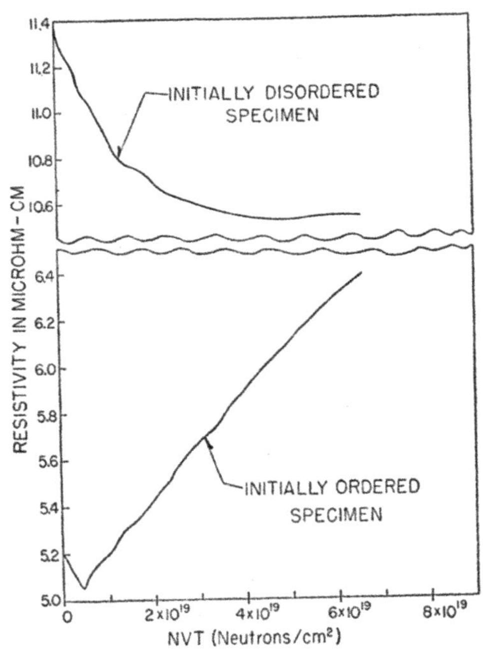

Figure 6.4. *Resistivity evolution under neutron irradiation at 80°C of initially ordered and disordered Cu_3Au samples (Glick et al. 1952). Fluence is expressed in nvt (neutron velocity time): fluence is equal to n the neutron density (neutrons/cm^3) multiplied by v the neutron velocity (cm/s) and by t the time (s). Surprisingly, neutron fluence (nvt) is the fluence of thermal neutrons*

This ability to order a metal alloy was confirmed by further reactor experiments at Oak Ridge (Figure 6.5) and by electron irradiation. Adam et al. (1953) at AERE-Harwell wrote:

> The experiment described here suggests that by introduction of artificially produced lattice defects it is possible to reach an equilibrium degree of order corresponding to a temperature lower than that obtainable by purely thermal treatment in a reasonable time.

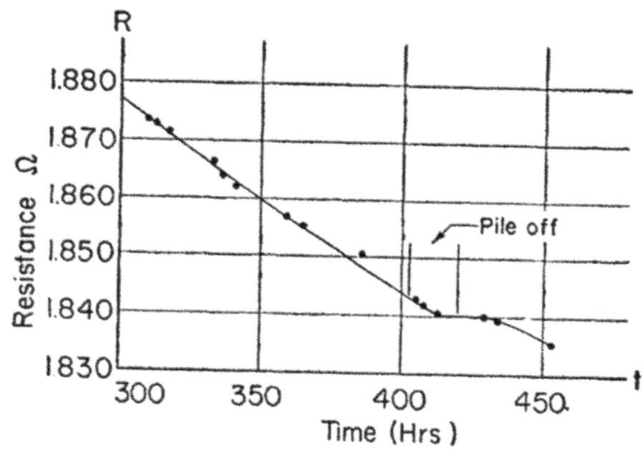

Figure 6.5. *Disordered Cu_3Au resistance versus time at 200°C and under a fast neutron flux of 10^{12} n/cm²/s. The decrease in electrical resistance demonstrates that neutron irradiation can help restore an ordered structure. The plateau at cell shutdown shows that temperature alone cannot restore the ordered state (Blewitt and Coltman 1952)*

6.3. Displacement threshold

The first measurement of displacement threshold energy was carried out in copper. For this, the variation in electrical resistivity was used to measure the creation of defects at liquid air temperature. The displacement threshold was thus evaluated as equal to 25 ± 1 eV (Eggen and Laubenstein 1953). The experiment by Corbett et al. (1957), who carried out the same type of measurement but at 10 K, followed. By varying the electron energy from 0.7 to 1.4 MeV, they obtained a threshold value of 22 ± 3 eV (Corbett et al. 1957). In 1961, Lucasson and Walker (General Electric Research Laboratory, New York) measured at low temperature (20.4 K) the evolution of the displacement cross-section σ_d as a function of electron energy (from 0.5 to 1.35 MeV). The defect concentration $c = n\sigma_d$ can be deduced from the resistivity measurements if the variation in resistivity $\Delta\rho_f$ per unit defect concentration has first been determined: $\sigma_d = \frac{\Delta\rho}{n\Delta\rho_f}$, n being the electron flux. Two assumptions are implied: the type of defects created is independent of energy and all defects are "frozen" at the irradiation temperature (Lucasson and Walker 1961). For comparison with theoretical models, they made two assumptions about the probability of displacements P_d of an atom, either a single threshold P_d (T) = 1 if T > Td, or P_d (T) varying linearly from 0 to 1 for T varying from T_d to $T_d + \Delta T_d$. The copper example in Figure 6.6 shows that the assumption of a single threshold at 22 eV seems the most likely.

The displacement thresholds for the various metals studied were as follows: Cu 22 eV, Ag 28 eV, Ni 24 eV and Fe 24 eV. Lucasson and Walker (1961) conclude with:

> In summary, the results support the validity of the simple displacement theory for low-energy collisions. The major limitation of the present work is the narrow energy range which was used. It is hoped that others who possess more energetic accelerators will extend this work in order to further test the above conclusions.

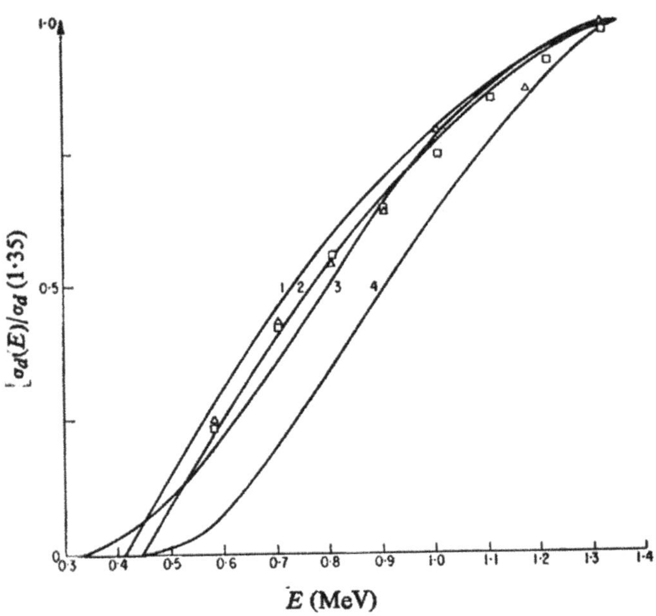

Figure 6.6. *Comparison of experimental and calculated displacement cross-sections normalized to 1.35 MeV for copper. 1: T_d = 20 eV, ΔT_d = 0; 2: T_d = 22 eV, ΔT_d = 0 eV; 3: T_d = 16 eV, ΔT_d = 16 eV; 4: T_d = 22 eV, ΔT_d = 22 eV (Lucasson and Walker 1961)*

Three years later, Bauer and Sosin (1964) of North American Aviation published a displacement threshold measurement for copper and gold. The damage production curve obtained by plotting the derivative of the variation in resistivity versus fluence for different electron energies showed the existence of two components, which they interpreted as arising from a contribution from direct displacements with a displacement threshold of T_d^{Cu} = 16–19 eV and T_d^{Au} = 33–36 eV and a second contribution from sub-threshold events associated with indirect displacements due to the presence of light impurities. The sequence $e^- \rightarrow A \rightarrow Cu$ may be more efficient than $e^- \rightarrow Cu$ if A is lighter than copper. More recently, these sub-threshold events have been remarkably highlighted by irradiation of binary compounds such as $Fe_{75}B_{25}$, TaS_2 or SmS (Lesueur et al. 1983). With regard to subthreshold events in copper, Wollenberger gave another interpretation based on the existence of anisotropy in the displacement threshold (Wollenberger and Wurm 1965). Several studies have been devoted to this energy threshold anisotropy. More or less sophisticated, they have shown, for example, that near the compact stacking directions of face-centered cubics $\langle 110 \rangle$ and $\langle 100 \rangle$, there are zones where the threshold is minimal and, on the contrary, maximal according to $\langle 111 \rangle$ (Jung et al. 1973). These results were confirmed and completed by King et al. (1981, 1983; Argonne). To appreciate the magnitude of the work, it is sufficient to note that 180 damage rate measurements were carried out at five energies between 0.5 and 1.1 MeV in ~ 40 directions distributed over the unit stereographic triangle (for further details, see section 5.2.1).

Conditions in which only isolated Frenkel pairs exist are relatively rare. The most favorable case is undoubtedly irradiation with electrons from a few hundred keV to MeV. The electrons' low mass favors low energy transfer, while maintaining a sufficiently long range. On the other hand, when irradiated with neutrons or ions, the primary atom can receive enough energy to become a projectile in its turn: a cascade of displacements is created. The number of vacancies and interstitials not annihilated in the cascade then represents only a few percent of the number of Frenkel pairs created. Numerical simulations have enabled us to gain a better understanding of the elementary processes taking place within a cascade, such as focused collisions, the formation of defect clusters and recombinations (Torrens and Robinson 1972; Robinson and Torrens 1974). Experimentally, it is more difficult to determine the number of free vacancies. One example is the experiment by Blewitt et al. (1971), whose aim was to evaluate the number of free vacancies in the case of neutron irradiation. Since diffusion is caused by the migration of vacancies, this property can be studied to assess the number of free vacancies. Diffusion can easily be monitored by electrical resistivity in a disordered alloy whose ordered state is stable under the conditions of the experiment. Vacancy jumps reestablish the ordered phase. The variation in resistivity gives the number of ordered atoms, and the number of vacant atoms can then be deduced. They find that between three and seven vacancies are released from a cascade created in Cu_3Au by a fast neutron. Based on Kinchin and Pease's (1955) estimates of the number of defects formed, they show that less than 2% of the vacancies created in a cascade remain isolated.

6.4. Description of defects

The description of the vacancy in a metal poses no real problem, apart from that of the relaxation of its first neighbors. On the other hand, the configuration of the interstitial has been the subject of much debate. Indeed, in the cfc lattice, symmetry considerations suggest the existence of six configurations for the interstitial: octahedral, quadratic (tetragonal), cubic and three configurations where two atoms share the same site along the directions $\langle 100 \rangle$, $\langle 111 \rangle$ and $\langle 110 \rangle$ directions (Figure 6.7). This is why, in a face-centered cubic metal like copper, the interstitial configuration was the subject of real controversy between the Americans and the Stuttgart group. The main issue was the migration energy of the interstitial in relation to its configuration. We will come back to this controversy in section 6.2.3.

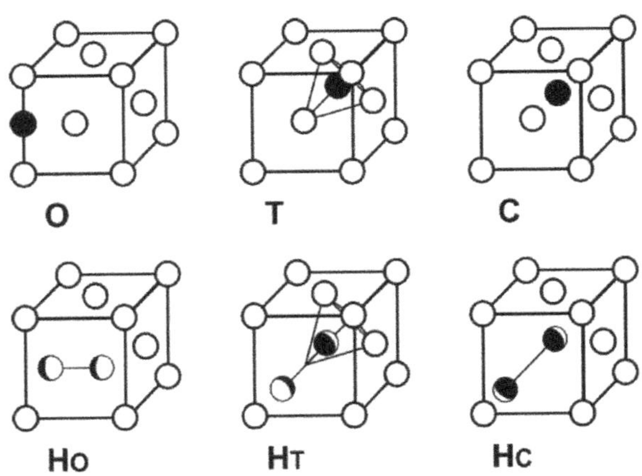

Figure 6.7. *The six possible interstitial configurations in a c.f.c. lattice: O: octahedral; T: quadratic; C: cubic; H_O, H_T, H_C: dumbbell configurations with the axis following the directions* $\langle 100 \rangle$, $\langle 111 \rangle$ *and* $\langle 110 \rangle$

But first, let us devote a few lines to describing experimental methods that provide information on defect configurations.

6.4.1. *Experimental observations of point defects*

The development of field emission electron microscopy *(Feldelektronenmikroskope)* by Müller (1937, 1938) in Berlin made it possible to "see" individual atoms for the first time. The principle of this microscope is based on a very fine metal tip applied at high voltage. Electrons emitted by the electric field at the apex of the tip follow field lines to a detector. As the field lines are representative of the local topology, the electron impacts on the detector reproduce an image of the tip structure. In reality, in the 1937 publication, the resolution was not sufficient to distinguish atoms. Müller (1943) showed that the blurring of the image was essentially determined by the electron velocity distribution.

Some 15 years later, Müller[11] made two major improvements to this technique, creating field ion microscopy (Müller 1951) and with tip cooling (Müller 1956). By reversing the polarity of the microscope and increasing the voltage, it became possible to extract the atoms adsorbed on the tip in the form of positive ions. As with electrons, these positive ions reproduce an image of the tip microstructure on the detector, with a resolution that reaches the interatomic distance. The atoms adsorbed were barium for the first experiments, then hydrogen atoms and finally helium atoms.

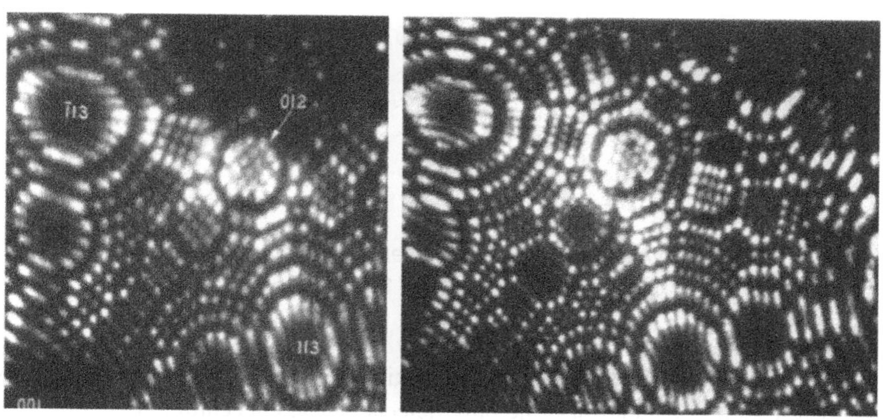

Figure 6.8. *Field ion microscopy of a platinum tip showing a vacancy in the [012] plane. On the left, the perfect plane; on the right, the same tip after removal of 15 atomic layers, the plane contains a vacancy. The following planes are again perfect (Müller 1959)*

11 By then, Müller had left Germany for Pennsylvania State University in the United States.

This technique enabled the first direct observation of a point defect. In 1959, Müller observed vacancies in platinum quenched from 1,500°C (0.85 T_f). By evaporation under field conditions at the temperature of liquid hydrogen, the crystalline layers were eliminated one after the other, sometimes revealing a vacancy like the one visible in the photograph on the right in Figure 6.8 (Müller 1959). The presence of this vacancy in a plane of atoms confirmed its preexistence, as evaporation under field conditions takes place from the edges of the planes. From the observation of five vacancies in 71 planes made up of 120 atoms, he deduced that the vacancy concentration was 5.9×10^{-4} and, as $c \propto \exp(E_f/kT)$, the vacancy formation energy E_f in platinum is equal to 1.2 eV. Müller concluded that the essential result was that the notion of vacancies, which had hitherto been based solely on a theoretical concept, was confirmed by direct observation of the atomic lattice using a low-temperature field ion microscope.

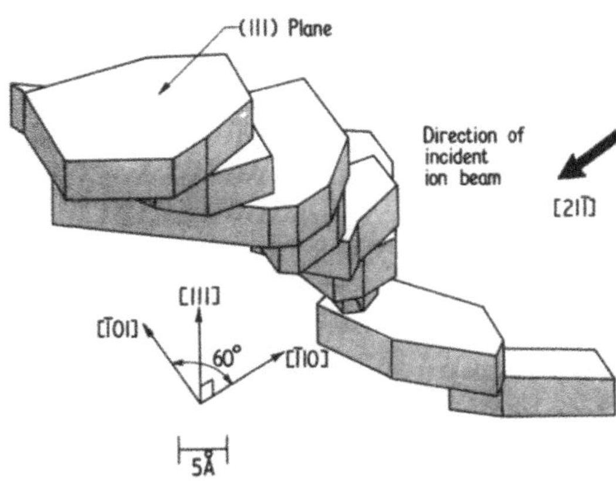

Figure 6.9. *Isometric drawing of a depleted zone constructed from the detection of vacancy clusters in [111] planes (Seidman 1973)*

Irradiating only with helium ions of maximum energy 1 keV, he could only observe vacancies or impurities, the only defects whose concentration is noticeable at thermodynamic equilibrium. Interstitials were only discovered 14 years later by Seidman (1973) of Cornell University, Ithaca, New York, in a paper entitled "The direct observation of point defect in irradiated or quenched metals by quantitative field ion microscopy". This excellent and difficult experiment on tungsten involved three stages in the field ion microscope: (1) irradiation with W^+ ions of around 30 keV at a temperature below the first annealing stage; (2) analysis of the point by pulsed field evaporation. This step enabled the detection of vacancy clusters and the reconstruction of atomically depleted zones (Figure 6.4); and (3) the study of self-interstitial diffusion during annealing by detecting their arrival at the tip surface. The aim of this study was to determine the temperature range in which self-interstitials migrate.

The discoveries of near-field microscopies, STM (Scanning Tunneling Microscope) and AFM (Atomic Force Microscope), have provided new tools for visualising individual atoms. It is hard to resist the pleasure of showing an AFM image of a vacancy in the surface of a graphite sample.

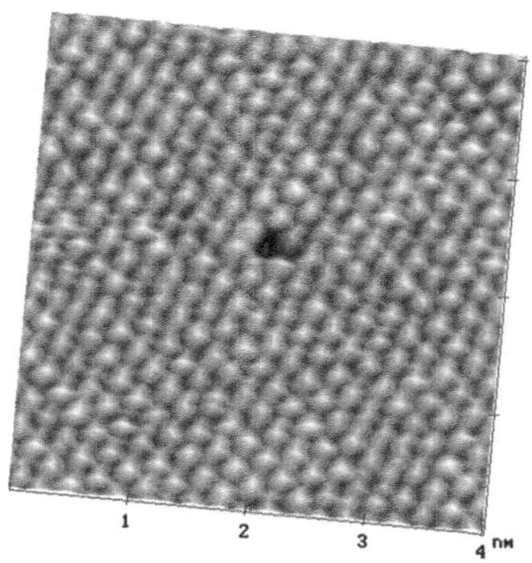

Figure 6.10. *Contact-mode atomic force microscopy of the surface of a graphite sample irradiated with high-energy heavy ions. Note the presence of a vacancy or divacancy (image: Serge Bouffard). For a color version of this figure, see www.iste.co.uk/bouffard/irradiation.zip*

Alongside these instruments, which allow us to see individual atoms, but without knowing which one is being visualized, there are statistical methods that often have much better resolution. High-resolution electron microscopy is perhaps the bridge between the two approaches. It is only very recently that the very high resolution of some transmission electron aberration-corrected microscopes has been demonstrated by the image of an interstitial in germanium (see Figure 7.8).

X-ray or neutron diffraction is an extremely powerful method for describing defects. Indeed, the distribution in reciprocal space of the intensity of diffuse scattering contains information about the structure of point defects. The finest method is certainly elastic diffuse scattering, which can be divided into three diffraction domains: small-angle diffraction; scattering between Bragg reflections; and Huang scattering (close to Bragg reflections). Ehrhart and Schilling (1973) (Institut für festkörperforschung der Kernforschungsanlage Jülich) used scattering near Bragg reflections (200), (400), (220) and (222) in directions parallel and perpendicular to the scattering vectors (Huang scattering) to study interstitials created by electron irradiation in aluminum and copper (Figure 6.11). They found that the interstitial was of cubic symmetry or slightly quadratic symmetry. They totally ruled out a crowdion-type configuration[12] (Ehrhart and Schilling 1973; Ehrhart et al. 1975). The intensity of diffuse scattering gives access to characteristic defect quantities: the volume change induced by an interstitial in copper is 0.45 atomic volume, and relaxation around a vacancy is –0.4 atomic volume.

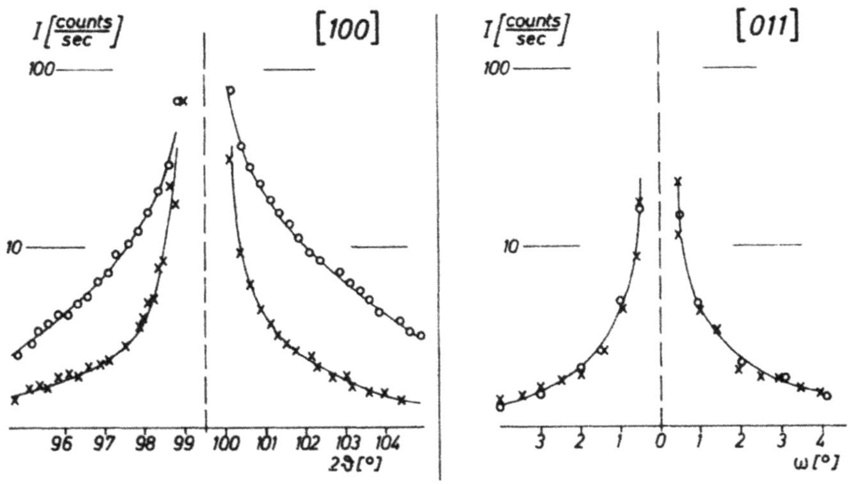

Figure 6.11. *Diffuse scattering intensity measured at 4.5°K for aluminum near $\vec{h}$ = (400) in the [100] and [011] directions. (×) and (o) are the measurement points before and after electron irradiation (Ehrhart and Schilling 1973)*

Exploration over a wider region of reciprocal space, typically along four circles on the Ewald sphere, enables possible interstitial configurations to be differentiated. In Figure 6.12, in the case of electron-irradiated aluminum[13], the scattered intensity profiles calculated for three configurations are compared with the measured profile.

12 The crowdion configuration for an interstitial consists of an extra atom in a dense row. The disturbance length is five to six cell parameters.

13 The irradiation dose is expressed as a resistivity increment at 4.5 K: $\Delta\rho_0 = 142$ nΩ. cm.

The dissociated interstitial configuration in the (100) direction is unambiguously the one in agreement with experimental measurements.

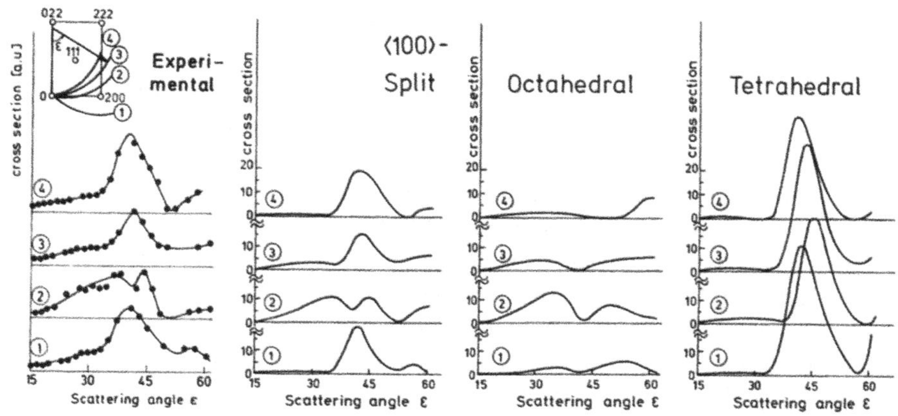

Figure 6.12. *Comparison of measured and calculated diffuse scattering intensity for three interstitial configurations in electron-irradiated aluminum (Ehrhart et al. 1974)*

Simmons and Balluffi (1960, University of Illinois) realized that the experiments used to characterize defects were unable to identify the type of defect, or to obtain the defect concentration directly, so they proposed an original method. The experiment involved measuring the difference between the relative variation of the sample dimension $\Delta L/L$ and its cell parameter $\Delta a/a$. The cell parameter is sensitive only to average expansion, whereas the macroscopic dimension is equally sensitive to the creation or destruction of atomic sites. So, if vacancies are generated, then new sites are created and $\Delta L/L - \Delta a/a$ is positive. An isotropic crystal of N atoms containing ΔN identical point defects has a defect concentration of $\Delta N/N = 3(\Delta L/L - \Delta a/a)$. For aluminum, near the melting point (660°C), they found that the dominant defect was, of course, the vacancy, with a concentration of 9.4×10^{-4} (Simmons and Balluffi 1960).

6.5. Defect annealing

Defect migration and recombination complicate the vision of electron irradiation, which is supposed to create only simple defects, that is, Frenkel pairs. But the mobility of these defects significantly increases the zoology of the defects: divacancy, di-interstitials and interstitials interacting with an impurity to remain in the simplest defect agglomerates. Defect diffusion also means that the number of displaced atoms can be very different from the number of stabilized defects. Post-irradiation annealing experiments have enabled us to gain some insight into the processes involved in the mobility of primary defects.

In 1952, Eggleston (1952, Atomic Energy Research Department, North American Aviation) published preliminary results on the annealing of copper worked at liquid helium temperature. Cold working consisted of twisting a copper wire. His results, obtained by measuring electrical resistivity, seem to show the existence of two annealing stages, one at −140°C and the other at 20°C. Assuming that the annealing process follows simple kinetics, he obtained activation energies of 10.2 ± 1.45 and 15.5 ± 2 kcal/mol for both stages (Eggleston 1952). However, exactly 46 days before that of Eggleston, the editor of *Physical Review* received a manuscript from Marx, Cooper and Henderson (Northwestern University, Illinois), also dealing with defects in copper but created by irradiation (Marx et al. 1952a). Defects created by 12 MeV deuterons present a thermal annealing stage around −165°C and an activation energy of 0.2 ± 0.05 eV/atom (4.6 ± 1.2 kcal/mol). There is only a factor of 2 between the two measurements, which is remarkable given the small temperature range over which these values are determined. Overhauser analyzed an experiment[14] of isothermal annealing after irradiation at −180°C with 12 MeV deuterons from the cyclotron of his group at the Department of Physics, University of Illinois. From each isotherm, which lasted around 2.5 h, the activation energy that could be extracted varied linearly from 0.2 to 0.7 eV between −180°C and –20°C. He concluded by pointing out that his hypotheses had varying degrees of certainty, which could be modified or rejected on the basis of experimental evidence or further theoretical examination. These included (1) annealing at low temperatures is due to recombination of interstitial–vacancy close pairs, (2) annealing above –40°C is due to volume diffusion of vacancies and their annihilation with interstitial atoms and (3) the energy of vacancy creation is 1.39 eV and the energy of migration is 0.68 eV (Overhauser 1953).

Also at the University of Illinois, Urbana, a team consisting of Cooper, Koelher and Marx made more precise measurements of the annealing stages of defects. They first observed that the initial slope of the damage curves at 10 K was twice as large as at liquid nitrogen temperature, and even slightly steeper than at 12 K. They interpreted this effect as an immediate annealing of the defects, the greater the temperature. After bombardment of the copper and silver samples, the evolution of resistivity was followed as a function of temperature from 10 K to room temperature. Annealing processes occurred in three temperature ranges: below

14 Made with the help of Henderson.

50 K, between 50 and 220 K and between 220 and 280 K (Figure 6.13). As to the physical origin of these annealing stages, they assumed that below 50 K, it was the vacancy–interstitial close pairs that recombined. On the other hand, no hypothesis was put forward as to the origin of annealing at higher temperatures (Cooper et al. 1954).

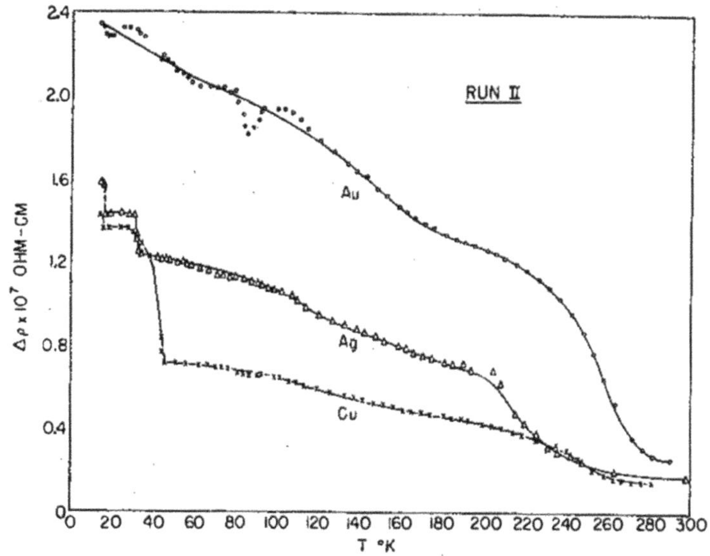

Figure 6.13. *Annealing of the irradiation-induced increase in resistivity in copper, silver and gold. The abrupt decrease in resistivity around 40 K for copper is irreversible (Cooper et al. 1955)*

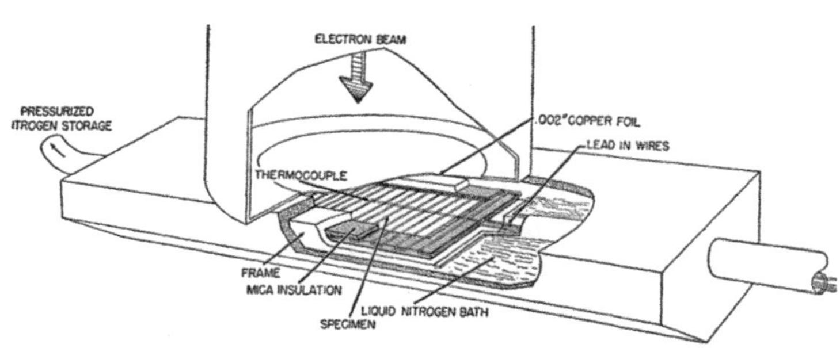

Figure 6.14. *Target box for low-temperature electron irradiation (liquid nitrogen). The window separating the accelerator vacuum from the liquid nitrogen is a 50 µm copper foil (Meechan and Brinkman 1956)*

Meechan and Brinkman (1956) of North American Aviation showed by electron irradiation at 80 K (see their irradiation device, Figure 6.14) that stage III was associated with a single process whose activation energy was equal to 0.60 ± 0.01 eV. Considering the amplitude of the annealing stages, they attributed stage III to interstitial migration. Stage IV resulted from the migration of trapped point defects released by temperature.

In 1958, Corbett and Koehler (1958) showed that stage I is composed of at least four distinct substages. The following year, Corbett, Smith and Walker of the General Electric Research Laboratory, New York, published two important papers on the annealing of electron-irradiated copper: the first on the annealing of close pairs (Corbett et al. 1959a) and the second on interstitial migration (Corbett et al. 1959b). Both papers were submitted to the editor on the same day, January 28, 1959. They began by stating that there was a consensus to define five stages in the thermal annealing of copper characterized by their temperature range: stage I: 14–65 K; stage II: 65–233 K; stage III: 233–373 K; stage IV: 373–473 K and stage V: T > 473 K. But although a great deal of experimental information existed, there was no general agreement on the defects and processes operating at each stage. One of the controversies was whether stage I involved the free migration of a defect over a long distance, or whether it only involved localized readjustments of the network. The first paper was devoted to stage I, which they showed was made up not of four but of five distinct substages named I_A, I_B, I_C, I_D and I_E (Figure 6.15).

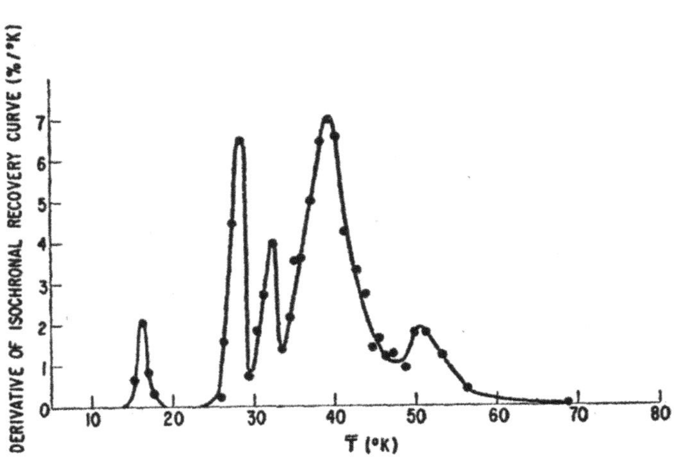

Figure 6.15. *Numerical derivative of the isochronous annealing curve. The various peaks correspond to the substages I_A, I_B,...I_E of the first annealing stage (Corbett et al. 1959a)*

Comparing the results of their electron experiments with those using deuterons (Magnuson et al. 1958) supported their hypothesis that the first three substages were intrinsic and could be attributed to the annealing of different types of close pairs.

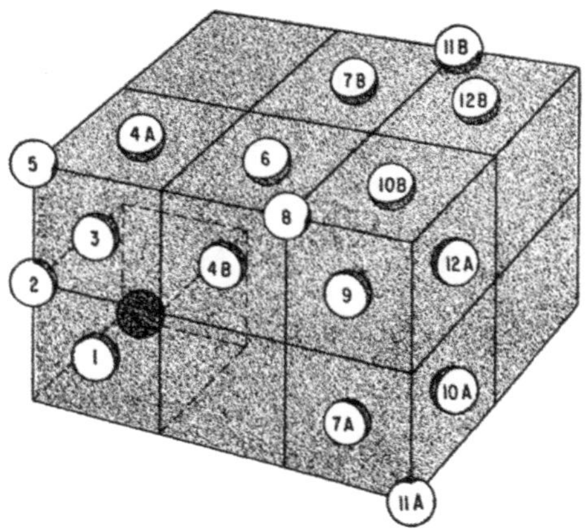

Figure 6.16. *Copper face-centered cubic lattice with a body-centered interstitial (dark atom). Atom sites are numbered according to their distance from the interstitial. Equivalent sites are not shown (Corbett et al. 1959a)*

Assuming that the interstitial was at the center of the unit cell, and based on Tewordt's (1958) calculation of lattice distortions for different configurations of the near pair, Corbett and his colleagues concluded that sites 1 and 3 in Figure 6.16 were unstable and that sites 2, 4, etc., were possible stable sites for close pairs[15]. They concluded their article with:

> [...] From this approach, we would conclude that I_A is associated with site 2, I_B with site 4 and I_C with site 5.
>
> Clearly at this point we must conclude that a final decision on the question of the defect configuration corresponding to the substages must await further calculation and experiment.

15 The method developed by Tewordt to study interstitials and vacancies in copper consists of defining three concentric regions in the metal: a region I centered around the defect where the atoms are discrete particles whose displacements can exceed the elastic limits, a region II also made up of discrete particles but where linear elasticity applies and a region III considered as a continuous medium. The formation energies calculated in this way are ≈ 0, 2.3 and 3.6 eV for the vacancy, centered interstitial and crowdion, respectively.

The second paper focuses on substages I_D and I_E, which the authors associated with free interstitial diffusion (Corbett et al. 1959b). The free diffusion of an interstitial ceases when it encounters a vacancy (annihilation), an impurity (trapping) or another interstitial (aggregate formation). Trapped interstitials are untrapped at higher temperatures. The substage I_D, which has the same activation energy as I_E, corresponds to the free migration of the interstitial before its annihilation with a vacancy.

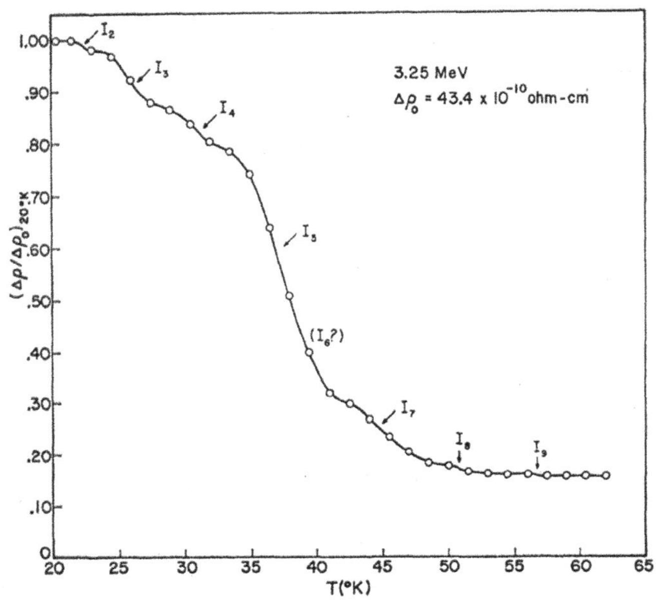

Figure 6.17. *Isochronous annealing of a copper sample irradiated with 3.25 MeV electrons at 20.4 K. Fractional recovery remaining after 10-min anneals versus annealing temperature (Tesk et al. 1964)*

A few years later, a group from Northwestern University, using the device developed by Corbett and Walker (1968), published a very precise study of the substages of copper stage I (Tesk et al. 1964). The experiment was carried out on eight different samples, with five electron energies ranging from 1.25 to 3.25 MeV and a total of 168 annealing temperatures. When you consider that temperature measurement and control were manual, you realize how cumbersome this experiment was: each annealing point required precise heating control so as not to exceed the annealing temperature, stabilization at the annealing temperature for 10 min, cooling to 20.4 K (in liquid hydrogen) and resistivity measurement. Since a point cannot be determined in less than 30 min, the total annealing time in stage I was at least 84 h (100 h is probably closer to the truth). This work enabled them to count nine substages called I_2 to I_9 (substage I_1 was at a lower temperature than their irradiation temperature) (Figure 6.17).

The phenomena involved in stage II were more difficult to characterize. Schilling and Sonnenberg (1973) of Forschungszentrum Jülich proposed two processes: the rearrangement of interstitial aggregates formed in stage I, and the removal of impurities from interstitials. The latter explains the sensitivity of stage II to sample purity, and hence the non-reproducibility of certain results.

For many years, the site properties of the interstitial remained an open question. In copper, for example, the main issue was the migration energy of the interstitial: was it ~ 0.1 eV or ~ 0.6 eV? In the former case, its annealing stage would be below 60 K, and in the latter around 250 K, stage III. In the absence of decisive experience, both points of view were based on the analysis of a large corpus of experimental results. Seitz was one of the proponents of an interstitial with low migration energy, while Seeger was one of the proponents of interstitials with high migration energy.

According to Corbett et al. (1959b), stage I_D corresponded to the free migration of the interstitial, which finally annihilated itself on its own vacancy. Stage I_E would still be the interstitial escaping its parent vacancy Granato and Nilan (1965), Urbana, attributed stage I_D to recombination of close pairs. The free interstitial would migrate to stage I_E. In contrast, the model proposed by Seeger et al. (1962) called for two types of interstitial: an extended defect, called a crowdion, consisting of an extra atom inserted in a dense direction. This defect migrates in a single direction at temperatures below those of stage I_E. Above this, it converts to a "classic" interstitial that migrates to stage III.

In 1968, a group from Jülich and Aachen raised the question of vacancy mobility in stage III (Dworschak et al. 1968). Following irradiation and quenching experiments, they came to following the conclusion:

> Summarizing, we can say that the results of the quenching experiments do not allow a decision pro or con vacancy migration in stage III. [...] There are, however, a number of indirect conclusions reported in which support vacancy migration in stage III just as our direct method of AI's does[16].

They therefore remained with hypotheses such as the formation of divacancy possibly complexed with an impurity. They concluded, however, that the results

16 At that time, AI did not mean artificial intelligence but analysis irradiation.

were best interpreted by assuming that the lacunae migrate to stage III, or more artificially by conversion to stage III of the interstitial.

Defects are characterized by well-targeted experimental techniques, including diffuse or small-angle X-ray scattering (Ehrhart and Schilling 1973), internal friction (Moser and Pichon 1973), positron annihilation (Seeger 1974) or perturbed $\gamma\gamma$ angular correlation (Wichert et al. 1978).

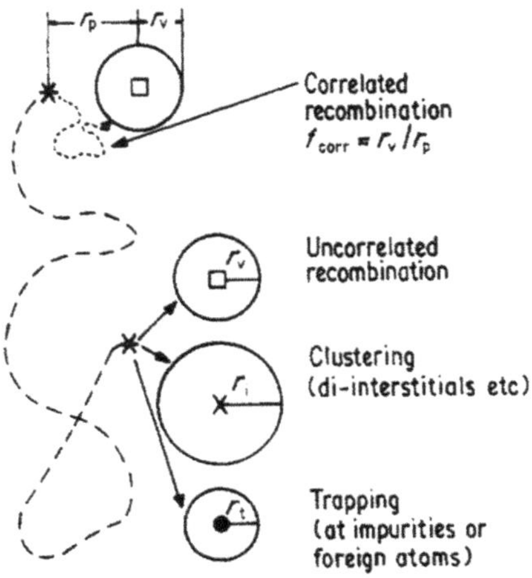

Figure 6.18. *Schematic representation of the different reaction pathways of a free interstitial in annealing stage I_D (correlated recombination) and I_E (noncorrelated recombination and aggregation with an interstitial or impurity) (Schilling and Sonnenberg 1973)*

In a review article published in 1973, Schilling and Sonnenberg recall the mechanisms attributed to the various stages and substages and schematically represent the different reactions that an interstitial may encounter (Schilling and Sonnenberg 1973). This drawing has been reproduced in numerous review articles and theses (Figure 6.18). The annealing stages and sub-stages are assigned as follows:

– I_A, I_B, I_C: close Frenkel pairs recombine;

– I_D: interstitials migrate freely before recombining with their parent vacancy;

– I_E: interstitials recombine with any other vacancy. These two substages are distinct only at low concentrations;

– II: interstitial clustering dominate;

– III: vacancies become mobile and annihilate at the interstitial clusters.

In the book *Electron Radiation Damage in Semiconductors and Metals* by Corbett (1966), we find a detailed description of studies of the annealing stages of copper (pp. 206–240) and gold (pp. 243–259).

Still driven by nuclear industry R&D, the effects of irradiation on metals are being studied from the atomic scale to the mechanical properties of nuclear reactor structural elements. It would therefore be necessary to write the history of knowledge on the formation of defect clusters, dislocation loops, bubbles and cavities, on defect-solute interaction and, on a larger scale, the effects on segregation processes, dislocation dynamics, mechanical and dimensional properties and corrosion. These are all subjects far from my expertise, and specialists are in better position than I am to describe their history.

The effects of irradiation on metals are described in greater detail in several works (e.g. Corbett 1966; Quéré 1967; Bullough 1972).

6.6. Chapter 6 references

6.6.1. *Biographies of some of the chapter's personalities*

Hillard Bell Huntington (1910, Wilkes-Barre–1992, Troy), a solid-state physicist, studied many important problems. He is best known for his work on electromigration, which he studied long before its impact on integrated circuit technology was recognized.

Figure 6.19. *Erwin Wilhelm Müller. Photo credit: Niels Bohr Library & Archives, Physic History Network*

Erwin Wilhelm Müller (1911, Berlin–1977, Washington, D.C.) invented the field emission electron microscope in 1936, while working at the Siemens Research Laboratory. In 1947, appointed to the Kaiser Wilhelm Institute for Physical Chemistry and Electrochemistry, he developed the field ion microscope, which achieved a resolution of 0.25 nm. This was the first instrument to "see" individual atoms. In 1952, he joined Pennsylvania State University, where he co-invented the tomographic atom probe.

James Stark Koehler (1914, Oshkosh–2006, Urbana). After defending his PhD in 1940, he completed a post-doctorate under the direction of Seitz. In 1944, he joined the Manhattan Engineer District to study the effects of radiation on solids. After the war, he worked on dislocation trapping and, in Illinois, on quenching point defects in metals. As a theorist, he supervised numerous PhD students who studied defects in materials using experimental approaches ranging from transmission electron microscopy to mechanical properties, X-ray diffraction, internal friction, electrical resistivity, channeling, etc. His scientific output spanned 57 years, during which he often arrived at the laboratory saying: "I've got a wild idea!".

6.7. References

Adam, J., Green, A., Dugdale, R.A. (1953). An effect of electron bombardment on order in Cu_3Au alloy. *Philosophical Magazine*, 43, 1216–1218.

Andrade, E.N.D.C. (1945). Effect of alpha-ray bombardment on glide in metal single crystals. *Nature*, 156(3952), 113–114.

Bauer, W. and Sosin, A. (1964). Threshold displacement energies and subthreshold displacements in copper and gold near 10°K. *Journal of Applied Physics*, 35(3), 703–709.

Billington, D.S. and Siegel, S. (1950). Effect of nuclear reactor radiation on metals. Report, AECD – 2810, Oak Ridge National Laboratory, U.S. Atomic Energy Commission, Oak Ridge.

Blewitt, T.H. and Coltman, R.R. (1951). The effect of pile irradiation on the stress-strain curve of copper. *Physical Review*, 82(2), 769.

Blewitt, T.H. and Coltman, R.R. (1952). The effect of neutron irradiation on metallic diffusion. *Physical Review*, 85, 384.

Blewitt, T.H., Coltman, R.R., Jamison, R.E., Redman, J.K. (1960). Radiation hardening of copper single crystals. *Journal of Nuclear Materials*, 2(4), 277–298.

Blewitt, T.H., Klank, A.C., Scott, T.L., Weber, W. (1971). Experiments to determine the number of "free vacancies" in a fission-neutron displacement cascade. In *International Conference on Radiation-Induced Voids in Metals*. State University of New York, Albany.

Bowen, D., Eggleston, R.R., Kropschot, R.H. (1952). A study of the annealing kinetics in cold worked copper. *Journal of Applied Physics*, 23(6), 630–634.

Brandsma, W.F. (1928). Thermokräfte in kaltbearbeiteten Metallen. *Zeitschrift für Physik*, 48(9), 703–705.

Broom, T. (1954). Lattice defects and the electrical resistivity of metals. *Advances in Physics*, 3(9), 26–83.

Bullough, P. (1972). Point defects and their aggregates in metals. *Journal of Physics F: Metal Physics*, 3, 233–496.

Cooper, H.G., Koehler, J.S., Marx, J.W. (1954). Resistivity changes in copper, silver, and gold produced by deuteron irradiation near 10°K. *Physical Review*, 94(2), 496.

Cooper, H.G., Koehler, J.S., Marx, J.W. (1955). Irradiation effects in Cu, Ag and Au near 10°K. *Physical Review*, 97(3), 599–607.

Corbett, J.W. (1966). *Electron Radiation Damage in Semiconductors and Metals*. Academic Press, New York and London.

Corbett, J.W. and Walker, R.M. (1958). Discrete recovery spectrum below 65°K in irradiated copper. *Physical Review*, 110, 767–768.

Corbett, J.W., Denney, J.M., Fiske, M.D., Walker, R.M. (1957). Electron irradiation of copper near 10°K. *Physical Review*, 108(4), 954.

Corbett, J.W., Smith, R.B., Walker, R.M. (1959a). Recovery of electron-irradiated copper. I. Close pair recovery. *Physical Review*, 114(6), 1452.

Corbett, J.W., Smith, R.B., Walker, R.M. (1959b). Recovery of electron-irradiated copper. II. Interstitial migration. *Physical Review*, 114(6), 1460.

Dienes, G.J. (1953). Radiation effects in solids. *Annual Review of Nuclear Science*, 2, 187–220.

Druyvesteyn, M.J. and Manintveld, J.A. (1951). Change of resistivity of gold by cold-working at liquid-air temperature and its recovery by temporary heating. *Nature*, 168(4281), 868–869.

Dworschak, F., Schuster, H., Wollenberger, H.J., Wurm, J. (1968). Analysis of point defect states in copper. 2. Effects of stage 1 and stage 3 annealing on electron-irradiation-induced Frenkel pairs. *Physica Status Solidi*, 29(1), 81.

Eggen, D.T. and Laubenstein, M.J. (1953). Displacement energy of face centered cubic alloy. *Physical Review*, 91, 238.

Eggleston, R.R. (1952). Cold work studies on copper at low temperature. *Journal of Applied Physics*, 23(12), 1400–1401.

Ehrhart, P. and Schilling, W. (1973). Investigation of interstitials in electron-irradiated aluminium by diffuse-X-ray scattering experiments. *Physical Review B (Condensed Matter and Materials Physics)*, 8(6), 2604–2621.

Ehrhart, P., Haubold, H.G., Schilling, W. (1974). Investigation of point defects and their agglomerates in irradiated metals by diffuse X-ray scattering. *Festörperprobleme*, 14, 87–110.

Ehrhart, P., Schlagheck, U., Schilling, W. (1975). Huang scattering from Frenkel defects in electron-irradiated f.c.c metals. *Journal of Applied Crystallography*, 8, 150.

Frenkel, J. (1926). Über die Wärmebewegung in festen und flüssigen Körpern. *Zeitschrift für Physik*, 35(8–9), 652–669.

Friedemann, H.C. (1956). Effects of irradiation on solids. In *The Effects of Radiation on Materials*, Harwood, J.J., Hausner, H.H., Morse, J.G. Rauch, W.G. (eds). Reinhold Publishing Corporation, New York.

Friedemann, H.C. (1957). Effects of irradiation on solids. A bibliography. Report, NP-5939, Penn-Texas Corporation, Houston.

Glen, J.W. (1955). A survey of irradiation effects in metals. *Advances in Physics*, 4(16), 381–477.

Glick, H.J., Brooks, F.C., Witzig, W.F., Johnson, W.E. (1952). The resistivity of Cu_3Au during neutron irradiation. *Physical Review*, 87, 1074.

Granato, A.V. and Nilan, T.G. (1965). Concentration dependence of stage i annealing in irradiated copper. *Physical Review*, 137(4A), A1250–A1263.

Gróh, J. and Hevesy, G. (1920). Die Selbstdiffusionsgeschwindigkeit des geschmolzenen Bleis. *Annalen der Physik*, 368(17), 85–92.

Gróh, J. and Hevesy, G. (1921). Die Selbstdiffusion in festem Blei. *Annalen der Physik*, 65(3), 216–222.

Henderson, J.W., Cooper, H.G., Marx, J.W. (1952). Analysis of the resistivity changes produced by deuteron irradiation of copper, silver and gold. *Physical Review*, 86(4), 642.

Hirth, J.P. (1985). A brief history of dislocation theory. *Metallurgical Transactions A*, 16(12), 2085–2090.

Huntington, H.B. (1942). Self-consistent treatment of the vacancy mechanism for metallic diffusion. *Physical Review*, 61(5–6), 325.

Huntington, H.B. and Seitz, F. (1942). Mechanism for self-diffusion in metallic copper. *Physical Review*, 61(5–6), 315.

Jeppson, M.R., Mather, R.L., Andrew, A., Yockey, H.P. (1955). Creep of aluminium under cyclotron irradiation. *Journal of Applied Physics*, 26, 365.

Jung, P., Chaplin, R.L., Fenzl, H.J., Reichelt, K., Wombacher, P. (1973). Anisotropy of threshold energy for production of Frenkel pairs in copper and platinum. *Physical Review B (Condensed Matter and Materials Physics)*, 8(2), 553–561.

Kinchin, G.H. and Pease, R.S. (1955). The displacement of atoms in solids by radiation. *Reports on Progress in Physics*, 18, 1–51.

King, W.E., Merkle, K.L., Meshii, M. (1981). Determination of the threshold-energy surface for copper using in-situ electrical-resistivity measurements in the high-voltage electron microscope. *Physical Review B (Condensed Matter and Materials Physics)*, 23(12), 6319–6334.

King, W.E., Merkle, K.L., Meshii, M. (1983). Threshold energy surface and Frenkel pair resistivity for Cu. *Journal of Nuclear Materials*, 117, 12–25.

Lesueur, D., Morillo, J., Mutka, H., Audouard, A., Jousset, J.C. (1983). Displacement threshold energies in binary compounds – Amorphous metallic alloy $Fe_{75}B_{25}$, low dimensional conductors TaS_2 and TaS_3, homogeneous intermediate valence compound SmS. *Radiation Effects and Defects in Solids*, 77(1–2), 125–144.

Lucasson, P. and Walker, R.M. (1961). Energy dependence of electron-induced atomic displacements in Al, Ag, Cu, Fe and Ni. *Discussions of the Faraday Society*, 31, 57–66.

Magnuson, G.D., Palmer, W., Koehler, J.S. (1958). Isothermal annealing below 60°K of deuteron irradiated noble metals. *Physical Review*, 109, 1990–2000.

Makin, M.J. (1954). Effect of alpha-particle bombardment on creep in cadmium single crystals. *Nature*, 174(4433), 752–753.

Martin, A.B., Austerman, S.B., Eggleston, R.R., McGee, J.F., Tarpinian, M. (1951). The effects of cyclotron bombardment on the physical properties of metals. *Physical Review*, 81, 664.

Marx, J.W., Cooper, H.G., Henderson, J.W. (1952a). Radiation damage and recovery in Cu, Ag, Au, Ni, and Ta. *Physical Review*, 88(1), 106.

Marx, J.W., Koehler, J.S., Wert, C.A. (1952b). Effect of deuteron bombardment on the electrical resistivity of copper, silver and gold. *Physical Review*, 86, 643.

Matthiessen, A., Holzmann, M., Wheatstone, C. (1860). On the effect of the presence of metals and metalloids upon the electric conducting power of pure copper. *Philosophical Transactions of the Royal Society of London*, 150, 85–92.

Meechan, C.J. and Brinkman, J.A. (1956). Electrical resistivity study of lattice defects introduced in copper by 1.25-Mev electron irradiation at 80°K. *Physical Review*, 103(5), 1193–1202.

Molenaar, J. and Aarts, W.H. (1950). Change of resistivity by cold working at liquid-air temperature. *Nature*, 166(4225), 690.

Moser, P. and Pichon, R. (1973). Internal friction phenomenon in relation to the properties of point defects in metals. *Journal of Physics F: Metal Physics*, 3, 363.

Müller, E.W. (1937). Elektronenmikroskopische Beobachtungen von Feldkathoden. *Zeitschrift für Physik*, 106(9–10), 541–550.

Müller, E.W. (1938). Weitere Beobachtungen mit dem Feldelektronenmikroskop. *Zeitschrift für Physik*, 108(9), 668–680.

Müller, E.W. (1943). Das Auflösungsvermögen des Feldelektronenmikroskops. *Zeitschrift für Physik*, 120(5), 270–282.

Müller, E.W. (1951). Das Feldionenmikroskop. *Zeitschrift für Physik*, 131(1), 136–142.

Müller, E.W. (1956). Resolution of the atomic structure of a metal surface by the field ion microscope. *Journal of Applied Physics*, 27(5), 474–476.

Müller, E.W. (1959). Beobachtung von nahezu fehlerfreien Metallkristallen und von Punktdefekten im Feldionenmikroskop. *Zeitschrift für Physik*, 156(3), 399–410.

Nix, F.C. and Shockley, W. (1938). Order-disorder transformation in alloys. *Reviews of Modern Physics*, 10, 1.

Orloff, N. (1904). Action of radium on metals. *Zhurnal Russkogo Fisiko-Khimicheskogo Obshchestva*, 36(2b), 41–45.

Orowan, E. (1934a). Zur Kristallplastizität. III Über den Mechanismus des Gleitvorganges. *Zeitschrift für Physik*, 89(9), 634–659.

Orowan, E. (1934b). Zur Kristallplastizität. II Die dynamishe Auffassung der Kristallplastizität. *Zeitschrift für Physik*, 89(9), 614–633.

Orowan, E. (1934c). Zur Kristallplastizität. I Tieftemperaturplastizität und Beckershe Formel. *Zeitschrift für Physik*, 89(9), 605–613.

Orowan, E. (1935a). Zur Kristallplastizität. *Zeitschrift für Physik*, 98(5), 382–387.

Orowan, E. (1935b). Zur Kristallplastizität. IV Wietere Begründung des dynamischen Platizitätsgesetzes. *Zeitschrift für Physik*, 97(9), 573–595.

Overhauser, A.W. (1953). Isothermal annealing effects in irradiated copper. *Physical Review*, 90(3), 393.

Polanyi, M. (1923). Über Strukturänderungen in Metallen durch Kaltbearbeitung. *Zeitschrift für Physik*, 17(1), 42–53.

Quéré, Y. (1967). *Défauts ponctuels dans les métaux*. Masson et Cie, Paris.

Roberts-Austen, W.C. (1896). Bakerian lecture – On the diffusion of metals. *Philosophical Transactions of the Royal Society of London, Series A: Mathematical and Physical Sciences*, 187, 383–415.

Robinson, M.T. and Torrens, I.M. (1974). Computer simulation of atomic-displacement cascades in solids in the binary-collision approximation. *Physical Review B (Condensed Matter and Materials Physics)*, 9, 5008–5024.

Rollin, B.V. (1939). Self-diffusion in copper. *Physical Review*, 55(2), 231.

Schilling, W. and Sonnenberg, K. (1973). Recovery of irradiated and quenched metals. *Journal of Physics F: Metal Physics*, 3, 322–350.

Seeger, A.K. (1974). The study of defects in crystals by positron annihilation *Applied Physics*, 4(3), 183–199.

Seeger, A.K., Mann, E., Jan, R. (1962). Zwischengitteratome in kubisch-flächenzentrierten kristallen, insbesondere in kupfer. *Journal of Physics and Chemistry of Solids*, 23(6), 639–658.

Seidman, D.N. (1973). The direct observation of point defects in irradiated or quenched metals by quantitative field ion microscopy. *Journal of Physics F: Metal Physics*, 3, 393–421.

Seitz, F. (1952). On the generation of vacancies by moving dislocations. *Advances in Physics*, 1(1), 43–90.

Seitz, F. and Koehler, J.S. (1956). Displacement of atoms during irradiation. *Solid State Physics*, 2, 305–448.

Siegel, S. (1949). Effect of neutrons bombardment on order in the alloy Cu_3Au. *Physical Review*, 75(12), 1823–1824.

Simmons, R.O. and Balluffi R.W. (1960). Measurements of equilibrium vacancy concentrations in aluminum. *Physical Review*, 117(1), 52.

Slater, J.C. (1951). The effects of irradiation on materials. *Journal of Applied Physics*, 22(3), 237–256.

Steigman, J., Shockley, W., Nix, F.C. (1939). The self-diffusion of copper. *Physical Review*, 56(1), 13–21.

Taylord, G.I. and Elam, C.F. (1923). Bakerian lecture: The distortion of an aluminium crystal during a tensile test. *Proceedings of the Royal Society of London A*, 102, 643–667.

Tesk, J.A., Jones, E.C., Kauffman, J.W. (1964). Stage I recovery spectrum of pure copper irradiated with electrons over the range 1.25 to 3.25 MeV. *Physical Review*, 133(1A), A288–A93.

Tewordt, L. (1958). Distortion of the lattice around an interstitial, a crowdion, and a vacancy in copper. *Physical Review*, 109(1), 61–68,

Torrens, I.M. and Robinson, M.T. (1972). Computer simulation of atomic displacement cascades in solids. In *Interatomic Potentials and Simulation of Lattice Defects*, Gehlen, P.C., Beeler, J.R. Jaffee, R.I. (eds). Plenum Press, New York.

Voltera, V. (1907). Sur l'équilibre des corps élastiques multiplement connexes. *Annales scientifiques de l'École Normale Supérieure*, 24, 401–517.

Wichert, T., Deicher, M., Echt, O., Recknagel, E. (1978). Defect annealing in copper around stage III. *Physical Review Letters*, 41(24), 1659–1662.

Wollenberger, H.J. and Wurm, J. (1965). Atomic displacement cross sections in copper for anisotropic threshold energy. *Physica Status Solidi*, 9, 601–609.

7

Semiconductors Under Irradiation

Knowledge of the specific properties of semiconductors is relatively recent, although Michael Faraday is said to have observed around 1833 that the electrical resistance of silver sulfate decreases with increasing temperature. This negative slope is typical of semiconductor behavior. The low number of carriers and the existence of two types of carrier are characteristics that explain the sensitivity of these materials to defects.

The difference in electrical conduction depending on the direction of the current in certain materials was observed as early as the end of the 19th century (Braun 1874; Schuster 1874). In reality, it is the metal–semiconductor junction that creates this rectification effect, a phenomenon used to create the first diodes by applying a metallic tip to the surface of a semiconductor. However, this technique, which represents an advance on vacuum tubes, would not have enabled the explosion of microelectronics if ion implantation had not paved the way to miniaturization. From this point onward, basic research, R&D and industrialization have been and are always closely linked. Technological interest is confirmed by the strong presence of industrial laboratories (Bell Telephone Laboratories, Radio Corporation of America, General Electric Research Laboratories, Fairchild Semiconductors, Ion Physics Corporation, etc.) in publications on semiconductor irradiation or implantation, the number of which has grown almost exponentially (Figure 7.1). The number of implanters has probably followed the same trend. Indeed, the International Atomic Energy Agency counted 12,000 implanters worldwide in May 2022.

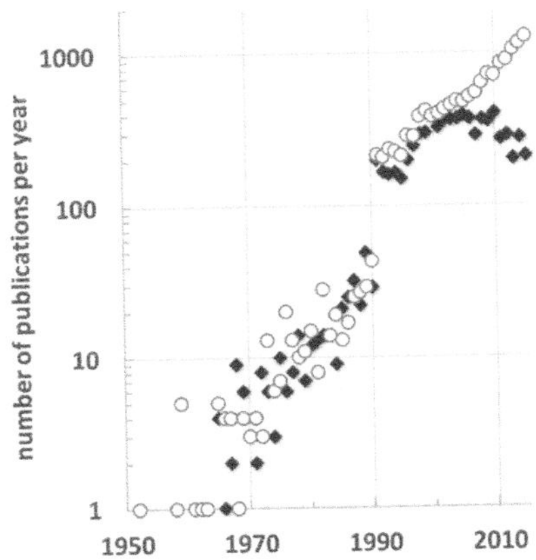

Figure 7.1. *Trend in the annual number of publications using the keywords "irradiation or bombardment – semiconductor" (solid symbols) and "implantation – semiconductor" (hollow circles). For a color version of this figure, see www.iste.co.uk/bouffard/irradiation.zip*

7.1. First irradiation of semiconductors

It is not easy to determine who first irradiated a semiconductor. If we exclude an irradiation of selenium by electrons of less than 100 eV (De Laer Kronig 1924), the oldest reference (1949) that we have found concerns neutron irradiation of germanium in the Oak Ridge reactor (Johnson and Lark-Horowitz 1949). However, the article states: "p-type samples show an increase in conductivity, as expected from the results of cyclotron irradiation". So this was not the first irradiation. The problem is that no reference is given. Furthermore, in the *Bibliography of cyclotron literature*, a collection of abstracts of work carried out with the Oak Ridge National Laboratories cyclotron from 1933 to 1955, there is no trace of irradiation of semiconductors (Livingston et al. 1956). Nevertheless, the idea of irradiation and its application was well established by the end of the 1940s. In 1949, William Shockley[1] filed a patent entitled "Semiconductor translating device", describing the formation of a p-n junction by ion implantation. (Shockley 1949). In 1954, he filed another patent entitled "Forming of Semiconductor Devices by Ionic Bombardment". The aim of this invention is to increase the frequency limit of a transistor junction by reducing the thickness of the base. The manufacturing process is based on monokinetic ion beam

1 William Shockley, John Bardeen and Walter Brattain invented the point-contact transistor at Bell Telephone Labs in 1947, for which they were awarded the Nobel Prize in Physics in 1956.

bombardment; the ions are considered as impurities after implantation. Ion implantation equipment featuring a mass separator, electrical and mechanical scanning, etc., is described. It also includes a post-implantation heating device to eliminate ion impact-induced damage (Shockley 1957). Russel S. Ohl also holds a patent entitled "Semiconductor translating device", which has many points in common with W. Schockley's patent[2]. Their purpose, however, is different. The object of R.S. Ohl's invention is to improve the characteristics of semiconductor signal conversion devices, and in particular:

> More specific objects of this invention are to increase the current carrying capacity and peak back voltage of semiconductor type rectifiers, to enhance the characteristics, such as sensitivity, speed of response and stability, of photocells, to obtain improved performance for point contact type semiconductor harmonic producers, detectors, frequency converters and the like, simplify and expedite the fabrication of very high frequency, i.e. microwave, semiconductor translating devices and facilitate the fabrication of amplifiers, of the type known as transistors, including a body of P conductivity type semiconductive material, such as silicon.

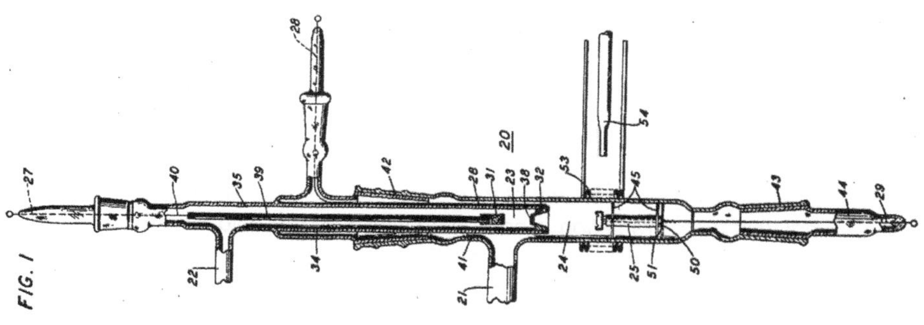

Figure 7.2. *The semiconductor bombardment device described in the patent by Ohl (1956) and, with a few exceptions, in the patent by Shockley (1957). The device consists of a closed glass envelope (20) pumped by (21), an ion gun (23) with anode (31) and cathode (32), and a bombardment chamber (24) with sample holder (25)*

Patents on implantation in the 1950s and 1960s are difficult to count, but their number is indicative of the technological interest in this technique. Regardless of date, a patent search on the United State Patent and Trademark Office database using the keywords "implantation" and "semiconductor" yields over 263,800 patents.

2 W. Schockley and R.S. Ohl both worked for Bell Telephone Laboratories, New York.

Let us go back to the first neutron irradiations at Oak Ridge National Laboratory. Under irradiation, conductivity increases monotonically for p-type samples. For n-type materials, on the other hand, conductivity initially decreases, passes through a minimum and then slowly increases. Hall effect measurements at the end of irradiation indicate that the material has changed from n- to p-type. This is interpreted as a change in electron mobility and the creation of acceptor sites by neutrons (Crawford and Lark-Horowitz 1950).

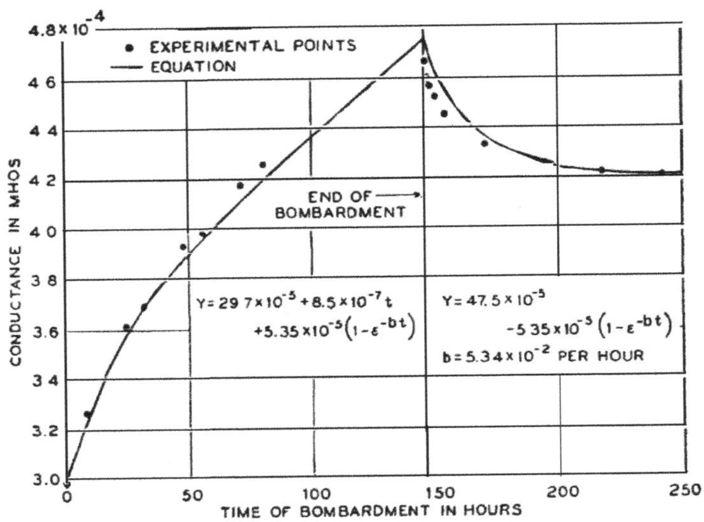

Figure 7.3. *Evolution of conductance expressed in mho[3] as a function of irradiation time. Irradiation is performed with 5.3 MeV polonium alphas (1.9 × 10^8 alphas per second) (Brattain and Pearson 1950)*

W. H. Brattain and G. L. Pearson of Bell Telephone Laboratories interpreted the evolution of the conductivity of a thin layer of n-type germanium subjected to 5.3 MeV polonium alphas as a succession of phases. The acceptor sites created by the bombardment begin by reducing the number of conduction electrons. Once the material has become p-type, bombardment continues to increase the concentration of acceptors and thus the number of conducting holes. From data analysis, they obtain the following quantitative results: (1) initially, when the bombarded layer is n-type, electrons are removed from the conduction band at a rate of 78 conducting electrons per alpha, and (2) after conversion, p-type conducting holes are introduced at a rate

3 The mho was replaced by the 1971 Siemens. 1 mho = 1 S = 1 $m^{-2}kg^{-1}s^3A^2$.

of 8.6 per alpha. These values seem reasonable; a calculation with SRIM-2013 giving 280 displacements per alpha. When irradiation stopped, 6.6 conducting holes disappeared, with a half-life of 13 h at room temperature (Figure 7.3). After bombardment and the relaxation phase, the depth distribution of holes was determined by removing thin layers of material and measuring the resulting change in conductance. The authors reported that the density of holes was greater at the end of the alphas' paths than near the surface, the ratio being 2.5 (Brattain and Pearson 1950).

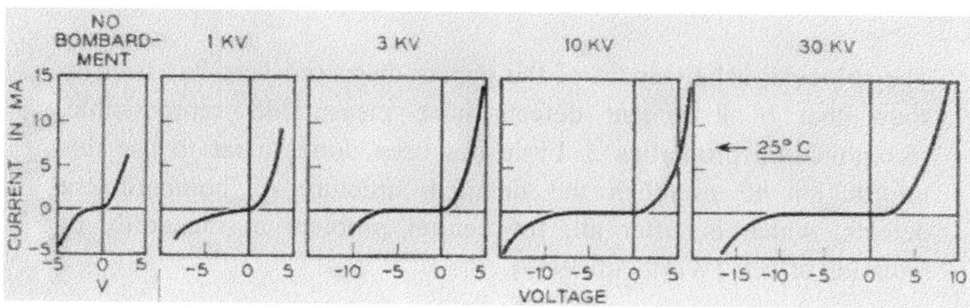

Figure 7.4. *Influence of helium ion energy on current-voltage characteristics (Ohl 1952)*

The first use of accelerated ions to modify an electrical property of a semiconductor dates back to 1950. Russel S. Ohl and W.J. Pietenpol of Bell Telephon Laboratories presented preliminary results on this subject at the Electronics Conference on June 22, 1950 at the University of Chicago. This was followed by a paper entitled *Properties of ionic bombarded silicon* (Ohl 1952). The 1952 paper demonstrates that the electrical properties of the silicon surface[4] can be greatly modified by bombardment with positive ions. The article concludes with an interesting property of irradiation:

> More particularly it is possible to contaminate silicon with impurities such as boron to the point where its rectifying properties are almost completely lost and by bombardment it is possible to convert the crystal into a very useful rectifier.

Also at Bell Telephone Laboratories, G.K. Wertheim was able to determine the energy position of the various levels created by electron irradiation in silicon, based

4 At no point does it define what the surface of silicon is, presumably the thickness corresponding to the average ion path.

on a number of more or less realistic assumptions, such as that mobility does not depend on the disorder introduced by bombardment. The author found, for example, an electron-trapping level at 0.16 eV below the conduction band and a hole-trapping level at 0.29 eV above the valence band (Wertheim 1957). Two years later, J.H. Crawford and J.W. Cleland of Oak Ridge National Laboratories found an electron-trapping level of 0.20 eV (Crawford and Cleland 1959). These bandgap states were early associated with charge carrier recombination mechanisms. G.K. Wertheim's review of experiments cites 11 papers on charge recombination in irradiated semiconductors. He concludes his 1959 article with:

> The principal acheivement of the papers discussed here has been to show that bombardment defects have measurable, reproducible, recombination properties ... Little has been done so far to use this information to establish the detailed structure of bombardment defects, which is, after all, the central problem in the study of radiation effects (Wertheim 1959).

Similar studies have been carried out on germanium. For example, a group at Oak Ridge National Laboratory, after irradiating germanium with neutrons, concluded that neutron bombardment introduces an average of 3.1 electron traps in the energy band gap per neutron. These traps could be due to the presence of vacancies. Hole traps must also be produced, but in germanium they are so close to the valence band that they are not observable (Crawford and Lark-Horowitz 1950; Cleland et al. 1951).

A similar series of experiments on germanium containing different amounts of impurity was carried out at Cambridge University by W.D. Cussins with different ions ranging from hydrogen to antimony and energies from 5 to 90 keV delivered by an accelerator designed specifically for this study. The junction consisted of a tungsten whisker applied to the germanium surface. In the discussion, he challenges the explanation given by K. Lark-Horovitz's explanation of conductivity evolution, based on the creation of point defects, and proposes a thermal spike approach (Cussins 1955).

In 1965, the Ion Physics Corporation, Burlington, Massachusetts, published an article showing everything they had successfully achieved with their accelerator–implanter: solar cells, radiation detectors, unipolar field-effect transistors and bipolar transistors (King et al. 1965).

7.2. Defect generation and counting

The proceedings of the "Radiation effects in semiconductors" conference[5] published in the *Journal of Applied Physics* give a good overview of the state of knowledge on semiconductors under irradiation at the end of the 1950s.

7.2.1. *Determining the displacement threshold*

E.E. Klontz and K. Lark-Horowitz of Purdue University in Indiana were the first to measure the minimum energy to be transmitted to a silicon atom in order to extract it from its site. They found, by measuring changes in conductivity at 78 K, that a minimum of 30.5 eV must be transmitted to a silicon atom for it to be ejected from its site (Klontz and Lark-Horowitz 1952). At the 1954 March meeting of the American Physical Society, Walter Kohn (Carnegie Mellon University, Pittsburgh, Pennsylvania) presented a calculation of the threshold energy in germanium. This determination was based on the assumption that the threshold energy is equal to the increase in lattice potential energy upon creation of the vacancy–interstitial pair, to which is added the energy required to break the four bonds with the nearest neighbors, giving an energy of between 7 and 15 eV. W. Kohn also found that it was easier to move an atom in one direction $(\bar{1}\bar{1}\bar{1})$ than in the (100) direction. It is difficult to give further details, as only the abstract of W. Kohn's presentation has been published. Information on the calculation comes from the articles (Brown and Augustyniak 1959; Loferski and Rappaport 1959). It should be noted that W. Kohn was awarded the Nobel Prize in Chemistry in 1998 for the development of density functional theory.

In 1956, W.L. Brown and W.M. Augustyniak (Bell Telephone Laboratories) published a paper containing a remarkable description of experimental conditions: sample set-up, electron flux measurement, beam energy calibration, determination of irradiated surface and electrical contacts. By irradiating single crystals of different orientations, they observed that crystallographic orientation had no effect on the displacement threshold. Whatever the orientation, the minimum electron energy required to create a defect is around 360 keV, giving a maximum transmitted energy of 15.3 eV. On the other hand, there is a difference in the efficiency with which the number of charge carriers is reduced: in the (111) direction, the efficiency is greater than in the (110) and (100) directions. They tried to explain their results, but were not very convincing. Aware of this, they wrote: "It seems remarkable that with *no*

5 Gatlinburg, Tennessee, May 1959.

detectable threshold energy difference there is found to be an orientation dependence in carrier removal rate at higher energy". They also observed an effect of temperature on threshold energy, but did not investigate the phenomenon further: "The precise determination of threshold energy is rather academic until these gross temperature effects are understood" (Brown and Augustyniak 1959).

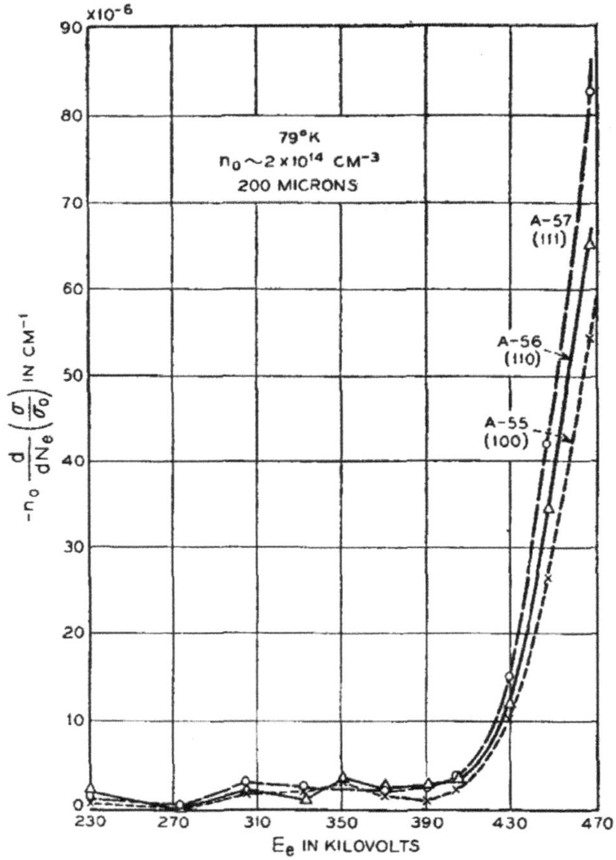

Figure 7.5. *Acceptor site creation rate as a function of electron energy. The three samples of different crystallographic orientation are irradiated at the same time (Brown and Augustyniak 1959)*

Figure 7.5 shows how difficult it is to accurately determine the threshold energy. The same applies to the effect of crystallographic orientation, which is very real, but only translates into an energy difference of the order of 3–4%. A few years later,

J.J. Loferski and P. Rappaport[6] again determined the displacement threshold energies of silicon and germanium, taking the lifetime τ minority carriers as the quantity sensitive to the presence of defects. The threshold energies measured were very low: 12.9 ± 0.6 eV for silicon and 14.5 ± 0.4 eV for germanium (Loferski and Rappaport 1958). In August 1959, in a paper entitled *Displacement thresholds in semiconductors*, J.J. Loferski and P. Rappaport summarized calculations and measurements of semiconductor displacement thresholds (Loferski and Rappaport 1959). Their table of results is shown in Figure 7.6.

TABLE I. Summary of threshold data.

Material	Temperature	Property measured	Sample thickness (cm)	E_{BO} (kev)	E_{LO} (ev)	Reference
n-type Ge	78°K	σ	0.05	630	30	6
n-type Ge	78°K	σ	0.006 to 0.002	360	14.5	12
n-type Ge	300°K	σ	0.005	500	22.3	15
n-type Ge	289°K	τ(EVE)	0.02	360	14.5	8
p-type Si	289°K	τ(EVE)	0.02	145	12.9	8
n-type InSb	78°K	σ	...	200	5.7	17

Figure 7.6. *Table taken from Loferski and Rappaport (1959). ELO: energy threshold, EBO: particle energy. References: 6: (Klontz and Lark-Horowitz 1952); 8: (Loferski and Rappaport 1958); 12: (Brown and Augustyniak 1959); 15: (Vavilov 1959); 17: (Eisen and Bickel 1959)*

7.2.2. *High-energy deposits*

Bombardments, as they were called at the time, were mainly carried out using ions or neutrons, that is, particles capable of transmitting significant energy to the target atoms. The creation of disordered zones was therefore a major concern.

Conversely, bombardment by Xe ions or fast neutrons of a thin sheet of germanium prepared by evaporation in a metastable amorphous state creates small crystallized regions. J.R. Parsons and R.W. Balluffi (University of Illinois) studied them by transmission electron microscopy in bright and dark fields: their number was proportional to the neutron fluence and their average size was equal to 190 Å for neutrons and varied from 260 to 360 Å when the energy of Xe ions increased

6 They worked at RCA Laboratories, Princeton, New Jersey. Laboratories were founded in 1942 with the aim of driving progress in the field of electrical engineering. This entity still exists under the name Sarnoff Corporation (David Sarnoff is the founder of RCA Laboratories).

from 40 to 90 keV[7]. Their interpretation was based on the displacement spike model. Crystallization occurs during the time of small displacement cascades (Parsons and Balluffi 1964).

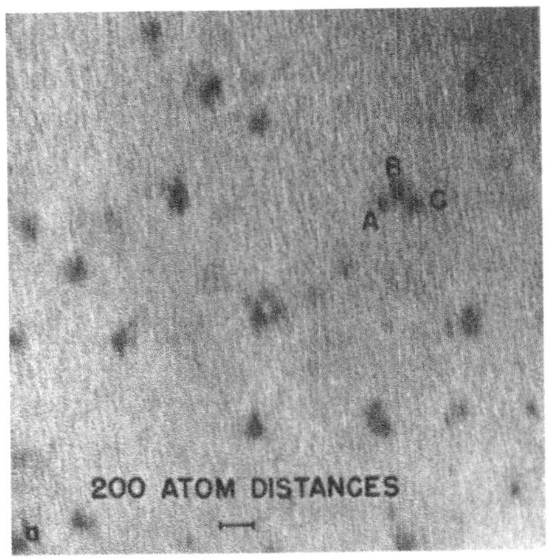

Figure 7.7. *Small recrystallized regions observed by transmission electron microscopy (Siemens Elmiskop I, 100 keV) in amorphous germanium irradiated with 2.5 × 10^{16} fast neutrons (Parsons and Balluffi 1964)*

Seitz and Koehler (1956, pp. 433–442) provide those interested with more details on the state of knowledge in the mid-1950s on the creation of defects in semiconductors. In addition, the review article by I.A. Abroyan (Kalinin Polytechnic Institute, Leningrad) gives the state of the art in the study of changes in the physical properties of semiconductors irradiated with ions, that is, in a field where defects are not created individually (Abroyan 1971).

7.2.3. *Description of defects*

Optical absorption, electrical conductivity, Hall effect[8] and electron paramagnetic resonance (EPR) were the first techniques used to characterize semiconductors, so they were naturally called upon to study the changes induced in

7 The ions were accelerated by the Cockroft Walton heavy ion accelerator at Argonne National Laboratory.

8 The first observation of the Hall effect in semiconductors was made in 1879 by the American physicist Edwin Herbert Hall (1855–1938).

semiconductors. Point defects are easily detected when their signature is a state in the energy band gap. In this case, spectroscopic methods such as EPR, infrared absorption spectroscopy or deep level transient spectroscopy can be very useful.

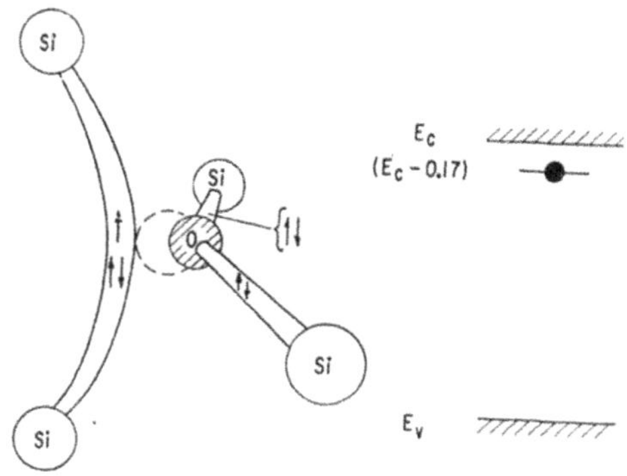

Figure 7.8. *Representation of the A center. The unpaired electron is at the origin of the spin resonance. On the right, the associated energy level is located at 0.17 eV from the bottom of the conduction band (Corbett et al. 1961)*

EPR was used to characterize impurities in semiconductors, so it was a natural progression to use this technique to study irradiation defects in these materials. The first experiment is said to have been carried out in 1955 by E. Schulz-Dubois, M. Nisenoff, H.Y. Fan and K. Lark-Horowitz on neutron-irradiated p-type silicon (presented to the 1955 March Meeting of the American Physical Society held at Baltimore, March 17–19, 1955). For n-type silicon, the situation was more complex. G. Bemski and colleagues (Bell Telephone Laboratories) observed two sets of absorption lines (a quadruplet and a doublet), which they attributed to ^{29}Si displaced from its site (Bemski et al. 1958). In contrast, researchers at General Electric Research Laboratory showed that the spectrum depended on the purity of the silicon. The EPR signal could be due to an impurity that has trapped a point defect, interstitial or vacancy. For silicon synthesized in a quartz crucible, the impurity could be oxygen, while for the purer silicon crystallizing in zone melting, the impurity could be the species used for doping, for example, potassium (Watkins et al. 1959). The defect stable at room temperature that has the strongest signal after electron irradiation is called A center or Si–A center. In 1961, J.W. Corbett and G.D. Watkins published two review articles on A center and, through correlations between several techniques, demonstrated that A center was a vacancy associated with an oxygen atom as shown in Figure 7.7 (Corbett et al. 1961; Watkins and Corbett 1961).

Among spectroscopic methods, a paragraph on deep level transcient spectroscopy (DLTS) would be in order, but I have not managed to go back in time to find the first experiment on an irradiated semiconductor. There is one publication, certainly not the first, which describes the technique and gives results for several semiconductors irradiated with neutrons in the research reactor at the University of Missouri (Farmer and Meese 1982).

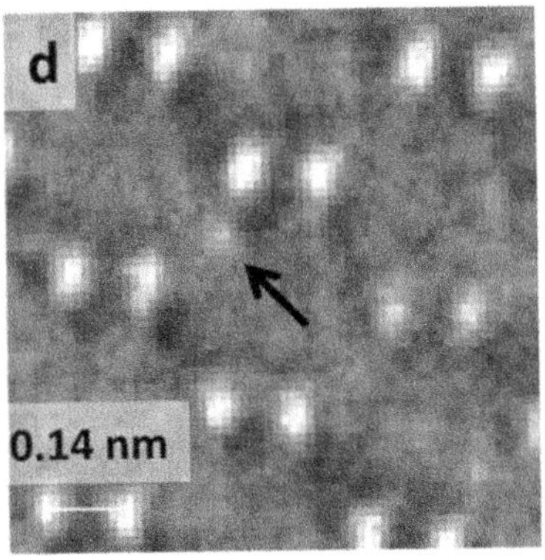

Figure 7.9. *High-resolution electron microscopy image of an oriented germanium crystal [110] in which an interstitial is well localized transversely (Alloyeau et al. 2009)*

With the development of aberration-corrected transmission electron microscopes, it has become almost possible to observe point defects (Figure 7.8). Unless there is an unfortunate omission, only one study of interstitial localization by transmission electron microscopy has been published (Alloyeau et al. 2009). The observation of 24 interstitials, created by electron irradiation, allows certain deductions to be made. Under the conditions of observation, thin plate under a continuous electron flux of 300 keV, interstitials were localized in hexagonal and quadratic sites; in contrast, in configuration ⟨110⟩ – *split* configuration, none were observed. These results, although in partial disagreement with density-functional theory calculations, show that this technique is now capable of visualizing an individual defect. However, it is important to bear in mind that, while transverse localization is excellent, longitudinal localization is only a few lattice parameters away. This work was preceded by a similar experiment involving the localization of an impurity in silicon (Oh et al. 2008).

It is clear that transmission electron microscopy was used much earlier to study defects in semiconductors. However, in the 1950s and 1960s, the resolution only allowed the study of extended defects and their stress fields. Figure 7.10 shows an image of extended defects with an average size of around 75 nm.

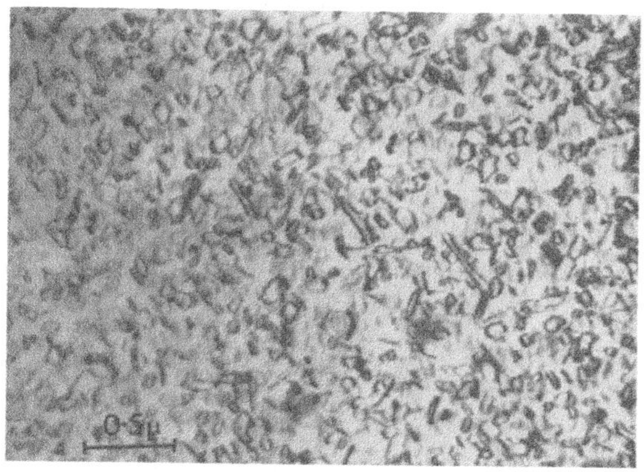

Figure 7.10. *Bright-field electron microscopy showing dislocations created by irradiation (3 × 10^{15} cm^{-2} ^{11}B de 40 keV) followed by 4 h annealing at 800°C (Mazey et al. 1968)*

7.3. Diffusion in semiconductors

As in most materials, the diffusion coefficient in a semiconductor at a given temperature is directly proportional to the vacancy concentration. Irradiation is a means of increasing vacancy concentration and thus accelerating diffusion. This was achieved as early as the 1950s in metals and a decade later in semiconductors. Thus, P. Baruch, C. Constantin, J.C. Pfister of the École Normale Supérieure de Paris, together with Ms R. Saintesprit of Lignes Télégraphiques et Téléphoniques de Conflans-Ste Honorine, studied the diffusion of gallium in silicon under irradiation. To do so, they observed the displacement of a p-n junction obtained at a depth of 6 μm by gallium diffusion. Under high-temperature bombardment (900 and 1,100°C) with 275 keV protons, the depth of the junction increases, whereas the proton range is only 3 μm. They assumed that the vacancies created near the surface diffused inward, thus increasing gallium diffusion (Baruch et al. 1961). This is just one example.

Defect diffusion in semiconductors and insulators is sensitive to the presence of excess electrons or holes, as well as to the charge state of the migrating species. In a

1978 paper, J.C. Bourgoin and J.W. Corbett[9] reviewed the mechanisms that can accelerate diffusion (field enhanced diffusion, defect enhanced diffusion, electro-striction enhanced diffusion, normal ionization enhanced diffusion, collision impact ionization enhanced diffusion) and detailed the mechanism they proposed in 1972. Since then, this mechanism has been known as the Bourgoin mechanism. This mechanism can operate if the potential minimum experienced by the migrating species in a charge state corresponds to the saddle point of the potential for another charge state, and vice versa. Matter is transported by a succession of charge state changes, without any thermally activated process. Figure 7.11 explains this mechanism, which can also be effective in insulators such as diamond (Bourgoin and Corbett 1972, 1978).

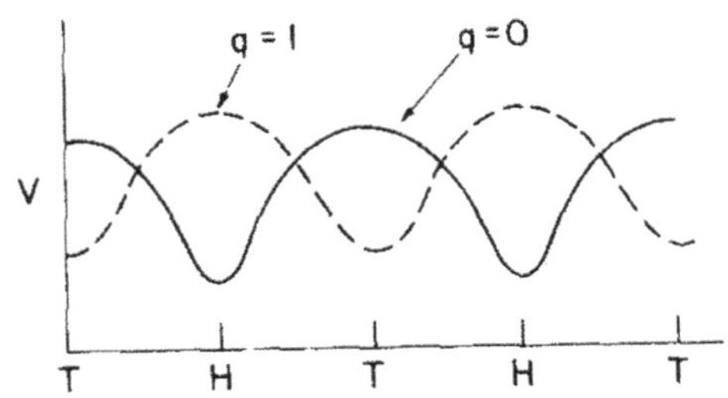

Figure 7.11. *Schematic representation of potential energy for two interstitial charge states allowing rapid movement between quadratic (T) and hexagonal (H) sites (Bourgoin and Corbett 1972)*

In 1965, P. Baruch and J.C. Pfister reported the presence of hydrogen bubbles obtained by precipitating hydrogen introduced by irradiation (Baruch and Pfister 1965). Controlling the formation of these bubbles is the basis of the Smart Cut process.

7.3.1. *Smart Cut process*

Implanting gases such as hydrogen or helium at a temperature at which diffusion mechanisms are active can lead to the formation of gas bubbles. Mastering the implantation of hydrogen in silicon and annealing enables the growth of hydrogen bubbles, the origin of the Smart Cut process. This process, invented in 1991 by

9 J.C. Bourgoin (École Normale Supérieure, Paris) and J.W. Corbett (State University of New York at Albany).

Michel Bruel of CEA/LETI[10] (Grenoble), has been operated since 1992 by SOITEC, a company set up by André-Jacques Auberton-Hervé and Jean-Michel Lamure of the same laboratory. The first application was the production of buried oxide layers for silicon-on-insulator (SOI) technology. Industrial production began in 1999 with 200 mm diameter wafers, rising to 300 mm in 2002. The Smart Cut process uses two technologies: wafer bonding and ion implantation combined with a heat treatment that induces a deep fracture parallel to the surface of the implanted wafer. Details of the Smart Cut process are described in Bruel et al. (1997).

The current state of the art in this technology is diverse, opening up to other semiconductors (SiC, GaN and InP) and other applications (see https://www.soitec.com/fr/produits/au-delà-du-silicium).

Research into the interaction of radiation with semiconductors therefore remains very active under the pressure of technological development.

7.4. Chapter 7 references

7.4.1. *Laboratories and personalities in this chapter*

Numerous academic and industrial researchers contributed to the advancement of knowledge about defects and their effects. The laboratories of the industries involved were very active, including two in the early days: Bell Telephone Laboratories and General Electric Research Laboratory. After briefly describing these two laboratories, I have chosen to present two researchers on a totally arbitrary basis.

Bell Telephone Laboratories was founded in 1876, about one year after the invention of the telephone, by Alexander Graham Bell under the name Bell Telephone Company. In 1889, this company came under the control of American Telephone & Telegraph Company (AT&T). In 1925, AT&T joined forces with Western Electric to create a new organization to strengthen their research: Bell Telephone Laboratories, Inc. This “laboratory” was particularly innovative, making decisive contributions to the transistor, fiber optics, solar cells and, more recently, satellite communications, the cell phone and even the Hubble telescope. In 2016, Nokia acquired 95% of the voting rights, and the company is now called Nokia Bell

10 CEA/LETI: Commissariat à l’Énergie Atomique/Laboratoire d’électronique et des technologies de l’information.

Labs. The list of researchers who have worked for this company includes 15 Nobel Prize winners.

General Electric Research Laboratory was founded in 1900 by the General Electric Company as a research laboratory for commercial applications based on new principles. At the time, the fields of physics, chemistry and mathematics were progressing so rapidly that university graduates from the early 1890s were outpaced by those who had graduated 5–10 years later, so management decided to hire young recruits. Research activity was intense, often culminating in the production of a prototype. Until the 1950s, management maintained a non-competitive, non-directed research atmosphere.

Karl Lark-Horovitz (1892, Vienna, Austro-Hungarian Empire–1958, West Lafayette, Indiana, USA), after starting to teach at the University of Vienna, Lark-Horovitz visited several laboratories in the United States in 1928 on an international research grant from the Rockefeller Foundation. During this tour, Purdue University offered him a position. Surprisingly, he accepted a position in a physics department with only five teachers, only one of whom had defended his thesis. He describes it thus:

> The Physics Department at the time of my arrival, existed mainly as a service department for the different schools. The training of the professional staff and the assistants had been only routine teaching, and with one exception, nobody was doing any research work or was interested in such work. The equipment was only suited for the different courses in undergraduate physics.

Two years later, he took over as head of the physics department, where he modernized teaching and research to make it an internationally recognized department. He ruled the department with an iron fist until his death in 1958.

He is best known for his work on structure determination by X-ray and electron diffraction.

James William Corbett (1928, Manhattann–1994, Albany, New York) began his career as a research associate at the General Electric Research Laboratory. His research was often pioneering in the field of electron irradiation of semiconductors (and metals). In 1968, he joined the Department of Physics at the University of Albany, where for over 20 years, he directed the Institute for the Study of Defects in Solids. For the period under review, his internationally recognized expertise focused on the characterization of irradiation defects in semiconductors and their dynamics.

Figure 7.12. *Right to left: Robert Walker, James Corbett and an unidentified person working on equipment to study low-temperature irradiation damage at the General Electric Research laboratory. Photo credit: General Electric Global Research, courtesy AIP Emilio Segrè Visual Archives, Physics Today Collection*

7.5. References

Abroyan, I.A. (1971). Effect of ion-bombardment on physical properties of semiconductors. *Soviet Physics Uspekhi-Ussr*, 14(3), 242–262.

Alloyeau, D., Freitag, B., Dag, S., Wang, L.W., Kisielowski, C. (2009). Atomic-resolution three-dimensional imaging of germanium self-interstitials near a surface: Aberration-corrected transmission electron microscopy. *Physical Review B (Condensed Matter and Materials Physics)*, 80, 014114.

Baruch, P. and Pfister, J.-C. (1965). Radiation induced diffusion in silicon. *Nuclear Instruments and Methods*, 38, 197.

Baruch, P., Constantin, C., Pfister, J.-C., Saintesprit, M.R. (1961). Non-metallic solids. Vacancy enhanced diffusion in silicon. Effects of irradiation and of chemical impurities. *Discussions of the Faraday Society*, 31, 76–85.

Bemski, G., Feher, G., Gere, E. (1958). Spin resonance in electron irradiated silicon. *Bukketin of the American Physical Society*, 3(2), 135.

Bourgoin, J.C. and Corbett, J.W. (1972). A new mechanism for interstistitial migration. *Physics Letters A*, 38(2), 135–137.

Bourgoin, J.C. and Corbett, J.W. (1978). Enhanced diffusion mechanisms. *Radiation Effects*, 36, 157–188.

Brattain, W.H. and Pearson, G.L. (1950). Changes in conductivity of germanium induced by alpha-particle bombardment. *Physical Review*, 80(5), 846–850.

Braun, F. (1874). Ueber die Stromleitung durch Schwefelmetalle (On the conduction of electricity through sulphur metals). *Annalen der Physik und Chemie*, 153(4), 556–563.

Brown, W.L. and Augustyniak, W.M. (1959). Energy, orientation, and temperature dependence of defect formation in electron irradiation of n-type germanium. *Journal of Applied Physics*, 30(8), 1300–1309.

Bruel, M., Aspar, B., Auberton-Hervé, A.-J. (1997). Smart-cut: A new silicon on insulator material technology based on hydrogen implantation and wafer bonding. *Japanese Journal of Applied Physics*, 36(3S), 1636.

Cleland, J.W., Crawford, J.H., Lark-Horowitz, K., Pigg, J.C., Young, F.W. (1951). The effect of fast neutron bombardment on the electrical properties of Germanium. *Physical Review*, 83(2), 312–319.

Corbett, J.W., Watkins, G.D., Chrenko, R.M., McDonald, R.S. (1961). Defects in irradiated silicon. II. Infrared absorption of the Si-*A* center. *Physical Review*, 121(4), 1015–1022.

Crawford, J.H. and Cleland, J.W. (1959). Nature of bombardment damage and energy levels in semiconductors. *Journal of Applied Physics*, 30, 1204-1213.

Crawford, J.H. and Lark-Horowitz, K. (1950). Fast neutron bombardment effects in germanium. *Physical Review*, 78, 815–816.

Cussins, W.D. (1955). Effects produced by the ionic bombardment of germanium. *Proceedings of the Physical Society. Section B*, 68(4), 213.

De Laer Kronig, R. (1924). Change of conductance of selenium due to electronic bombardment. *Physical Review*, 24(4), 377–382.

Eisen F.H. and Bickel P.W. (1959). Electron damage thresholds in InSb. *Physical Review*, 115(2), 345–346.

Farmer, J.W. and Meese, J.M. (1982). Deep level transient spectroscopy of neutron irradiated semiconductors. *Journal of Nuclear Materials*, 108–109, 700–708.

Johnson, W.E. and Lark-Horowitz, K. (1949). Neutron irradiated semiconductors. *Physical Review*, 76, 442–443.

King, W.J., Burrill, J.T., Harrison, S., Martin, F., Kellett, C. (1965). Experience in fabricating semiconductor devices using ion implantation techniques. *Nuclear Instruments & Methods*, 38, 178–179.

Klontz, E.E. and Lark-Horowitz, K. (1952). Displacements produced by electron bombardment of Germanium (abstract). *Physical Review*, 86(4), 643.

Livingston, R.S., Howard, F.T., Graham, R. (1956). Bibliography of cyclotron literature. Report, ONRL-2023, Oak Ridge National Laboratory.

Loferski, J.J. and Rappaport, P. (1958). Radiation damage in Ge and Si detected by carrier lifetime changes: Damage thresholds. *Physical Review*, 111(2), 432.

Loferski, J.J. and Rappaport, P. (1959). Displacement thresholds in semiconductors. *Journal of Applied Physics*, 30(8), 1296-1299.

Mazey, D.J., Nelson, R.S., Barnes, R.S. (1968). Observation of ion bombardment damage in silicon. *The Philosophical Magazine: A Journal of Theoretical Experimental and Applied Physics*, 17(150), 1145–1161.

Oh, S.H., Benthem, K.V., Molina, S.I., Borisevich, A.Y., Luo, W., Werner, P., Zakharov, N.D., Kumar, D., Pantelides, S.T., Pennycook, S.J. (2008). Point defect configurations of supersaturated Au atoms inside Si nanowires. *Nano Letters*, 8(4), 1016–1019.

Ohl, R.S. (1952). Properties of ionic bombarded silicon. *The Bell System Technical Journal*, XXXI, 104–121.

Ohl, R.S. (1956). Semiconductor translating device. US Patent, 2 750 541.

Parsons, J.R. and Balluffi, R.W. (1964). Displacement spike crystallization of amorphous germanium during irradiation. *Journal of Physics and Chemistry of Solids*, 25(3), 263–272.

Schuster, A. (1874). On unilateral conductivity. *The London, Edinburgh, and Dublin Philosophical Magazine and Journal of Science*, 48(318), 251–258.

Seitz, F. and Koehler, J.S. (1956). Displacement of atoms during irradiation. *Solid State Physics*, 2, 305–448.

Shockley, W. (1949). Semiconductor translating device. US Patent, 2 666 814.

Shockley, W. (1957). Forming of semiconductor devices by ionic bombardment. US Patent, 2 787 564.

Vavilov V.S. (1959). Radiative recombination in semiconductor. *Soviet Physics Uspekhi*, 2(3), 455–464.

Watkins, G.D. and Corbett, J.W. (1961). Defects in irradiated silicon. I. Electron spin resonance of the Si-A center. *Physical Review*, 121(4), 1001–1014.

Watkins, G.D., Corbett, J.W., Walker, R.M. (1959). Spin resonance in electron irradiated silicon. *Journal of Applied Physics*, 30(8), 1198–1203.

Wertheim, G.K. (1957). Energy levels in electron-bombarded silicon. *Physical Review*, 105(6), 1730–1735.

Wertheim, G.K. (1959). Recombination properties of bombardment defects in semiconductors. *Journal of Applied Physics*, 30(8), 1166–1174.

8

Iono-covalent Insulators Under Irradiation

The "inorganic insulators" category is heterogeneous and can be divided into two groups: ionic crystals and iono-covalent materials. We have already seen the intense activity surrounding the fluorescence or phosphorescence of ionic materials, so we will not go into that again. This chapter will therefore be devoted to iono-covalent materials.

8.1. Iono-covalent materials under irradiation

We saw in section 2.1.1 that the notion of disorder appeared in geology long before the effects of radiation were discovered. It was personalities such as J.J. Berzelius, R.J. Haüy, A. Des Cloizeaux, W.C. Broegger, A. Hamberg, etc., who advanced the understanding of the metamict state in iono-covalent materials.

Among the first to make observations of color change by irradiation of iono-covalent materials were S. Meyer and K. Przibram. They described the various situations in which they observed color changes. Their experimental conditions were rather complicated: "Different glasses exposed to beta rays take on either a brown or a violet color". But carefully heated, brown glass turns violet, while exposed to ultraviolet radiation, it turns pale yellow! (Meyer and Przibram 1912, author's translation). Eight years later, S.C. Lind took stock of this knowledge. After discussing the disappearance of color by heating or beta irradiation, S.C. Lind turned his attention to changes in mechanical properties. E. Rutherford is quoted as having noted that striations appear on silica tubes containing radium, leading in time to tube disintegration (Lind 1920).

At the same time, the enigma of pleochroic halos was explained. A correlation was first established between these colored zones observed in zircons and biotites (of the mica family) and the path of alphas (Joly 1907). The use of a photometer to digitize the radial profile of the halos greatly improved precision and enabled a finer comparison with Bragg curves (Henderson and Bateson 1934; Henderson and Turnbull 1934).

From the 1920s to the 1940s, studies of the effects of irradiation on insulators gradually moved from color changes in absorption and emission to experiments on the mechanisms of defect creation and their characterization. From the 1920s onward, structural approaches gained in importance. However, as early as 1916, Lars Vegard[1] wrote: "The 'metamic' crystal of thorite gave no X-ray reflexion. The lattice is completely destroyed" (Vegard 1916). The conditions for a material to undergo a metamic transformation were defined by Victor Moritz Goldschmidt[2]:

> A. The original structure must be only weakly ionic and possibly susceptible to hydrolysis. B. The structure must contain one or more kinds of ions that are readily susceptible to changes in the state of ionization. C. In many cases it may additionally be necessary that the crystal is subjected to relatively strong radiation, either from radioactive material within the crystal itself or from outside sources. No one of these conditions is alone sufficient for isotropization (Goldschmidt 1924, author's translation).

In the years that followed, numerous studies focused on the reconstitution of the structure of the metamictized mineral by heating. The first experiments used powder diffraction. Peak intensities in powder diffraction give information on the diffracting volume, but do not indicate whether it is an "epitaxial" recrystallization[3]. The use of single crystals shows that partially metamictized minerals reform a single crystal after heating. Fully amorphized zircons, on the other hand, do not recrystallize in a single-crystal form (Stackelberg and Chudoba 1937; Stackelberg and Rottenbach 1939).

1 Lars Vegard (1880–1963): Norwegian physicist, pioneer of crystallography. Physical Institute, Christiania, Denmark.

2 Victor Moritz Goldschmidt (1888–1947): chemist at the University of Oslo and from 1929 to 1935 at the University of Göttingen. Co-founder of modern geochemistry and crystallochemistry.

3 The term epitaxy was coined around this time.

In 1952, Adolf Pabst (University of California), at the end of his journal article *The metamic state*, expressed surprise at the low number of metamictization experiments:

> Since the metamict state is now generally agreed to arise from alpha particle bombardment it is surprising that so few attempts have been made to produce it artificially.
>
> More powerful sources of radiation are now available. They might prove useful in the study of metamictization (Pabst 1952).

Developments in characterization techniques opened up new perspectives on defects. X-ray diffraction has made a significant contribution to characterizing the effects of irradiation. For example, in zircon ($ZrSiO_4$), the diffraction angle 2θ of the (112) planes varied from 35.635° to 35.1° for a dose of 6×10^{15} alphas/mg, and for the same dose, the intensity decreased to zero. From these values, the authors[4] deduced that each alpha displaced around 4,500 atoms. Furthermore, this relationship between the 2θ angle and the number of alphas/mg makes it possible to date a zircon sample as soon as the value of its $2\theta_{(112)}$ angle and its alpha activity are known, and provided it is homogeneous and not too damaged (Hurley and Fairbairn 1952).

8.1.1. *Defects in iono-covalent materials*

Among irradiated materials, nuclear fuel is undoubtedly subjected to the highest doses of irradiation during its time in operation in a reactor[5]. At full power, a pressurized water reactor core is the site of 1.4×10^{13} fissions/cm^3/s, which release around 3.6×10^{13} neutrons/cm^3/s and around 2.8×10^{13} fission fragments/cm^3/s. All these particles emitted within the material are sources of irradiation, to which must be added significant chemical modifications to fission events and the creation of fission fragments. Research into uranium dioxide under irradiation is therefore very active, but is mainly devoted to the technological problems of high burnup rates (swelling, fracturing, release of fission gases, etc.). In the 1950s and 1970s, faced with the industrial challenge, knowledge of the basic mechanisms was of less interest to the nuclear industry. Nevertheless, there are publications on the fundamental phenomena of UO_2 irradiation. For example, tracks of fission

4 Patrick M. Hurlay and Harold W. Fairbairn are from the Massachusetts Institute of Technology.

5 In future fusion reactors, materials could be subjected to higher irradiation stresses (first wall, tritiated blanket, etc.).

fragments in the fuel were observed by transmission electron microscopy (Bierlein and Mastel 1960). The creation of point defects was postulated on the basis of the restoration of the crystalline parameter during isochronous annealing of samples irradiated in the EL2 pile at Saclay. Jacqueline Bloch highlighted two annealing stages beginning at 200 and 500°C, which she attributed to the presence of point defects (Bloch 1961). Twenty years later, W.J. Weber described three annealing stages: the first at 300°C was associated with the migration of oxygen vacancies, the second at 575°C with uranium vacancies and the third at 925°C with the migration or dissociation of the vacancy–helium complex (Weber 1983). Understanding fission gas migration mechanisms was also of fundamental and practical interest. H.E. Evans and colleagues at the Department of Physics, University of London, Surrey, demonstrated the potential of positron annihilation spectroscopy to study the dynamics of vacancy-type defects and their interaction with noble gases (Evans et al. 1992). The emission of uranium atoms induced by fissions near the surface of UO_2 disks was measured as early as the 1960s using the "catcher foil" method (Nilsson 1966).

Among the multitude of iono-covalent materials, quartz, alumina and magnesia are also among those whose behavior under irradiation has been the subject of numerous studies. Point defects in these materials can generally be counted and characterized by optical, magnetic or nuclear methods.

Electron spin resonance provides information on the configuration of paramagnetic defects or defects interacting with a paramagnetic impurity. The first paramagnetic resonance signal was observed in the insulator MgO irradiated in a reactor. This signal was attributed to the F center, an anionic vacancy that had captured an electron (Wertz et al. 1957). In fact, it was a team from the Department of Physics, University of California, who first observed an electron spin resonance line from an F-center in KCl irradiated with 2.5 MeV electrons or 250 keV X-rays. However, this material is not iono-covalent (Kip et al. 1953).

Positron annihilation is also a technique that enables us to follow the production of vacancy-type defects, their recombination and the early stages of their aggregation. Quartz was one of the first inorganic insulators to be studied using this technique. The angular correlation of annihilation photons is different between fused and crystalline quartz (Page et al. 1955; Colombino et al. 1960). However, it was not until the late 1960s that an experiment on irradiation defects in this material was found. Still on quartz, a narrow component of angular correlation disappears under electron irradiation and is restored by annealing (Coussot 1969).

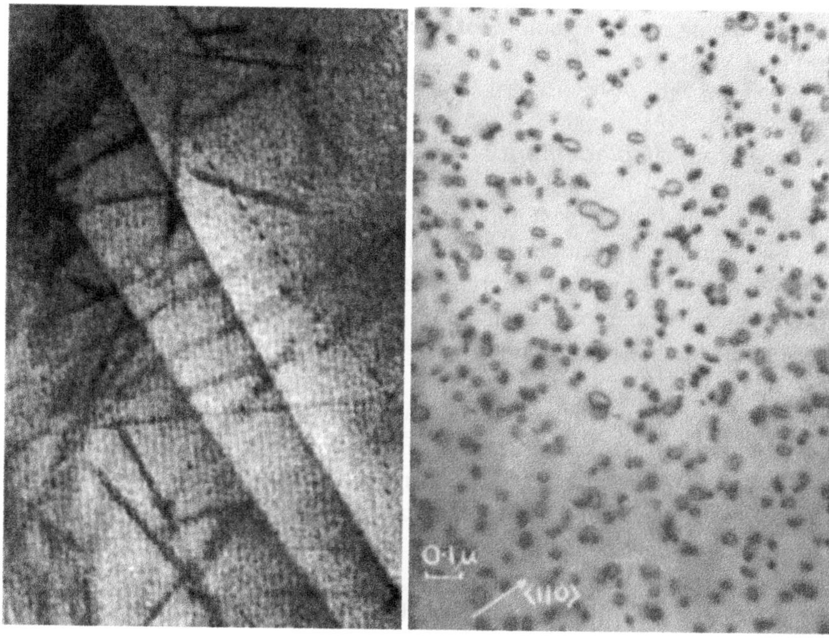

Figure 8.1. *Left: tracks due to fission fragments in mica, the two long oblique lines are the result of changes in thickness (Silk and Barnes 1959). Right: dislocation loops {110} in MgO irradiated at 650°C and annealed at 1,200°C for 1 h (Groves and Kelly 1963)*

In the 1960s, transmission electron microscopy became an important tool for studying extended defects. The principle of this microscopy was invented in Berlin by Max Knoll and Ernst Ruska in 1931 (Knoll and Ruska 1932). With a resolution of around 50 Å for the best microscopes, the objects that can be studied are typically dislocations, dislocation loops, latent traces of high-energy heavy ions and large clusters of defects. Images of these defects are often due to the presence of a stress field. Figure 8.1 shows latent tracks created by fission fragments in muscovite mica. This is probably one of the earliest images of latent tracks. The tracks are generally straight and less than 300 Å in diameter. The article is signed by E.C.H. Silk and R.S. Barnes of the Atomic Energy Research Establishment, Harwell, but it states that the microscopy was carried out by C.J. Jackson and B.C. Jenkins (Silk and Barnes 1959). In the same figure, on the right, dislocation loops are observed in MgO irradiated and annealed at 1,200°C. The loops have an average diameter of 280 Å (Groves and Kelly 1963). The authors are from the Department of Metallurgy at Cambridge University. The emergence of high-resolution electron microscopy made it possible to measure the diameter of latent traces and their evolution with electronic stopping power. Examples of track images in various oxides are shown in Houpert et al. (1988).

8.1.2. *Threshold displacement energy in inorganic insulators*

The notion of displacement threshold energy is not relevant in cases where damage by electronic excitation is the dominant process. This is the case for ionic crystals. Whatever the material, determining the value of the displacement threshold energy is a cumbersome experiment. As a result, few people bother to determine these displacement threshold energies.

For uranium dioxide, the reference values were obtained very indirectly by Jacques Soullard of the Commissariat à l'Énergie Atomique, Fontenay-aux-Roses. The publication by J. Soullard and A. Alamo in *Radiation Effects* is unanimously cited to justify the values of 20 and 40 eV for the displacement threshold energies of oxygen and uranium, respectively (Soullard et al. 1978). However, the methods used to extract these values are not described. For details of the experiment, Jacques Soullard's thesis is the best place to read it. In it, the hypotheses are detailed, such as:

> The displacement threshold energy is mainly determined by the number of bonds that need to be broken to eject an atom from its site, the energy required for the defect once created to leave the recombination volume being only a few eV. In this case, the threshold energy of the oxygen atom is 2 times smaller than that of the uranium atom, since only 4 U-O bonds need to be broken to create an oxygen defect, instead of 8 to create a uranium defect (Soullard 1977, author's translation).

The value of the threshold energy is determined in such a way that the number of displacements calculated in the case of an indirect process $e^- \rightarrow O \rightarrow U$ corresponds to the number of vacancies required to create the vacancy loops observed in a high-voltage electron microscope[6]. Twenty-eight years later, Costantin Meis and Alain Chartier of the Commissariat à l'Énergie Atomique, Saclay, estimated threshold energies in eight different crystallographic directions using the sudden approximation approach. Their results are close to those of J. Soullard: $E_d[O]$~20 eV and $E_d[U]$~50 eV (Meis and Chartier 2005).

6 The comparison is more subtle than that. It takes into account more parameters such as interstitial migration energy.

Displacement threshold energies have been measured for a number of oxides, with the difficulty that, unlike resistivity for metals and semiconductors, the measured quantity, generally optical absorption in the uv-visible, is more cumbersome to measure in situ. Furthermore, as shown by the first publication on the determination of a displacement threshold for a non-conductive material, α-Al_2O_3, tracing the absorption line at 205 nm did not reveal which threshold was being measured. George W. Arnold and W. Dale Compton of the United States Naval Research Laboratory (Washington, D.C.) indicated both possibilities: a threshold at 50 ± 5 eV for aluminum or 90 ± 5 eV for oxygen (Figure 8.1). But they wrote that aluminum displacements might be more likely because they have greater freedom of movement than oxygen atoms, whose stacking is more compact (Arnold and Compton 1960).

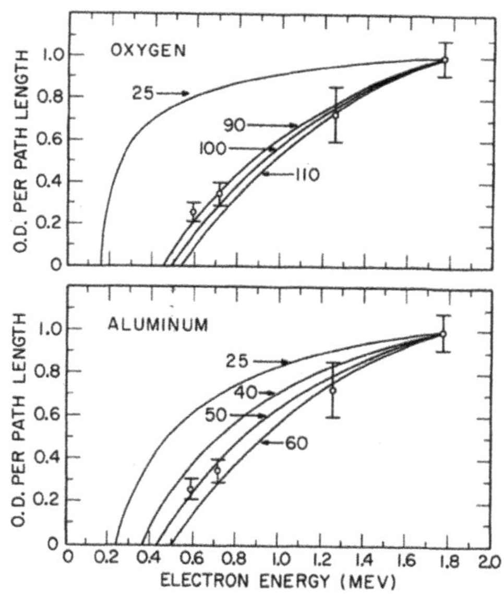

Figure 8.2. *Optical density, normalized by electron path length, compared with total cross-section calculated for different values of oxygen and aluminum displacement threshold energy. All curves are normalized by their value at 1.77 MeV (Arnold and Compton 1960)*

It took almost 20 years before a new measurement was published for alumina. The approach was different, and involved detecting the minimum energy required to observe the appearance of extended defects (for example, dislocation loops). The experiment was carried out in the AEI-EM7 high-voltage electron microscope at Harwell in a single-tilt heated stage. At a given temperature, G.P. Pells and

D.C. Phillips increased the accelerating voltage in 20 keV steps until extended defects appeared. The operation was repeated for eight temperatures between 300 K and 1,120 K. This is an indirect approach that only works when the defects are mobile. At room temperature, defects were observed from an energy of 390 keV, giving a threshold at 44 eV for aluminum or 74 eV for oxygen (Pells and Phillips 1979).

8.1.3. *Phase transformation under irradiation*

The accumulation of defects can lead to the loss of crystalline order or, in some cases, to a transition to another crystalline phase.

The first phase transitions induced by irradiation were, as we have seen, transitions to the metamict state, amorphization. With this in mind, William Primak (Chemistry Division, Argonne National Laboratory) studied the behavior of vitreous silica and quartz. He explained his results on the irradiation of vitreous silica and quartz using a thermal spike approach. Up to a fluence of 2×10^{18} neutrons/cm², the density of silica increased by 0.18% per unit dose[7] and that of quartz decreased by 0.07% per unit dose. At higher fluences, the rate of change in quartz density accelerated to 0.2% per dose unit, and then reached saturation. This more rapid variation in density was explained by a decrease in thermal conductivity under irradiation, which amplified the effect of thermal spike (Primak 1955).

The thermal spike model was also put forward to explain the transformation from the low-temperature monoclinic phase of zirconia (ZrO_2) to the high-temperature cubic phase. However, the authors, M.C. Wittels and F.A. Sherrill of Oak Ridge National Laboratory, believe that their hypothesis is questionable, as the high-temperature phase (>1,900°C) could not be stabilized by laboratory quenching. However, they point out that the quenching speed of a thermal spike has been calculated at between 10^{-12} and 10^{-10} s. They therefore suggest another mechanism. The phase transformation would be due to the stress field associated with the presence of interstitials (Wittels and Sherrill 1956). A few months earlier, but only published in an internal report, G.E. Klein of the same laboratory had observed that in stabilized zirconia (cubic plus monoclinic), the monoclinic phase disappeared under neutron irradiation (Klein 1955).

Working at Harwell on the corrosion resistance of zirconium alloy claddings for future water-cooled reactors, J. Adam and B. Cox investigated the role of phase transformation on the strength of the protective oxide layer. Their conclusions were slightly different: pure zirconia does not transform under irradiation with fast

7 one dose unit = 10^{18} neutrons/cm².

neutrons, whereas baddeleyite (natural zirconia) transforms into a quadratic phase (Adam and Cox 1959).

8.2. Biographies of some of the chapter's personalities

Many people have contributed to the advancement of knowledge of iono-covalent materials under irradiation. However, there will be no summary biography for this chapter, either because no biography is available or because it was the subject of a short biography in one of the previous chapters.

8.3. References

Adam, J. and Cox, B. (1959). The irradiation-induced phase transformation in zirconia solid solutions. *Journal of Nuclear Energy. Part A. Reactor Science*, 11, 31–33.

Arnold, G.W. and Compton, W.D. (1960). Threshold energy for lattice displacement in a-Al_2O_3. *Physical Review Letters*, 4(2), 66–68.

Bierlein, T.K. and Mastel, B. (1960). Damage in UO_2 films and particles during reactor irradiation. *Journal of Applied Physics*, 31(12), 2314–2315.

Bloch, J. (1961). Restauration thermique du paramêtre de l'UO2 faiblement irradié. *Journal of Nuclear Materials*, 3(2), 237–238.

Colombino, P., Degregori, I., Mayrone, L., Trossi, L., DeBenedetti, S. (1960). Angular correlation of annihilation radiation in silica and aqueous solutions. *Nuovo Cimento*, 18(4), 632–643.

Coussot, G. (1969). Influence of the electron irradiation on the narrow component in the angular correlation of positron annihilation in quartz crystal. *Physics Letters*, 30A(2), 138–139.

Evans, H.E., Evans, J.H., Rice-Evans, P., Smith, D.L., Smith, C. (1992). A slow positron study into the recovery behaviour of as-polished and krypton-implanted uranium dioxide. *Journal of Nuclear Materials*, 199(1), 79–83.

Goldschmidt, V.M. (1924). Über die Umwandlung krystallisierter Minerale in den metamik ten Zustand (Isotropisierung) (On the transformation of crystallised minerals into the metamictic state (isotropisation)). *Geochemische Verteilungsgesetze der Elemente*, III, 51–58.

Groves, G.W. and Kelly, A. (1963). Neutron damage in MgO. *The Philosophical Magazine: A Journal of Theoretical Experimental and Applied Physics*, 8(93), 1437–1454.

Henderson, G.H. and Bateson, S. (1934). A quantitative study of pleochroic haloes – I. *Proceedings of the Royal Society of London, Series A Mathematical and Physical Sciences*, 145(855), 563–581.

Henderson, G.H. and Turnbull, L.G. (1934). A quantitative study of pleochroic haloes – II. *Proceedings of the Royal Society of London, Series A Mathematical and Physical Sciences*, 145(855), 582–591.

Houpert, C., Hervieu, M., Groult, D., Studer, F., Toulemonde, M. (1988). HREM investigation of GeV heavy ion latent tracks in ferrites. *Nuclear Instruments and Methods in Physics Research Section B: Beam Interactions with Materials and Atoms*, 32(1), 393–396.

Hurley, P.M. and Fairbairn, H.W. (1952). Alpha-radiation damage in zircon. *Journal of Applied Physics*, 23(12), 1408.

Joly, J. (1907). Pleochroic halos. *The London, Edinburgh and Dublin Philosophical Magazine and Journal of Science (6th series)*, XIII(LXXV), 381–383.

Kip, A.F., Kittel, C., Levy, R.A., Portis, A.M. (1953). Electronic structure of F centers: Hyperfine interactions in electron spin resonance. *Physical Review*, 91(5), 1066–1071.

Klein, G.E. (1955). X-ray examination of ceramic samples in solid state division semiannual progress report. Report, ORNL-1852, Oak Ridge National Laboratory, Tennessee.

Knoll, M. and Ruska, E. (1932). Das Elektronenmikroskop (The electronic microscope). *Zeitschrift für Physik*, 78(5), 318–339.

Lind, S.C. (1920). Notes on the coloring and thermoluminescence of glass produced by radium radiation. *Journal of Physical Chemistry*, 24(6), 437–443.

Meis, C. and Chartier, A. (2005). Calculation of the threshold displacement energies in UO_2 using ionic potentials. *Journal of Nuclear Materials*, 341(1), 25–30.

Meyer, S. and Przibram, K. (1912). Influence of radium radiations on glass and certain minerals. *Sitzungsberichte der Akademie der Wissenschaften in Wien, Mathematisch-Naturwissenschaftliche Klasse*, 121(2a), 1413–1418.

Nilsson, G. (1966). Ejection of uranium atoms from sintered UO_2 by fission fragments in different gases and at different gas pressures. *Journal of Nuclear Materials*, 20, 215–230.

Pabst, A. (1952). The metamict state. *American Mineralogist*, 37(3–4), 137–157.

Page, L.A., Heinberg, M., Wallace, J., Trout, T. (1955). Angular correlation in two-photon annihilation in quartz. *Physical Review*, 98(1), 206–208.

Pells, G.P. and Phillips, D.C. (1979). Radiation damage of α-Al_2O_3 in the HVEM: I. Temperature dependence of the displacement threshold. *Journal of Nuclear Materials*, 80(2), 207–214.

Primak, W. (1955). Experimental evidence for thermal spikes in radiation damage. *Physical Review*, 98(6), 1854.

Silk, E.C.H. and Barnes, R.S. (1959). Examination of fission fragment tracks with an electron microscope. *Philosophical Magazine*, 4, 970–972.

Soullard, J. (1977). Contribution to the study of structural defects in uranium dioxide. Report, CEA-R–4882, Commissariat à l'Energie Atomique.

Soullard, J., Leteurtre, J., Genthon, J.P., Cance, M. (1978). Etude du ralentissement des ions lourds dans une cible diatomique. I – Calcul de la distance parcourue et du nombre d'atomes primaires créés. *Radiation Effects*, 38(1–2), 119–123.

Stackelberg, M. and Chudoba, K. (1937). Dichte und Struktur des Zirkons II (Density and structure of the zircon II). *Zeitschrift für Kristallographic, Kristallgeometrie, Kristallphysiks, Kristallchemie*, 97, 252–262.

Stackelberg, M. and Rottenbach, E. (1939). Dichte und Struktur des Zirkons IV. Die Ursache des Isotropisierung des Zirkons (Density and structure of the zircon IV. The cause of the isotropisation of the zircon). *Zeitschrift für Kristallographic, Kristallgeometrie, Kristallphysiks, Kristallchemie*, 102, 207–208.

Vegard, L. (1916). VI. Results of crystal analysis. *The London, Edinburgh, and Dublin Philosophical Magazine and Journal of Science*, 32(187), 65–96.

Weber, W.J. (1983). Thermal recovery of lattice defects in alpha-irradiated UO_2 crystals. *Journal of Materials Research*, 114, 213–221.

Wertz, J.E., Auzins, P., Weeks, R.A., Silsbee, R.H. (1957). Electron spin resonance of F centers in magnesium oxide; confirmation of the spin of magnesium-25. *Physical Review*, 107(6), 1535–1537.

Wittels, M.C. and Sherrill, F.A. (1956). Irradiation-induced phase transformation in zirconia. *Journal of Applied Physics*, 27(6), 643–644.

9

Polymers Under Irradiation

9.1. First irradiations of polymers

The first intentional irradiations of organic materials were carried out at virtually the same time in Kiel and Paris. Publications by August Becker and Henry Becquerel are dated May 16 and 18, 1903 respectively (Becker 1903; Becquerel 1903). In both studies, the material concerned was paraffin[1]. Paraffins are linear alkanes with the formula C_nH_{2n+2} with n between 18 and 32. But it is possible that J.J. Thomson had precedence. In fact, Becquerel began his note to the May 18, 1903 session of the Académie des Sciences by pointing out that Thomson had observed that solid paraffin through which X-rays were passed seemed to have acquired a kind of electrolytic conductivity. As was often the case at the time, he gave no bibliographical references enabling the source of his assertion to be traced. Becker also begins his article with a reference to Thomson, but on the subject of gaseous insulators. That said, they both pointed out that paraffin becomes slightly conductive under the influence of radium radiation, and that this conductivity slowly disappears once the radioactive source is removed. It should be noted that in his article Becker supported his results with a remarkable analysis of possible causes of error. In the same year, he carried out the same experiments using cathode radiation as the source of irradiation (Becker 1904). His set-up is shown in Figure 9.1.

This work on paraffin conductivity does not really fit into the framework of materials under irradiation: no permanent effect is observed.

1 Becker reported that he had carried out identical experiments on mica, shellac and rubber.

A Historical Approach to Materials Under Irradiation,
Serge BOUFFARD.

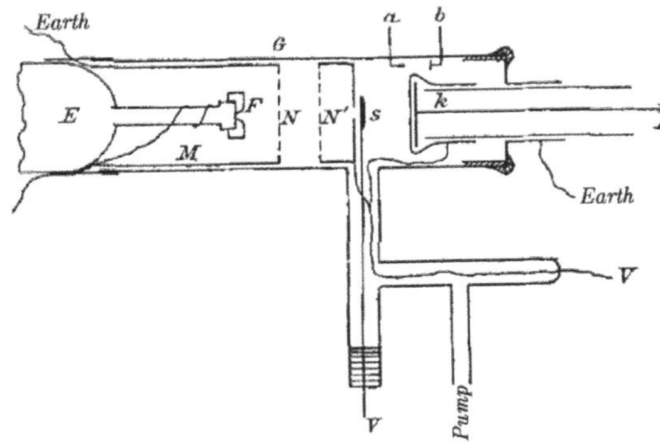

Figure 9.1. *Cathode ray irradiation device (Becker 1904). E: discharge tube, F: aluminum window, S: sample*

A few years later, Samuel Colville Lind of the Laboratory of Physical Chemistry, University of Michigan, studied the chemical effects of α rays in the gas phase. He was the first to suggest a close relationship between the ionization of a gas and the chemical action produced in that gas. He provided quantitative evidence in support of this idea. When studying the amount of ozone produced in oxygen by alpha particles, he found that the total number of ozone molecules formed in a given time was almost equal to the calculated number of ion pairs produced during the same time, on the assumption that one ion pair is involved in the production of one ozone molecule. He pointed out that this correlation implied that ions are the primary actors in the reaction. He published this result in the *American Chemical Journal* (Lind 1912a), in French in *Le Radium* (Lind 1912b), and in German in *Monatshefte für Chemie* (Lind 1912c). A first step toward the condensed phase was to show that reactions in liquid systems obey the same laws as those in the gas phase (Lind 1912d).

Twelve years later, working for the US Bureau of Mines, Lind demonstrated for the first time that irradiation of a low-mass hydrocarbon could create higher mass hydrocarbons. As with his earlier work, he studied the gas-phase behavior of ethane (C_2H_6) under irradiation by alpha particles. He observed the appearance of a liquid phase, which he explained as the formation of octane (C_8H_{18}), the lower-mass paraffin remaining liquid at low vapor pressure at room temperature (Lind and Bardwell 1924). This was the first evidence of a cross-linking process between organic molecules, albeit in the gas phase[2]. This research was still in the realm of molecular collision physics, but in an endnote, the authors pointed out:

> Professor Boltwood[3] (private communication) informs us that in work on the production of helium from radium he and Professor Rutherford had occasion to confine radon by a solid plug of paraffin. After several days large quantities of hydrogen were found to have been given off and the paraffin plug could no longer be melted, even by heating almost to the softening point of the glass.

The extensive correspondence between these two figures has been compiled in a book (Badash 1969), but I have not found any letters relating to this result. However, it may be considered the first evidence of irradiation-induced structural modification in a polymer. Paraffin reaches its gel point by cross-linking.

More than 20 years after the publications of Becker and Becquerel, studies on the effects of radiation on this material really began (or at least were published). At the October 25, 1926 meeting of the Cambridge Philosophical Society, Rutherford presented a note by William T. Richards on the effects of alpha particles on paraffin (Richards 1927). The aim this time was to generalize a result obtained in water to another material. Gaseous emission due to the action of α particles is about 20 times lower in ice than in liquid water (Duane and Scheuer 1913). This liquid/solid effect exists in paraffin, but only to a limited extent.

It seems that this work on paraffin was the only research carried out on the irradiation of organic matter before 1945, with the exception of the rubber industry, which was expecting radiation-induced cross-linking to provide an alternative to the chemical cross-linking of natural rubber. In 1929, BF Goodrich society filed a patent entitled *"Method of vulcanizing rubber"*. The invention involved vulcanizing rubber by exposure to high-speed electrons. The conditions for complete vulcanization were given: a thin film of rubber placed in a nitrogen atmosphere 2 inches from the window of a 250 kV cathode-ray tube delivering a current of 1 mA for 20–25 s was completely vulcanized (Newton 1929). Almost 20 years later, two researchers, engineers at the same company[4], published a paper on the behavior under pile

2 Lind has been called the father of modern radiochemistry.

3 Bertram B. Boltwood, chemist and professor at Yale University, collaborated extensively in letters with his friend Ernest Rutherford.

4 One of them, W.L. Davidson, was also attached to the Clinton National Laboratory in Oak Ridge, which explains his access to irradiation in nuclear reactors.

irradiation (in a nuclear reactor) of three basic rubbers (natural, butyl and polyisobutyl) containing different additives (Davidson and Geib 1948). They observed that the molecular weight of natural rubber increased under pile bombardment, while that of polyisobutylene decreased. They suggested that the first two stages, ionization and neutralization, were similar in both materials, and then:

> If the ionization occurs at the expense of a valence electron belonging to a chain carbon, cleavage of the chain probably takes place. Upon neutralization free radicals are generated. In natural rubber these radicals are able to initiate chain reactions involving double bonds. Apparently the cross linking established in this way more than compensates for the broken chains. With PIB no chain reaction can be effected and so the end result has to be degradation.

Neutralization by electron capture of the ions initially formed leads to the formation of radicals by breaking covalent bonds. This process leads to a change in the molecular weight of polymer chains: backbone scission.

In parallel with these rubber studies, several groups were working on the effects of different sources of radiation. Such was the case of F.L. Hopwood and J.T. Phillips of St. Bartholomew's Hospital, London, who published a very short article in *Nature* entitled "*Polymerization of liquids by irradiation with neutrons and other rays*" (Hopwood and Phillips 1939).

Ten years later, we find a similar approach in a publication by Michel Magat's group at the Physical Chemistry Laboratory of the Paris Faculty of Science (Chapiro et al. 1949). This time, however, the chemical reactions were induced by slow neutrons. The mechanism triggering the creation of free radicals was the Swilard-Chalmer effect: the recoil energy upon capture of a neutron by an atom could be sufficient to eject the atom from its site. Secondary reactions producing β and γ radiation were also responsible for a significant proportion of radicals. The feasibility of initiating styrene polymerization[5] by slow neutrons was shown in the article by Chapiro et al., which concluded with a lengthy discussion of the chemical mechanisms and associated yields.

Thanks to these studies on low-mass molecules, it was established that one of the initial actions of ionizing radiation on organic matter was the creation of ions whose neutralization led to the formation of radicals by breaking bonds.

5 Common name for phenylethylene $C_6H_5{-}CH{=}CH_2$.

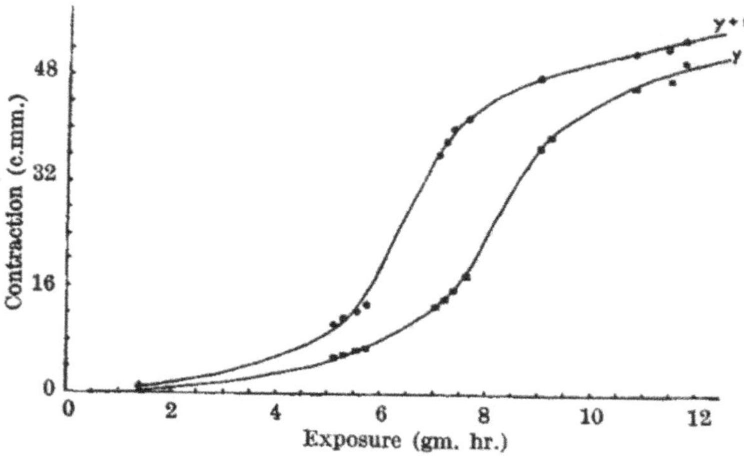

Figure 9.2. *Polymerization of methyl acrylate by irradiation with neutrons and γ rays (top curve) and γ rays alone (bottom curve). Neutrons come from the reaction* $^{9}_{4}Be + ^{4}_{2}He \rightarrow ^{12}_{6}C + ^{1}_{0}n$*. Exposure is expressed by the product of radium mass and time, and polymerization by density measurement using a dilatometer (Hopwood and Phillips 1939)*

9.2. Research into degradation mechanisms

These polymerization studies have taken us slightly away from the subject of polymers under irradiation. However, cross-linking and the notion of gel have already been discussed. The theoretical aspects of cross-linking were first proposed by Paul John Flory[6] in three linked papers (Flory 1941a, 1941b, 1941c) and a fourth the following year (Flory 1942) in which he established a correlation between gel formation and the formation of branched macromolecules. Moreover, assuming a uniform initial distribution of chain lengths, he proposed a formula for calculating the gel point and the average molecular mass at that point. One of the simplifying assumptions made was that no intramolecular reactions were taken into account. Walter Hugo Stockmayer of Colombia University, New York, proposed a different method, allowing chains of any length with branches of any functionality to be taken into account (Stockmayer 1943). The following year, he extended his theory to take account of the initial distribution of chain lengths. His results can be used for certain copolymers, such as vinyl-divinyl copolymers (Stockmayer 1944).

6 Flory was a member of Esso Laboratories of the Standard Oil Development Company. In 1974, he was awarded the Nobel Prize in Chemistry for his work on macromolecules.

As in the case of many materials, it was the availability of irradiation facilities that enabled the development of this research, essentially from the 1940s onwards. In 1943, J.G. Burr and W.M. Garrison of Argonne National Laboratory reviewed the state of knowledge on the effects of radiation on the physical properties of plastics in a report[7] for the United State Atomic Energy Commission (Burr and Garrison 1943). This compilation brought together electrical and mechanical data on a large number of commercial plastics. Irradiations were carried out with 2.5 MeV electrons from the Van de Graaff accelerator at Notre Dame University at a dose rate of 10^5 Röntgen/mn (i.e. 60 kGy/h) and with 3 MeV gammas at the Massachusetts Institute of Technology (600 kGy/h). Working with industrial polymers, they were well aware of the difficulties associated with not being familiar with the many added constituents these could contain:

> In considering the effect of radiation on the specific resistance of plastics, considerable difficulties are encountered in attempting to correlate the results closely with their chemical and physical properties. The difficulty is in the fact that the plastics as supplied to us, although basically composed of one or two molecular species, contain, in addition, large and often variable amounts of plasticizer, filler, etc. Frequently plastics of the same type, e.g., Koroseal, vary tremendously in the relative amounts of these added constituents. Therefore any detailed complete correlation of data is practically impossible.

Thus, for 22 different plastics, the evolution of specific resistance was measured, and when mechanical properties were important for their use (seals, for example), the evolution of breaking strength was also given. In their concluding remarks, they stated that in a few cases they observed a dose rate effect, which they attributed to heating under irradiation. The experimental results were only supported by a few vague ideas about the behavior of these materials under irradiation.

In 1951, O. Sisman and C.D. Bopp presented a compilation of the physical properties of 33 plastics irradiated in Hole 19 of the Oak Ridge National Laboratory reactor. They pointed out that the irradiations were carried out in air, but that the effects of oxygen and the post-irradiation storage period would have to be studied in greater depth. The number of characterizations is impressive: stress–strain curves, tensile strength, impact strength, shear strength, elastic modulus, percent elongation, hardness, weight loss, density, water absorption, light transmission and scattering, resistivity, dielectric strength and resistance to electrical arc. This document is a

7 This report was declassified on June 25, 1948.

good basis for choosing a polymer for a technological application. In terms of scientific explanations, however, the remarks remain very general (Sisman and Bopp 1951).

Studies of the structural effects of irradiation began in earnest in the early 1950s with work carried out at the Atomic Energy Research Establishment in Harwell. K. Little's work on the irradiation of high molecular weight linear polymers is noteworthy (Little 1952). In the first paragraph, he stated:

> The initial effect is to break down bonds, some ionic, the majority covalent. The radicals formed can react in a number of ways. In the presence of oxygen they can form low-molecular weight compounds and carboxyl end-groups; different end-groups and low-molecular weight compounds can be formed in the absence of oxygen; they can reform the original bonds; or they can recombine in other ways.

He used X-ray diffraction to show that three factors play an important role:

– *Presence or absence of air*: in the presence of air, samples are rapidly decomposed to form volatile compounds, whereas in a vacuum, free-end reactions create new bonds and stabilize the polymer.

– *Irradiation intensity*: taking into account heating by the beam, he asserted that the effect of irradiation intensity was more easily interpreted in terms of an effect on the number of defects than in terms of an evolution in their type.

– *The chemical nature of polymers*: the presence of unsaturated rings is an important factor in limiting rapid polymer degradation. Polystyrene is more resistant to irradiation than polyethylene.

Arthur Charlesby joined Harwell in 1949 to take charge of the group studying the effects of radiation. He quickly established a serious reputation in the field of polymer irradiation. One of his first publications on polymer irradiation dates from 1952, and is entitled *"Cross-linking of polythene by pile irradiation"* (Charlesby 1952). This article provides quantitative data. For example, polythene becomes insoluble when 1.1%–1.4% of the carbon atoms are involved in cross-linking, or an energy of the order of 25 eV is required to break a C-H bond. The evolution of sample weights suggests that two mechanisms are in competition: (1) bond breaks release hydrogen and low-mass hydrocarbons and (2) an oxidation process. The first mechanism is proportional to sample mass and the second to sample surface (oxygen diffuses slowly).

Charlesby continued this work on aliphatic compounds by irradiating alkanes from C_7H_{16} to $C_{36}H_{74}$ again in Harwell's BEPO[8] pile. The advantage of alkanes lies in their perfectly controlled chain length. It shows that the energy required to create cross-linking is virtually independent of chain length. The radiation dose at which the polymer becomes non-meltable can be used to calculate the cross-linking energy (Charlesby 1954a). In the years 1953–1954, Charlesby and his colleagues published several papers on the effects of cross-linking on various properties such as elastic modulus, density and melting. (Charlesby and Ross 1953; Charlesby and Hancock 1953) Although Charlesby (1954b) did not neglect the degradation of long polymer chains, he treated this mechanism independently of cross-linking.

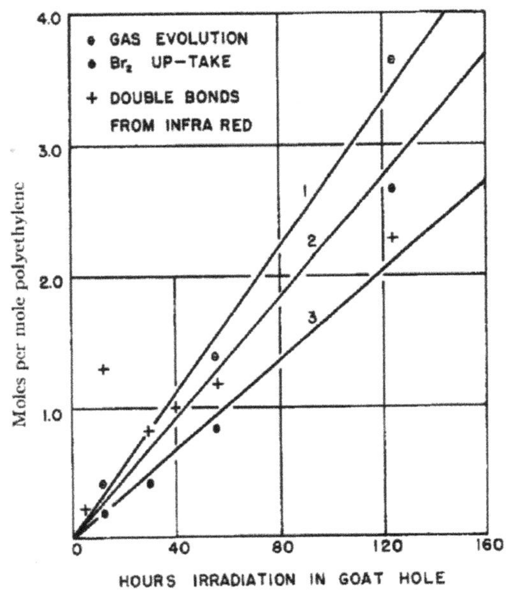

Figure 9.3. *Number of moles of gas emitted per mole of polyethylene (curve 1), number of moles of double bonds determined by bromine vapor absorption (curve 2) and infrared absorption (curve 3)*[9] *(Dole et al. 1954)*

One of the conclusions of this work on pile irradiation of solid alkanes was that unsaturation production is negligible, and that all the effects observed are due to

8 BEPO: British Experimental Pile 0: a 6 MW pile that operated at Harwell from 1948 to 1968.

9 In the core of an atomic pile, the place where samples are irradiated is known as the "goat hole". I have not found the origin of this term.

cross-linking. This assertion was contested by Malcolm Dole (Department of Chemistry, Northwestern University), who believed that the hypothesis of no olefinic group production was not supported by any experimental evidence. He also pointed out that high-energy irradiation in the absence of air produced both unsaturation and cross-linking. These modifications were compounded by significant oxidation in the presence of air (Dole et al. 1954). His relatively direct measurements of the number of double bonds by bromination, combined with the weighing and quantification of emitted gases and supplemented by infra-red absorption spectroscopy analyses, enabled him to extract some noteworthy results. He concluded that 70–80% of the hydrogen atoms emitted were due to the formation of unsaturations, with the remaining 20–30% resulting from inter-chain bonds. He also established that C-C bonds were cut half as much as C-H bonds.

Furthermore, in interpreting his results, Dole proposed the chemical equations for the formation of molecular hydrogen. It all starts with ionization followed by neutralization, which releases an atomic hydrogen that reacts with another hydrogen to form the H_2 molecule.

$$-CH_2CH_2CH_2CH_2 - +\gamma \text{ rays} \rightarrow -CH_2C^+H_2CH_2CH_2 - +e^-$$

$$-CH_2C^+H_2CH_2CH_2 - +e^- \rightarrow -CH_2C^\bullet HCH_2CH_2 - +H$$

$$H^\bullet + -CH_2CH_2CH_2CH_2- \rightarrow H_2 - CH_2C^\bullet HCH_2CH_2 -$$

He concluded: "Thus the production of large amounts of hydrogen is readily understood". But he said nothing about the fate of the two radicals created on the chains.

Of all the modifications induced by irradiation, vinyl groups ($-HC{=}CH_2$) display surprising behavior. They disappear rapidly, whereas for the polymers studied the probability of vinyl groups being impacted by incident radiation is low, on the order of one in a thousand. There are two vinylidenes for every 2,300 CH_2 groups. To explain this non-random effect, Dole proposed the long-distance migration of a reactive species until it interacted with the vinyl groups. These reactive species could be free radicals, which by a random walk mechanism moved along carbon chains (Dole and Keeling 1953). A few years later, the author proposed that electronic excitation could be the migrating species. The transfer of an electronic excitation created on a CH_2 group to the unsaturation produced a triplet state from which new reactions were initiated (Dole et al. 1958).

Atomic piles were used extensively in early experiments on the effects of radiation on polymers. However, their use had its drawbacks: a rather high ambient temperature, a long exposure time and, often, the samples could only be removed when the pile was opened for other reasons. On these points, electron beams, which benefitted from this period of technological development, offered serious advantages, especially if energy and flux were high. The development of linear accelerators for radiotherapy opened up opportunities for polymer research. This was the case for Metropolitan-Vickers Electrical Company's collaborations with the Atomic Energy Research Establishment and the Medical Research Council for the accelerator, and with Imperial Chemical Industries Ltd for polymers (Miller 1953; Baskett and Miller 1954).

The General Electric Company team, using the 800 keV electron accelerator they had developed in-house, published a paper in 1953 giving extensive quantitative data on the behavior of poyethylene (Lawton et al. 1953). Hydrogen emission dominated (85%) the gaseous emissions, accompanied by volatile compounds with two, three or four carbon atoms (C_2, C_3 and C_4). Based on Charlesby's hypothesis that each hydrogen molecule released was associated with the creation of an inter-chain bond, they found that 3.33×10^{18} cross-links per cubic centimeter were created for an irradiation dose of 10^6 Roentgen, giving an efficiency of 1.87 cross-links per ionization[10].

Also in 1955, Chapiro studied the effects of γ radiation from a ^{60}Co source on polymers (Chapiro 1955). For vacuum irradiation, he followed Charlesby's results. The increase in melting point was directly linked to the increase in average molecular weight. Chapiro's article ended with a calculation of some radiochemical yields. For polyethylene, the cross-linking yield was 1.9 under vacuum and 0.75 under air (for 100 eV).

Osamu Saito (Faculty of Engineering, Chuo University, Tokyo) published two papers in 1958 on the effects of high-energy radiation on polymers (Saito 1958a, Saito 1958b). The first paper dealt with cross-linking and degradation, while the second dealt with end-bonding and gel fraction. With these two articles, he tackled the most general case in which the effects of radiation were classified into cross-linking, including intra-chain cross-linking (cyclization), degradation and end-linking. In the same year, Charlesby studied the influence of polyethylene crystallinity on its evolution under irradiation (Charlesby and Callaghan 1958).

By the end of the 1950s, most of the concepts for irradiating polymers in an inert atmosphere had been established, although some were still intuitive.

10 This is surprisingly high efficiency. Defect creation efficiency was defined as the ratio of the number of defects to the number of ionizations, until Milton Burton's proposal to define G efficiency as the number of entities created or destroyed per 100 eV absorbed by the material.

9.3. Radio-oxidation of polymers

The presence of oxygen in a polymer's environment profoundly alters its evolution under irradiation. As industrial plastic materials are generally used in the presence of air, the degradation of industrial polymer properties under irradiation in real-life conditions has been the subject of ongoing research. Of all the physicochemical processes involved, one point that is essential to understand is the potential heterogeneity of degradation. Early research on polymers under irradiation in the presence of air revealed heterogeneities. As we have seen, Charlesby (1952) introduced a surface effect and hence heterogeneity in his early publications:

> When irradiation is carried out in air, a very thin wax-like surface film is produced, which is only visible after fairly prolonged radiation (10 to 20 units). The transition from this oxidation surface film to the remainder of the specimen is very sharp. No film is formed if the polythene is irradiated in a vacuum, nor does it form to any appreciable extent after radiation when the specimen is left in the air for long periods.

In 1954, using a ^{60}Co source, researchers at Brookhaven National Laboratory and the Polytechnic Institute of Brooklyn used infrared absorption to show the progression of unsaturation and the formation of oxygen-containing functional groups with the deposited dose. They assumed that oxidation processes were localized near the surface and must play only a minor role in the volume: "About 80% of the carbonyl absorption at 60 Mrep[11] can be attributed to the surface while only about 30% of the unsaturation is associated with the surface layers at this irradiation" (Ballentine et al. 1954). The diffusion of oxygen from the surface to the volume was clearly demonstrated by Alexander and Toms[12]. They observed that, for irradiation in air, the gel fraction decreased with the thickness of the polythene film. This result was interpreted as meaning that the amount of dissolved oxygen was too small to influence detectable reactions, and that the effect of film thickness was due to oxygen diffusion during irradiation (Alexander and Toms 1956).

---

11 Note: The rep ("Röntgen equivalent physical") was a legal unit of absorbed dose until the 1960s, when it was gradually replaced by the rad (100 erg/g), then by the Gray (1 J/kg). The value of the rep fluctuated widely between 83 and 93 erg/g, until it was fixed by the US National Bureau of Standard at 9.3 mGy. Note that 60 Mrep is therefore equivalent to 560 kGy.

12 P. Alexander worked at the Chester Beatty Research Institute, Royal Cancer Hospital, London and D. Toms at the Chemistry Department of Imperial College, London.

9.4. Research and development, an active field

From the earliest experiments on the effects of radiation on organic matter, BFGoodrich investigated the possibility of replacing rubber vulcanization with high-energy radiation-induced cross-linking. BFGoodrich is an American company founded in 1870. It specializes in aircraft, tires and military equipment. Its tire business was acquired by Michelin in 1990.

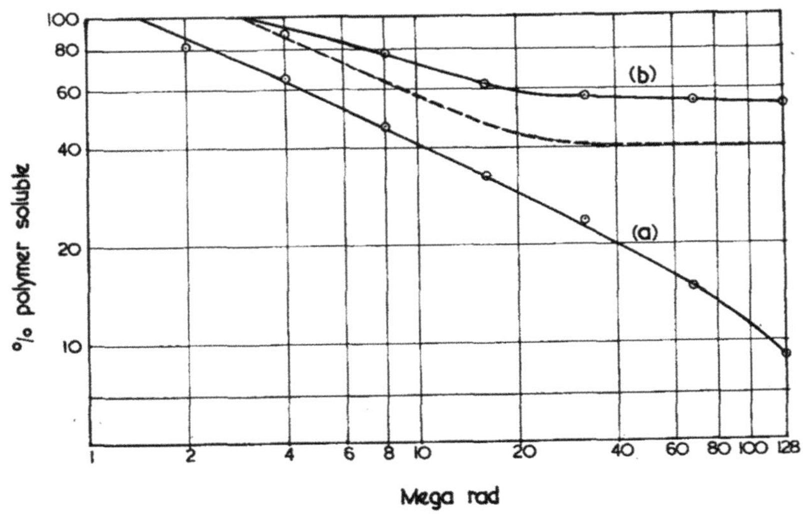

Figure 9.4. *Percentage of insoluble polythene as a function of irradiation dose (2 MeV electrons) for a 1-µm-thick film. Curve (a): irradiation in vacuum, curve (b) irradiation in air, dashed curve: calculation considering that each cross-linking is associated with a break in the main chain (Alexander and Toms 1956)*

Polymers are closely linked to the oil industry. So it is only natural that the oil industry should be interested in modifying these materials. Flory worked for the Standard Oil Development Company. Raychem Corporation was founded in 1957. It was a pioneer in the marketing of products made using radiation chemistry. The first of these products was electric cables whose insulator had been hardened by cross-linking under irradiation for military applications. Raychem then invented heat-shrink tubing. General Electric carried out extensive research into the irradiation of polyethylene by high-energy electrons. As a result, both the accelerator and the irradiated polymer were commercialized. The polymer was marketed under the name Irrathene. It is still marketed as heat-shrinkable film for insulating electric motor coils, for example.

The industrial applications of irradiating polymers with ions of high electronic stopping power are developed in section 12.6: 'Ion track technology'.

9.5. Chapter 9 references

Among the many researchers who have explored the effects of irradiation on polymers – Charlesby, Dole, Chapiro, Flory, Lind, Saito, etc. – I will only mention the first two for a short biography.

9.5.1. *Biographies of some of the chapter's personalities*

Arthur Charlesby (1915, London–1996, Swindon): Although a trained physicist, Charlesby can be considered the father of radiation chemistry. He graduated from the Imperial College of Science in London, and curiously was responsible for air traffic planning in post-war Europe. He then joined the Atomic Energy Research Establishment (AERE) in Harwell to work on the effects of radiation on materials. He made a significant contribution to the understanding of cross-linking mechanisms.

Malcolm Dole (1930, Melrose Massachusetts–1990, Los Gatos California): During his career, most of which was spent at Northwestern University, Illinois, Dole made important contributions to the field of physical chemistry (particularly electrochemistry and polymer chemistry). Dole is also known for the discovery of the "Dole effect" that the atomic mass of oxygen in air is greater than that of oxygen in water due to a difference in the concentration of ^{18}O, as well as for his work on ionization in electrosprays.

9.6. References

Alexander, P. and Toms, D. (1956). The effect of oxygen on the changes produced by ionizing radiations in polymers. *Journal of Polymer Science*, 22(101), 343–348.

Badash, L. (1969). *Rutherford and Boltwood, Lettres on Radioactivity.* Yale University Press, New Haven and London.

Ballentine, D.S., Dienes, G.J., Manowitz, B., Ander, P., Mesrobian, R.B. (1954). Evaluation of polyethylene crosslinked by ionizing radiation. *Journal of Polymer Science*, 13(70), 410–414.

Baskett, A.C. and Miller, C.W. (1954). Electron irradiation of polythene. *Nature*, 174, 364–365.

Becker, A. (1903). Über die Leitfähigkeit fester Isolatoren unter dem Einfluß von Radiumstrahlen (On the conductivity of solid insulators under the influence of radium rays). *Annalen der Physik*, 317(9), 124–143.

Becker, A. (1904). Über den Einfluß von Kathodenstrahlen auf feste Isolatoren (The influence of cathode rays on solid insulators). *Annalen der Physik*, 318(2), 394–421.

Becquerel, H. (1903). Conductivité et ionisation résiduelle de la paraffine solide sous l'influence du rayonnement du radium. *Comptes Rendus Hebdomadaires des Séances de l'Academie des Sciences*, 136, 1173–1176.

Burr, J.G. and Garrison, W.M. (1943). The effect of radiation on the physical properties of plastics. Report, AECD-2078, United States Atomic Energy Commission, Argonne National Laboratory.

Chapiro, A. (1955). Effets de l'irradiation g sur les polymères dans l'état solide I. Réticulation du polyéthylène. *Journal de Chimie Physique*, 52, 246–258.

Chapiro, A., Cousin, C.H., Landler, Y., Magat, M. (1949). Contribution à l'étude des polymérisations amorcées par des rayonnements nucléaires. *Recueil des Travaux Chimiques des Pays-Bas*, 68(11), 1037–1068.

Charlesby, A. (1952). Cross-linking of polythene by pile radiation. *Proceedings of the Royal Society of London, Series A Mathematical and Physical Sciences*, 215, 187–214.

Charlesby, A. (1954a). Crosslinking and degradation of paraffin chains by high energy radiation. *Proceedings of the Royal Society of London, Series A Mathematical and Physical Sciences*, 222(1148), 60.

Charlesby, A. (1954b). Molecular-weight changes in the degradation of long-chain polymers. *Proceedings of the Royal Society of London. Series A. Mathematical and Physical Sciences*, 224(1156), 120–128.

Charlesby, A. and Callaghan, L. (1958). Crystallinity changes in irradiated polyethylenes. *Journal of Physics and Chemistry of Solids*, 4(4), 306–314.

Charlesby, A. and Hancock, N.H. (1953). The effect of cross-linking on the elastic modulus of polythene. *Proceedings of the Royal Society of London. Series A. Mathematical and Physical Sciences*, 218(1133), 245–255.

Charlesby, A. and Ross, M. (1953). The effect of cross-linking on the density and melting of polythene. *Proceedings of the Royal Society of London. Series A. Mathematical and Physical Sciences*, 217(1128), 122–135.

Davidson, W.L. and Geib, I.G. (1948). The effects of pile bombardment on uncured elastomers. *Journal of Applied Physics*, 19, 427–433.

Dole, M. and Keeling, C.D. (1953). Long range migration of chemical activity in the solid state. *Journal of the American Chemical Society*, 75(23), 6082–6083.

Dole, M., Keeling, C.D., Rose, D.G. (1954). The pile irradiation of polyethylene. *Journal of the American Chemical Society*, 76(17), 4304–4311.

Dole, M., Milner, D.C., Williams, T.F. (1958). Irradiation of polyethylene. II. Kinetics of unsaturation effects. *Journal of the American Chemical Society*, 80(7), 1580–1588.

Duane, W. and Scheuer, O. (1913). Recherches sur la décomposition de l'eau par les rayons a. *Le Radium*, 10(2), 33–46.

Flory, P.J. (1941a). Molecule size distribution in three dimensional polymers. I. Gelation. *Journal of the American Chemical Society*, 63, 3083–3090.

Flory, P.J. (1941b). Molecule size distribution in three dimensional polymers. II. Trifunctional branching units. *Journal of the American Chemical Society*, 63, 3091–3096.

Flory, P.J. (1941c). Molecule size distribution in three dimensional polymers. III. Tetrafunctional branching units. *Journal of the American Chemical Society*, 63, 3099–4100.

Flory, P.J. (1942). Constitution of three-dimensional polymers and the theory of gelation. *The Journal of Physical Chemistry*, 46(1), 132–140.

Hopwood, F.L. and Phillips, J.T. (1939). Polymerization of liquids by irradiation with neutrons and other rays. *Nature*, 143(3624), 640.

Lawton, E.J., Bueche, A.M., Balwit, J.S. (1953). Irradiation of polymers by high-energy electrons. *Nature*, 172(4367), 76–77.

Lind, S.C. (1912a). The ozonization of oxygen by α-rays. *American Chemical Journal*, 47, 397–415.

Lind, S.C. (1912b). Ozonisation de l'oxygène par les rayons α. *Radium (Paris)*, 9(3), 104–106.

Lind, S.C. (1912c). Ozonisierung des Sauerstoffes durch a-Strahlen (Ozonisation of the oxygen by α-rays). *Monatshefte für Chemie und verwandte Teile anderer Wissenschaften*, 33(3), 295–310.

Lind, S.C. (1912d). On the nature of the chemical action produced by α particles and the probable role played by ions. *Journal of Physical Chemistry*, 16(7), 564–613.

Lind, S.C. and Bardwell, D.C. (1924). Direct synthesis of higher from lower hydrocarbons. *Science*, 60, 364–365.

Little, K. (1952). Irradiation of linear high polymers. *Nature*, 170, 1075.

Miller, C.W. (1953). Travelling-wave linear accelerator for X-ray therapy. *Nature*, 171(4346), 297–298.

Newton, E.B. (1929). Method of vulcanizing rubber. US Patent 1,906,402.

Richards, W.T. (1927). Note on the effect of alpha particles on paraffin. *Proceedings of the Cambridge Philosophical Society*, 23(5), 516–522.

Saito, O. (1958a). On the effect of high energy radiation to polymers: I. Cross-linking and degradation. *Journal of the Physical Society of Japan*, 13(2), 198–206.

Saito, O. (1958b). On the effect of high energy radiation to polymers: II. End-linking and gel fraction. *Journal of the Physical Society of Japan*, 13(12), 1451–1464.

Sisman, O. and Bopp, C.D. (1951). Physical properties of irradiated plastics. Report, ORNL-928, Oak Ridge National Laboratory.

Stockmayer, W.H. (1943). Theory of molecular size distribution and gel formation in branched-chain polymers. *The Journal of Chemical Physics*, 11(2), 45–55.

Stockmayer, W.H. (1944). Theory of molecular size distribution and gel formation in branched polymers II. General cross linking. *Journal of Chemical Physics*, 12(4), 125–131.

10

Radiolysis of Liquids

The radiolysis of liquids has been the subject of several historical approaches. I particularly recommend reading the articles on the history of radiolysis written by Christiane Ferradini and Jean-Paul Jay-Gerin (1996, 1997, 1999) as well as the articles "A short history of the radiation chemistry of water" by Charles D. Jonah (1995) and "Historic landmarks in radiation chemistry since early observations by Marie Sklodowska-Curie and Pierre Curie" by Jacqueline Belloni (2011). These texts can be complemented by John D. Zimbrick's (2002) article "radiation chemistry and the Radiation Research Society: A history from the beginning". The book *Early Developments in Radiation Chemistry* edited by Jerzy Kroh (1989) gathers the recollections of important players in radiation chemistry, mainly during and after the Second World War. Not wishing to reproduce what has already been well described, I shall confine myself in this chapter to the major advances in the radiolysis of water.

10.1. Upstream of the notion of radiolysis

One of the first observations of radiolysis was reported in a short article by Pierre and Marie Curie: "the rays emitted by highly active radiferous[1] barium salts are capable of transforming oxygen into ozone" (Curie and Curie 1899, author's translation). But if we limit ourselves to water, most of the actors of the time and the authors of magazine articles attribute the first observation of the decomposition of water by irradiation to Friedrich Otto Giesel (Braunschweig, German Empire). While they cite a publication, it is a 1902 article entitled *"Ueber Radium und radioactive Stoffe"*[2] (Giesel 1902a). In this article, Giesel does indeed speak of a

1 Barium salts which emit radiation.

2 About radium and radioactive substances.

A Historical Approach to Materials Under Irradiation,
Serge Bouffard.

colorless gas constantly being released, which could be a mixture of hydrogen and oxygen. But this paper is above all dedicated to the search for the radioactive bodies that accompany radium, including the mysterious emanium, a radioactive gas that emanates from radium. His experiments led him to discover actinium in 1902, independently of André-Louis Debienne who had discovered it in 1899. In a presentation on his work, Giesel noted that radium rays made paper brown and brittle, and described their physiological effects: 2 h application of 0.3 g radium bromide to the arm left a trace that was still visible a year and a half later (Giesel 1900, 1902b). Becquerel and Curie (1901) voluntarily or accidentally made the same observations.

Studying the gaseous emanations from radium bromide solutions, William Ramsay and Frederick Soddy of University College London concluded that radium produced helium and hydrogen, the origin of which was the action of rays on water. They explained that the significant excess of hydrogen over the composition required to form water was due to the greater solubility of oxygen (Ramsay and Soddy 1904). A few years later, Ramsay (1907) published an article on the chemical action of radium emanation, and more specifically its action on distilled water. Two-thirds of the article is devoted to the study of radium emanation. From these experiments on the effect of emanium on water, he extracted that emanium decomposes water and produces an excess of hydrogen. The formation of hydrogen peroxide was the only possible cause of this excess, but he ruled it out as he had not detected any hydrogen peroxide.

André-Louis Debierne, collaborator and friend of Marie and Pierre Curie and best known for his discovery of actinium, confirmed this decomposition of water. At the March 15, 1909 session of the Académie des Sciences de Paris, he presented a paper on the decomposition of water by radium salts (Debierne 1909). This note, which is a model of scientific quality in terms of the description of experimental procedures, the provision of raw results and their interpretation, deserves a closer look. The main question is the origin of the gaseous emission: "the decomposition of water in a solution of radium salt could be attributed either to a direct action of radium, emanation or their transformation products, or to the action of the various radiations they emit, α, β and γ rays" (author's translation). By isolating the radioactive substance from the water, he first demonstrated that the latter action appeared to be the correct one. He then made an initial hypothesis: the violent collisions of particles with water molecules induce movements equivalent to a thermal agitation of several billion degrees at the origin of molecular decomposition. Pointing out that violent collisions are infrequent, he put forward a second hypothesis in which collisions produce positive and negative ions, which in turn transfer kinetic energy to neighboring molecules, so that as the particle passes by, a kind of local heating must occur, capable of dissociating the molecules. He listed four recombinations between the ions formed, which explain the formation of hydrogen peroxide and oxygen. This note was followed by a second on the study of off-gassing by Miroslaw Kernbaum of the same laboratory (Kernbaum 1909). In the fall of 1908, Kernbaum, from Poland, became assistant to Marie Sklodowska-Curie at the age of 26. He worked on the decomposition of water under the influence of various types of radiation. He published five or six articles on the subject. Unfortunately, his career was short-lived, and when he returned to Poland, he committed suicide in Krakow in 1911. In the above-mentioned note, he concluded with a meticulous approach: "penetrating radium rays, acting on distilled water through glass, release hydrogen and simultaneously form hydrogen peroxide. The reaction takes place according to the formula[3] $2H_2O = H_2O_2 + H_2$". However, Kernbaum concluded: "it is conceivable that this mode of decomposition is not the primary reaction" (author's translation). At the same time, he demonstrated that Röntgen rays do not give off gas and hydrogen peroxide in the absence of air. Although this observation went unnoticed, it was rediscovered some 20 years later by Risse (1929).

During the same period in America, more specifically at the Laboratory of Physical Chemistry at the University of Michigan, Samuel Colville Lind (1912) carried out important work on gases and liquids, showing that there was a strong relationship between the number of ionizations and the chemical reactions induced by radioactivity. He observed this relationship in all of the gases he studied, and indicated that reactions in liquid systems were apparently of the same nature and obeyed the same general laws as in gases (Lind 1912).

Also at Sklodowska-Curie's laboratory, in 1912–1913, William Duane and Otto Scheuer showed in a note published in "Le Radium" and in the *Comptes rendus hebdomadaires des séances de l'Acamédie des Sciences* that the results already published on the decomposition of water by α rays depended very much on the radioactive source used and the experimental method (Duane and Scheuer 1913a, 1913b). At the request of Sklodowska-Curie, they undertook a study of the decomposition of water in the three solid, liquid and gaseous states, under controlled experimental conditions. They observed that:

> α rays decompose water, whatever its state, into hydrogen and oxygen. [...] in the liquid state there is first of all excess hydrogen, the oxygen formed at the same time combining in part with the water to form hydrogen peroxide. This in turn is decomposed, releasing excess oxygen (author's translation).

3 In the article, the formula is written as $2H^2O = H^2O^2 + H^2$.

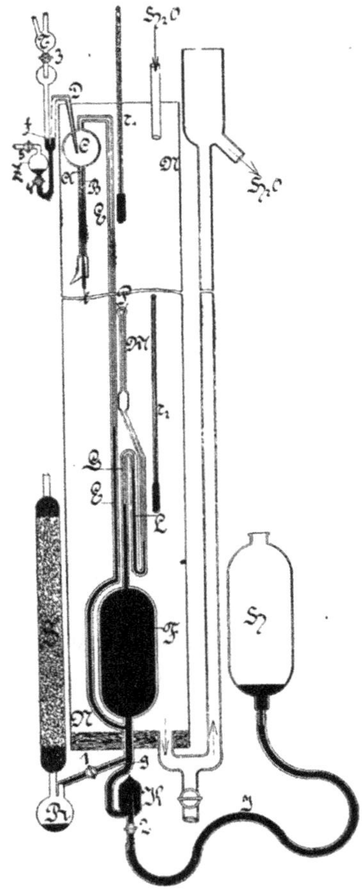

Figure 10.1. *Device developed by Duane and Scheuer (1913a) for precise measurement of the quantity of gas produced during radiolysis of water by emanium*

In the same article, they give a "yield" of gas formation: 0.825 cm^3 per Curie.h (Duane and Scheuer 1913a, 1913b). This yield, converted to SI units, corresponds to a yield of around 0.7×10^{-7} mole/Joule[4]. In 1914, Debierne stated in the journal *Annales de Physique* that the presence of helium gas in the gases released established "with certainty that the phenomena of radioactivity are those that accompany atomic transformations, which had only been considered

4 Assuming that the gas is at normal pressure and temperature and that only the first four short-lived alphas in the decay chain of ^{226}Ra are involved in radiolysis.

hypothetically". Concerning the radiolysis of water, he assumed that, under the action of α rays, water must contain four species of ions OH^-, H^+, H_2O^- and H_2O^+. Their recombinations give the following equations:

$$H^+ + OH^- \rightarrow H_2O$$
$$H_2O^+ + H_2O^- \rightarrow 2H_2O$$
$$H_2O^+ + OH^- \rightarrow H_2O + OH \text{ with } OH + OH \rightarrow H_2O_2$$
$$H_2O^- + H^+ \rightarrow H_2O + H \text{ with } H + H \rightarrow H_2$$

Water decomposition is therefore simply the result of charge neutralization (Debierne 1914). However, this hypothesis seemed to contradict the low radiolysis yield under X or β irradiation. The accumulation of radicals favored their interactions and the creation of radiolysis products. As a result, this explanation lost popularity in favor of the activated water model.

10.2. Activated water

Radiation creates an activated state of water that gives back a molecule of water: $H_2O \xrightarrow{\text{radiat.}} H_2O^{\text{activated}} \longrightarrow H_2O$ except in the presence of a solute on which the activated state can react to form a new product. This explains why deaerated water is not decomposed by X-rays (Risse 1929). This observation was confirmed by a more precise study by Fricke[5] and Brownscombe (1933). In the absence of the solutes on which H and OH radicals can react, the free radicals formed during the decomposition of water will disappear through an inverse reaction (back-reaction). In reality, the apparent absence of reaction in pure water is simply due to a steady-state concentration too low to be detected by conventional analytical methods. This was demonstrated much later by Allen (1948).

Fricke was one of the driving forces behind this model, demonstrating that over a very wide concentration range, the effect of radiation on the reagent depends only on the absorbed dose. The number of ferrous ions oxidized for a given dose is within a certain limit independent of the ferrous ion concentration (Fricke and Hart 1935). This remarkable property can be used in dosimetry: Fricke dosimetry. In this chemical dosimetry, the sensitive element is the ferrous ion (Fe^{2+}) which, by radiolysis, is oxidized to the ferric ion (Fe^{3+}). The ferric ion concentration is measured by spectrophotometry of the absorption band at 304 nm (Fricke and Morse 1929). In 1936, Nurnberger (1936) of Peiping Union Medical College in Peiping, China, set up one of the first facilities for studying the radiolysis of deaerated water, with which he unambiguously demonstrated that, unlike X-rays, alphas decompose

5 The Walter B. James Laboratory for Biophysics, Cold Springs Harbor.

deaerated water. Without naming it, he described a linear energy transfer (LET) effect.

10.3. Free radicals

This era of activated water, as Jonah calls it, led to the conclusion that irradiation acts first on the solvent (Jonah 1995). As water was likely to be used as a coolant in the nuclear reactors built during the Second World War, research into the radiolysis of water made significant progress, leading in particular to the introduction of free radicals and the solvated electron. It was at the University of Chicago, around Milton Burton, that the hypothesis of the creation of the two radicals H and OH and their interactions were reintroduced.

$$
\begin{aligned}
&H_2O \xrightarrow{\text{radiation}} H + OH \\
&H + H \rightarrow H_2 \\
&H + OH \rightarrow H_2O \\
&OH + OH \rightarrow H_2O_2 \\
&H + \text{reactant} \rightarrow \text{product} \\
&OH + \text{reactant} \rightarrow \text{product}
\end{aligned}
$$

But this model which remained classified for a long time, could not explain the lack of reactivity of pure water when irradiated with X-rays, or why water decomposition is inhibited in the presence of hydrogen, whereas it is amplified in the presence of oxygen or hydrogen peroxide. It was Augustine O. Allen, a specialist in free radical chemistry, who advanced the understanding of water's radiation chemistry. First, he proposed two reactions that restore the water molecule and lead to a very low steady-state concentration that can be interpreted as a lack of reactivity

$$
\begin{aligned}
&H_2 + OH \rightarrow H_2O + H \\
&H + H_2O_2 \rightarrow H_2O + OH
\end{aligned}
$$

by introducing ionic species.

$$
\begin{aligned}
&H_2O \xrightarrow{\text{radiation}} H_2O^+ + e^- \\
&e^- + H_2O \rightarrow H^\bullet + OH^- \\
&H^+ + OH^- \rightarrow H_2O \\
&H^\bullet + OH^\bullet \rightarrow H_2O \\
&H^\bullet + H^\bullet \rightarrow H_2 \\
&H_2 + OH^\bullet \rightarrow H_2O + H^\bullet \\
&OH^\bullet + OH^\bullet \rightarrow H_2O_2
\end{aligned}
$$

This series of chemical reactions was published on February 22, 1944 by Allen in Report No. 1321 from the Metallurgical Laboratory of the University of Chicago, based on the recollections of Gordon (1989). It should be noted that in 1944, the role of the hydrated electron as a reactive intermediate had not yet been imagined. In 1952, Hochanadel (1952) of the Allen group, in a paper on the equations of water radiolysis and the determination of their k constant, described a reaction between the electron and a water molecule in an aqueous medium. $e^- + H_2O + aq \rightarrow OH^-aq + H$. In the same year, Stein (1952) of Hebrew University, Jerusalem, more explicit about the role of electrons, included in his article a chapter entitled "Chemical actions due to the ejected electrons". These were the first studies of an entity, the hydrated or solvated electron, which was to become the focus of research into the radiolysis of water.

This research was supported by work on nuclear energy, but at the same time studies on the effects of irradiation on biological systems led to the same results (see, for example, Weiss 1944).

Improvements in measurements and concepts revealed that the yields of radical reactions depend on solute concentration, dose rate and the nature of the particle. Thus, as early as 1947, it was known that the yield of hydrogen molecules was low for γ rays and increased with the LET of other particles. Allen (1948) interpreted this in a presentation at the *Symposium on Radiation Chemistry and Photochemistry* at the University of Notre Dame, Indiana (June 24–27, 1947) on the basis of inhomogeneous kinetics. The yield depends on radical recombinations, which are favored in regions of greatest deposited energy density, such as the electron path ends (Allen 1948).

10.4. Solvated electrons

As the concentration of a solute increases, its disappearance rate rises, demonstrating that it is the fraction of energy absorbed by the water which, through an indirect effect, causes the modification of the solute. Specialists in the field agree that Dale (1943) of the Christie Hospital and Holt Radium Institute, Manchester, is credited with the discovery of this dilution effect, for his work on the deactivation of enzymes in aqueous solution. It was therefore necessary to involve reaction intermediates from the radiolysis of water (Dale 1943). The free radical model is particularly well-suited to designate the OH radical as the oxidizing species, in line with experimental results. On the other hand, the reducing character of the H$^\bullet$ radical is more complex to explain. To account for all experimental results, it is necessary to introduce two types of atom, H and H'. Readers interested in experimental research into the nature of these two entities will find all of the information they need in previous studies (Jonah 1995; Ferradini and Jay-Gerin 1999; Zimbrick 2002; Belloni 2021).

In a discussion[6] between Allen, Burton, Magee, Onsager and Pollard during the informal conference entitled *Basic mechanisms in radiobiology. Physical and chemical aspects*[7], Robert Leroy Platzman (1953) described the fate of electrons from the moment they reach thermal energy:

> For this reason the electron becomes hydrated. One must, of course, be very cautious about using the word "hydrated" because it means so many different things to different people. I mean here that the electron polarizes the dielectric and is bound in a stable quantum state to it. Then, finally, the chemical reaction can proceed as I have suggested, and the OH^- ion and H atom are formed. In between there is the time for the hydration to take place, which must, as Dr. Onsager said, be a minimum of the relaxation time 10^{-11} s. It may be greater; I have been trying to calculate this time by applying the theory suggested by Landau for the polarization of a crystal which turned out to be wrong for the crystal case but, I think, might apply here. I don't have any results to report on this, as yet. It does not concern the main picture because we are certain that the time cannot be shorter. The energy of the hydrated electrons, that is, the total energy of solution, by analogy to the case of ammonia, turns out to be about 2 electron volts, and the mobility of the "trapped" electron turns out to correspond to an effective electron mass of about 1/5 of an ionic mass (Platzman 1953).

Platzman joined the Manhattan Project in 1942, working at the Metallurgical Laboratory, then at Purdue University (Indiana).

10.4.1. *Solvated electrons, an old story*

As Belloni (2021) reminds us in *La fabuleuse histoire de l'électron solvaté* (Belloni 2021), the notion of the solvated electron was introduced long before it was used to explain the radiolysis of water. Between 1904 and 1908, Kraus (1908) of the Massachusetts Institute of Technology, measuring the electrical conductivity of various ions in liquid ammonia solutions, discovered that, for alkali metals, conductivity did not depend on the nature of the metal. He concluded that alkali metals are dissociated into ions in ammonia according to the formula $Me = Me^{+} + \theta$ (θbeing

6 Now, this discussion would be called a round table.

7 Highland Park, Illinois, May 7–9, 1953.

an electron), the cation and anion being more or less combined with solvent molecules. Moreover, he asserted that the anion responsible for most of the conductivity and blue color can only be an electron surrounded by solvent molecules: the solvated electron (Kraus 1908). Some eight years later, Gilson and Argo (1916) of the Chemical Laboratory of the University of California, studying the blue color of these metallic solutions, first attributed it to the optical absorption of the non-solvated electron. Two years later, more systematic measurements led them to attribute this absorption to the solvated electron (Gibson and Argo 1918).

Studies on the solvated electron and the understanding of radiolytic phenomena owe much to time-resolved experiments. The first experiments were carried out using continuous radiation sources, in this case sources of ^{60}Co, and interposing a rotating sector in front of the water cell. In fact, given the difficulty of rotating a 16.2-cm-thick lead disk, the Argonne group chose to rotate cells filled with the liquid under study in front of a slit in the source shielding (Hart and Matheson 1952). The group from the University of Saskatchewen in Canada built a rotating sector consisting of a steel cylinder 15.2 cm in diameter and 30.5 cm thick. Driven by a variable-speed motor, the 60° hollow sectors enable the time range to be covered from milliseconds to tens of seconds (Hummel et al. 1954). An important step was taken with linear accelerators and the temporal structure of their beams, which initially consisted of microsecond pulses. In 1960, three groups, including an English group from *Christie Hospital and Holt Radium Institute, Manchester, and two American groups from Du Pont Experimental Station, Wilmington and Argonne National Laboratory*, published results on pulsed radiolysis (Keene 1960; Matheson and Dorfman 1960; McCarthy and MacLachlan 1960). The advent of accelerators, with electrons injected by field emission and the high voltage provided by a Marx generator, brought a two- to three-order gain in temporal resolution, reaching tens of nanoseconds, or even a few nanoseconds (Lesigne and Sauneuf 1976). The most widely used of these generators were the Febetron from Field Emission Corporation (see section 11.1.4).

When pulsed radiolysis experiments became actually operational, the notion of a solvated electron was widely accepted. It was Keene (1960, 1962) of Christie Hospital and Holt Radium Institute, Manchester, who observed transient absorptions in water with a 4 MeV linac and 2 μs pulses from the Linear Accelerator Facility of Associated Electrical Industries in Manchester. The light source was continuous and, at a given wavelength, absorption was measured as a function of time using a photoelectric cell and oscilloscope. However, the presence of Čerenkov light meant that measurements could only be made after the beam had stopped, and the lack of sensitivity of the photoelectric cell beyond 6,000 Å meant that the position of the peak could not be accurately determined.

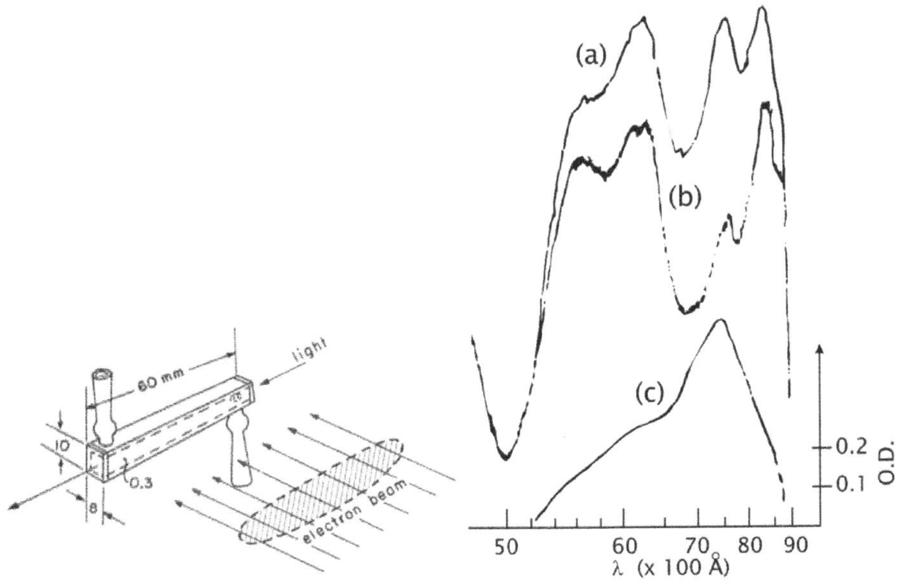

Figure 10.2. *Left: device used by Hart and Boag to measure the transient optical absorption of the solvated electron; right: transient absorption in pure deaerated water, (a) unirradiated, (b) simultaneously with a 2 µs (10 krad) electron pulse and (c) the spectrum of the solvated electron is obtained by difference between (a) and (b) (Boag and Hart 1963)*

It was also Hart of Argonne National Laboratory with Boag of the Research Unit in Radiobiology[8], Mount Vernon Hospital, UK, who confirmed the existence of the solvated electron by demonstrating long-lived optical absorption (Figure 10.2). This pulsed radiolysis experiment was carried out using a linear accelerator delivering pulses of 1.8 MeV electrons lasting around 2 µs, with a peak current of up to 0.5 A, giving a dose per pulse of between 4 and 10 krad (40–100 Gy). The optical absorption of water contained in a quartz cell was recorded on a photographic plate behind an Adam Hilger Ltd prism spectrograph. Illumination was provided by a spark light source with a duration of 4 µs. The transient absorption obtained by subtracting the spectra under and without irradiation showed an absorption peak at around 7,000 Å, which they attributed to the hydrated electron. These two experiments were presented at the *2nd International Congress of Radiation Research* in 1962 in Harrogate (UK).

8 A laboratory created by L.H. Gray.

Boag and Hart (1963) and Keene (1963) agreed to publish their results simultaneously in the same issue of *Nature*.

Solvated electrons were soon a great success. The *Weyl colloquia*[9] held in 1963 were largely devoted to this subject. The real breakthrough then came experimentally with the development of pulses with a duration of around one hundred femtoseconds. This temporal resolution, obtained using lasers that generate biphotonic excitations in the ultraviolet, makes it possible to follow the dynamics of electron solvation in water.

It should be noted that Boag (1989) tried unsuccessfully in the 1960s to use Čerenkov light as a light source synchronized with the electron pulse. This idea, and above all the development of accelerators delivering picosecond pulses, were at the origin of the development in Toronto (Bronskill et al. 1970) and Argonne (Jonah 1975) of picosecond radiolysis. In Toronto, the 40 MeV electron linear accelerator delivered pulses of less than 10 ps duration with a recurrence of 350 ps. Flashes of Čerenkov light were produced by passing the electron beam through 10 cm of air. The length of the optical path made it possible to adjust their time offset and thus describe the signal decay point by point: this being a stroboscopic method. The Argonne device is identical in principle, but differs in several respects: electron energy is 20 MeV, Čerenkov light is generated in xenon and acquisition is different. These accelerators enabled time-resolved observation of initial radiolysis processes a few picoseconds after energy deposition. The disadvantage of this stroboscopic method was that the accessible time range was reduced to around a hundred picoseconds. The advent of photo-triggered linear accelerators removed this constraint.

The discovery of the hydrated electron made it possible to describe the mechanisms of radiolysis of water in terms of radical chemistry e^-_{aq}, $H^\bullet$ and $^\bullet OH$. Radical and molecular yields can be calculated and measured. These yields are linked by the following equations:

$$G_{e^-_{aq}} = G_{H^\bullet}$$

$$G_{e^-_{aq}} + G_{H^\bullet} + 2G_{H_2} = G_{^\bullet OH} + 2G_{H_2O_2} = 2G_{H_2O}$$

The notation G_x represents the number of creation or disappearance of species X per 100 eV absorbed in the solution.

9 The Weyl colloquia were created to commemorate the centenary of Theodor Weyl's discovery of metal solutions in ammonia.

10.5. Effects of the spatial structure of energy deposits

But this radical chemistry associated with the solvated electron does not explain the differences in behavior between the different particle types and energies. Indeed, it was shown that the yield of hydrogen molecule production was low for gammas and increased with the LET of the particle. In June 1947, Allen presented an explanation at a meeting that was published in the *Journal of Physical and Colloid Chemistry* (Allen 1948). In a paragraph entitled "The proportion of radicals available for reaction with solutes", he explained that, unlike fast electrons, for which the initial distribution of radicals can be considered homogeneous, the more strongly ionizing particles create, with high probability, high concentrations of radicals in small areas along the particle's path. A significant fraction of these radicals reacts with each other, increasing the yield of hydrogen and hydrogen peroxide. Allen (1989) himself declared: "it certainly constitutes my major contribution to science" and indeed, this idea had a major impact on the radiation chemistry community.

In 1953, Magee and Samuel (1953) of Notre Dame University, Illinois, published a paper in which they presented a model describing the effects of the geometric distribution of energy deposition on the radiolysis of water. They did so as a consequence of the diffusion of H and OH free radicals alone. They considered that primary ionizations are isolated by an average distance that depends on the particle's velocity, and that secondary ionizations a few angstroms away from the primary events create small clusters called spurs. At 10^{-13} s after the particle's passage, the spurs are made up of an equal number of H and OH radicals. They assumed that these radicals have a finite probability of interacting with each other to give the molecules H_2, H_2O_2 and H_2O. The rest of the radicals diffuse out of the spurs. This description helps explain the results observed for electron irradiation. For alphas, Magee and Samuel (1953) considered a diffusion model in cylindrical geometry. This theoretical approach formed the basis of subsequent developments.

In the same year, in a long article, Magee (1953) reviewed the state of knowledge in chemistry under radiation from physics to chemistry. With an approach closer to physics than to chemistry, he described 27 elementary processes grouped under six items: impact processes, reactions of vibrationally excited ions, reactions producing additional electrons, reactions exchanging charge and excitation, neutralization reactions and radical reactions. The article also gives a table of time scales for the various phenomena and is a source of references, with 161 references cited.

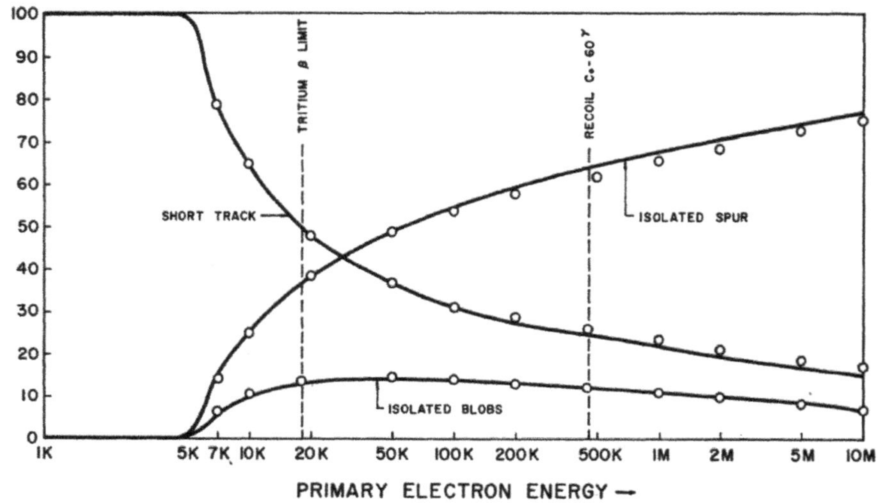

Figure 10.3. *Energy distribution between different ionization configurations: short tracks, spurs and blobs (Mozumder and Magee 1966a)*

In the years that followed, the Notre Dame group published increasingly sophisticated theories to describe radical diffusion (Monchick et al. 1957, Mozumder and Magee 1966c, 1966a, 1966b).

Figure 10.3 shows the distribution of ionizations between the three cluster geometries as a function of initial electron energy.

The codes incorporated more and more concepts for the physical and chemical phases. For example, there is an article in which physicists and chemists, specialists in irradiation (Turner, Magee, Wright, Chatterjee, Hamm and Ritchie), work together to calculate the evolution of chemical species around the trajectory of an electron (Turner et al. 1983).

This Monte Carlo simulation code is structured in three phases: (1) electron and secondary electron transport giving the initial distribution of H_2O^+ and H_2O^*, (2) a first chemical phase for $t \sim 10^{-11}$ s, which leads to the formation of OH, H_3O^+, H, H_2, O and the hydrated electron and (3) then comes a phase where the reactions are controlled by diffusion. This phase is followed until $t = 10^{-6}$ s. From 10^{-11} to 10^{-6} s, the distribution of chemical elements, initially heterogeneous, becomes virtually homogeneous (Figure 10.4).

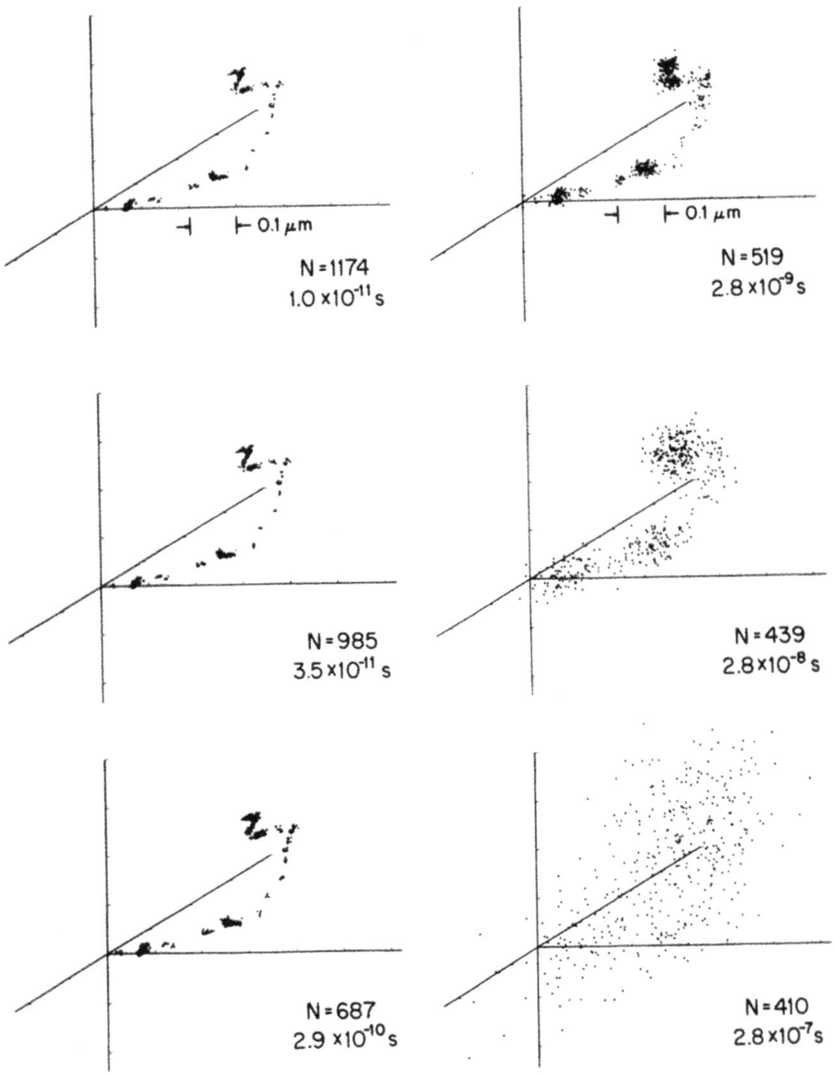

Figure 10.4. *Spatial evolution of chemical elements created by a 5 keV electron in water from 10^{-11} to 2.8×10^{-7} s (Turner et al. 1983)*

Subsequently, Monte Carlo codes increasingly incorporated variables such as temperature (Sanguanmith et al. 2012) or multi-ionizations (Gervais et al. 2005).

10.6. Radiolysis yields

Influenced by results obtained in gases, the most commonly used way of quantifying the effects of radiolysis in liquids was the ratio of the number of species created (M) to the number of ionizations (N). Faced with the difficulty of determining N in liquids, Milton Burton proposed using the number of molecules created or destroyed per 100 eV of energy deposited, during a debate at the meeting "A general discussion on radiation chemistry" held at Leeds University in April 1952 (Dewhurst et al. 1952):

> [...] I have discussed this matter with several participants in the Discussion and we have jointly arrived at the conclusion which I have summarized as follows: G: this notation (i.e. 100 eV yield) should be used whenever molecules produced or converted and actual energy input are actually measured. Of course, errors may occur in both determinations but such values are subject to check and the methods (if adequately described) are subject to criticism.

The origin of symbol G is reported in a note on page 241 in Kimura et al. (1993)[10] by Mr. Inokuti, who worked with Mr. Burton:

> The origin of symbol G does not seem to have been documented. One of us (MI) was fortunate to learn it from the inventor, the late Milton Burton. During the work on the Manhattan Project (1942–1945), Burton felt the need for a symbol for the radiation-chemical yield and used γ in association with γ rays, which were the radiations then employed in experiments. However, a security officer objected to the use of γ on the ground that it might indicate the connection of the Manhattan Project work with radiation. Thereafter, Burton used G instead of γ.

10.7. Chapter 10 references

10.7.1. *Biographies of some of the chapter's personalities*

Many people have contributed to the advancement of knowledge in the field of liquid radiolysis. Below are brief biographies of four of them, arbitrarily chosen.

---

10 This reference was given to me by Jean-Paul Jay-Gerin of the Université de Sherbrooke, Canada, for which I thank him.

Augustine Oliver Allen (1910–1988) was one of the founders of modern radiation chemistry. In particular, he developed a theory of radical recombination in dense ionization traces. He joined Brookhaven National Laboratory in 1949, where he spent his entire career. He worked on the radiolysis of hydrocarbons and the radiolysis of liquids absorbed on inert solids.

Milton Burton (1902, Stapleton–1985, Mishawaka) played a key role in radiolysis research. After working in industry for 10 years, he joined New York University in 1935. In 1942, he joined the Manhattan Project. He became head of radiation chemistry at the Metallurgical Laboratory. At the end of the war, he moved to the Department of Chemistry at Notre Dame University where, with the support of the U.S. Atomic Energy Commission, he set up the Radiation Laboratory in 1948. He was its director until 1971, when he retired.

Figure 10.5. *Hugo Fricke (1923). For a color version of this figure, see www.iste.co.uk/bouffard/irradiation.zip*

Hugo Fricke (1892, Aarhus–1972, New York) invented chemical dosimetry based on the oxidation of Fe^{2+} to Fe^{3+} by radiolysis products. The production of Fe^{3+} can be measured by its absorption at 304 nm. He studied at the University of Copenhagen, where he became Niels Bohr's assistant. After a short spell at Lund University, he emigrated to the United States to work at Colombia University and then Harvard University with Theodore Lyman. It was during his stay (1921–1928) at the Cleveland Clinic, where he ran a biophysics laboratory, that he discovered the effect of radiation on iron sulphate. From 1928 to 1955, he worked at Cold Spring Harbor Laboratory on the north shore of Long Island, New York. From 1955, he worked at Argonne National Laboratory and, from 1966, in parallel, at the Danish Atomic Energy Laboratory.

John Lafayette Magee (1914, Franklinton–2005, Moraga), began his career in 1943, after a post doc at Princeton University, at Los Alamos National Laboratory on the Manhattan Project. From 1946 to 1948, he worked at Argonne National Laboratory, then joined the Chemistry Department at Notre Dame University. After directing the Radiation Laboratory, he moved to Lawrence Berkeley National Laboratory, where he studied the biological effects of ionizing radiation. He was president of the Radiation Research Society.

10.8. References

Allen, A.O. (1948). Radiation chemistry of aqueous solutions. *Journal of Physical and Colloid Chemistry*, 52(3), 479–490.

Allen, A.O. (1989). The story of the radiation chemistry of water. In *Early Developments in Radiation Chemistry*, Kroh, J. (ed.). Royal Society of Chemistry, Cambridge.

Becquerel, H. and Curie, P. (1901). *Action physiologique des rayons de radium*. Institut de France, Académie des Sciences, Paris.

Belloni, J. (2011). Historic landmarks in radiation chemistry since early observations by Marie Skłodowska-Curie and Pierre Curie. *Nukleonika*, 56(3), 203–211.

Belloni, J. (2021). La fabuleuse histoire de l'électron solvaté. *L'Actualité Chimique*, 460–461, 17–22.

Boag, J.W. (1989). A historical account of the discovery of the optical absorption spectrum of the hydrated electron. In *Early Developments in Radiation Chemistry*, Kroh, J. (ed.). Royal Society of Chemistry, Cambridge.

Boag, J.W. and Hart, E.J. (1963). Absorption spectra in irradiated water and some solutions: Absorption spectra of 'hydrated' electron. *Nature*, 197(4862), 45–47.

Bronskill, M.J., Taylor, W.B., Wolff, R.K., Hunt, J.W. (1970). Design and performance of a pulse radiolysis system capable of picosecond time resolution. *Review of Scientific Instruments*, 41(3), 333–340.

Curie, P. and Curie, M. (1899). Effets chimiques produits par les rayons de Becquerel. *Comptes rendus hebdomadaires des séances de l'Academie des sciences*, 129, 823–825.

Dale, W.M. (1943). Effects of X-rays on acetylcholine solutions showing the dilution and protection phenomena, found for enzymes. *The Journal of Physiology*, 102(1), 50–54.

Debierne, A. (1909). Sur la décomposition de l'eau par les sels de radium. *Comptes rendus hebdomadaires des séances de l'Academie des sciences*, 148, 703–705.

Debierne, A. (1914). Recherches sur les gaz produits par les substances radioactives. Décomposition de l'eau. *Annales de Physique*, 9(2), 97–127.

Dewhurst, H.A., Dale, W.M., Bacq, Z.M., Swallow, A.J., Weiss, J., Magat, M., Minder, W., Suttle, J.F., Schulte, J.W., Burton, M. et al. (1952). General discussion. *Discussions of the Faraday Society*, 7, 312–318.

Duane, W. and Scheuer, O. (1913a). Recherches sur la décomposition de l'eau par les rayons α. *Le Radium*, 10(2), 33–46.

Duane, W. and Scheuer, O. (1913b). Décomposition de l'eau par les rayons α. *Comptes rendus hebdomadaires des séances de l'Académie des sciences*, 156, 466–467.

Ferradini, C. and Jay-Gerin, J.-P. (1996). La radiolyse de l'eau : historique de la découverte de l'électron hydraté jusqu'à la fin du siècle. *Cahiers de radiobiologie*, 4, 23.

Ferradini, C. and Jay-Gerin, J.-P. (1997). La radiolyse de l'eau : historique de la découverte de l'électron hydraté jusqu'à la fin du siècle (suite). *Cahiers de radiobiologie*, 6, 29.

Ferradini, C. and Jay-Gerin, J.-P. (1999). La radiolyse de l'eau et des solutions aqueuses : historique et actualité. *Canadian Journal of Chemistry*, 77, 1542–1575.

Fricke, H. and Brownscombe, E.R. (1933). Inability of X-rays to decompose water. *Physical Review*, 44(3), 240.

Fricke, H. and Hart, E.J. (1935). The oxidation of Fe^{++} to Fe^{+++} by the irradiation with X-rays of solutions of ferrous sulfate in sulfuric acid. *The Journal of Chemical Physics*, 3(1), 60–61.

Fricke, H. and Morse, S. (1929). The action of X-rays on ferrous sulphate solutions. *The London, Edinburgh and Dublin Philosophical Magazine and Journal of Science*, 7(41), 129–141.

Gervais, B., Beuve, M., Olivera, G.H., Galassi, M.E., Rivarola, R.D. (2005). Production of HO_2 and O_2 by multiple ionization in water radiolysis by swift carbon ions. *Chemical Physics Letters*, 410(4), 330–334.

Gibson, G.E. and Argo, W.L. (1916). The absorption spectra of the blue solutions of sodium and magnesium in liquid ammonia. *Physical Review*, 7(1), 33–48.

Gibson, G.E. and Argo, W.L. (1918). The absoprtion spectra of the blue solutions of certain alkali and alkali earth metals in liquid ammonia and in methylamine. *Journal of the American Chemical Society*, 40(9), 1327–1361.

Giesel, F. (1900). Ueber radioactive Stoffe. *Berichte der deutschen chemischen Gesellschaft*, 33(3), 3569–3571.

Giesel, F. (1902a). Ueber Radium und radioactive Stoffe. *Berichte der deutschen chemischen Gesellschaft*, 35(3), 3608–36011.

Giesel, F. (1902b). Über Bequerelstrahlen und die radioaktiven Substanzen. *Zeitschrift für Elektrochemie*, 8, 579–585.

Gordon, S. (1989). Radiation chemistry at the Metallurgical Laboratory, Manhattan Project, University of Chicago (1942–1947) and the Argonne National Laboratory, Argonne, IL (1947–1984). In *Early Developments in Radiation Chemistry*, Kroh, J. (ed.). Royal Society of Chemistry, Cambridge.

Hart, E.J. and Matheson, M.S. (1952). Mechanism and rate constants of the γ-ray induced decomposition of hydrogen peroxide in aqueous solutions. *Discussions of the Faraday Society*, 12, 169–188.

Hochanadel, C.J. (1952). Effects of cobalt γ-radiation on water and aqueous solutions. *Journal of Physical Chemistry*, 56(5), 587–594.

Hummel, R.W., Freeman, G.R., Van Cleave, A.B., Spinks, J.T.W. (1954). Rotating sector method applied to reactions induced by Co^{60} gamma rays. *Science*, 119(3083), 159–160.

Jonah, C.D. (1975). A wide–time range pulse radiolysis system of picosecond time resolution. *Review of Scientific Instruments*, 46(1), 62–66.

Jonah, C.D. (1995). A short history of the radiation chemistry of water. *Radiation Research*, 144(2), 141–147.

Keene, J.P. (1960). Kinetics of radiation-induced chemical reactions. *Nature*, 188(4753), 843–844.

Keene, J.P. (1962). Pulse radiolysis with 4 MeV electrons. *The British Journal of Radiology*, 35, 650.

Keene, J.P. (1963). Absorption spectra in irradiated water and some solutions: Optical absorptions in irradiated water. *Nature*, 197(4862), 47–48.

Kernbaum, M. (1909). Action chimique sur l'eau des rayons pénétrants de radium. *Comptes rendus hebdomadaires des séances de l'Academie des sciences*, 149, 126–127.

Kimura, M., Inokuti, M., Dillon, M.A. (1993). Electron degradation in molecular substances. *Advances in Chemical Physics*, 84, 193–291.

Kraus, C.A. (1908). Solution of metals in non-metallic solvents. II. On the formation of compounds between metals and ammonia. *Journal of the American Chemical Society*, 30(5), 653–668.

Kroh, J. (1989). *Early Developments in Radiation Chemistry*. Royal Society of Chemistry, Cambridge.

Lesigne, B. and Sauneuf, R. (1976). Reduction of the pulse duration from a Febetron 707. *Review of Scientific Instruments*, 47(9), 1063–1064.

Lind, S.C. (1912). On the nature of the chemical action produced by α particles and the probable role played by ions. *Journal of Physical Chemistry*, 16(7), 564–613.

Magee, J.L. (1953). Radiation chemistry. *Annual Review of Nuclear Science*, 3, 171–192.

Matheson, M.S. and Dorfman, L.M. (1960). Detection of short-lived transients in radiation chemistry. *The Journal of Chemical Physics*, 32(6), 1870–1871.

McCarthy, R.L. and MacLachlan, A. (1960). Transient benzyl radical reactions produced by high-energy radiation. *Transactions of the Faraday Society*, 56(0), 1187–1200.

Monchick, L., Magee, J.L., Samuel, A.H. (1957). Theory of radiation chemistry. 4. Chemical reactions in the general track composed of n-particles. *Journal of Chemical Physics*, 26(4), 935–941.

Mozumder, A. and Magee, J.L. (1966a). Model of tracks of ionizing radiations for radical reaction mechanisms. *Radiation Research*, 28(2), 203–214.

Mozumder, A. and Magee, J.L. (1966b). Theory of radiation chemistry. VII. Structure and reactions in low LET tracks. *The Journal of Chemical Physics*, 45(9), 3332–3341.

Mozumder, A. and Magee, J.L. (1966c). A simplified approach to diffusion-controlled radical reactions in tracks of ionizing radiations. *Radiation Research*, 28(2), 215.

Nurnberger, C.E. (1936). The decomposition of air-free water by alpha rays. *Journal of Physical Chemistry*, 41, 431–435.

Platzman, R.L. (1953). Energy transfer from secondary electrons to matter. In *Proceedings of the Informal Conference "Basic Mechanisms in Radiobiology. II. Physical and Chemical Aspects"*, 7–9 May. Committee on Nuclear Science National Research Council, Washington, D.C.

Ramsay, W. (1907). LXXXVI.–The chemical action of the radium emanation. Part I. Action on distilled water. *Journal of the Chemical Society, Faraday Transactions*, 91, 931–942.

Ramsay, W. and Soddy, F. (1904). Experiments in radioactivity, and the production of helium from radium. *Proceedings of the Royal Society of London*, 72(477–486), 204–207.

Risse, O. (1929). Über die Röntgenphotolyse des Hydroperoxyds. *Zeitschrift für Physikalische Chemie*, 140A(1), 133–157.

Samuel, A.H. and Magee, J.L. (1953). Theory of radiation chemistry. 2. Track effects in radiolysis of water. *Journal of Chemical Physics*, 21(6), 1080–1087.

Sanguanmith, S., Meesungnoen, J., Muroya, Y., Lin, M., Katsumura, Y., Jay-Gerin, J.-P. (2012). On the spur lifetime and its temperature dependence in the low linear energy transfer radiolysis of water. *Physical Chemistry Chemical Physics*, 14(48), 16731–16736.

Stein, G. (1952). Some aspects of the radiation chemistry of organic solutes. *Discussions of the Faraday Society*, 12, 227–234.

Turner, J.E., Magee, J.L., Wright, H.A., Chatterjee, A., Hamm, R.N., Ritchie, R.H. (1983). Physical and chemical development of electron tracks in liquid water. *Radiation Research*, 96(3), 437–449.

Weiss, J. (1944). Radiochemistry of aqueous solutions. *Nature*, 153(3894), 748–750.

Zimbrick, J.D. (2002). Radiation chemistry and the radiation research society: A history from the beginning. *Radiation Research*, 158(2), 127–140.

11
Irradiation Tools

The study of materials under irradiation uses different irradiation sources, the oldest of which is of course natural irradiation. These are mainly emitters α and β for terrestrial sources, and all kinds of particles for cosmic radiation. It should be remembered that we are exposed to radiation from our own bodies (~120 Bq/kg, including ~64 and ~53 Bq/kg, respectively, for ^{40}K and ^{14}C) and from radon in the air (~100 Bq/m^3), to which we must add the contribution of cosmic radiation (0.08 µSv/h at sea level)[1].

Three irradiation sources were used for the first studies of defects. Two of these sources, cathode ray (electron) and X-ray generators, originated in the Crookes tube and can be likened to accelerators. The third source is the natural radioactivity of radium, polonium, etc.

We saw in section 2.2.1 that the foundations of accelerators date back to the 17th century, with the invention in 1663 of the first electrostatic generator by De Guericke (1672). A few arbitrarily chosen milestones include the endless-belt electrostatic machine by Charles Louis Walckiers de St Amand, forerunner of the Van de Graaff accelerators (Walckiers de St Amand 1784), the Crookes tube (Crookes 1879) and the discoveries of X-rays, radioactivity and electrons between 1895 and 1897. The first real accelerator was not designed until 1925. During these

1 The Sievert (Sv) is the unit for measuring the impact of radiation on living matter. It is the quantity of energy absorbed per unit of mass (Gray J/kg) multiplied by two factors, one depending on the effectiveness of the radiation and the other on the sensitivity of the organ concerned. The Sievert is therefore expressed like the Gray in J/kg. The name of this unit is a tribute to the Swedish physicist Rolf Sievert (1896–1966), who worked on dose measurement and the biological effects of radiation.

25 years, the use of sources having become widespread, numerous results were obtained on the effects of α-matter interaction.

For neutron irradiation, we had to wait for the discovery of the neutron by James Chadwick in 1932, followed six years later by the discovery of fission by Otto Hahn and Fritz Strassman, and its interpretation by Lise Meitner, before we could consider building a nuclear reactor. This was achieved in fall 1942 with the divergence of the Chicago pile.

11.1. Accelerators

Researchers working to understand the internal structure of the atom in the 1920s–1930s reached the limit of what was feasible with radioactive matter alone. They intuited the need for higher energy ions. In 1928, Rutherford told the Royal Society of the need for a source of positive particles with higher energy than the alphas of radioactive substances.

A major scientific and technological effort was therefore undertaken in the 1930s to define an accelerator structure. Two approaches were developed in parallel: acceleration by a radiofrequency electric field and acceleration by high voltage.

11.1.1. *Radio-frequency cavity accelerators*

Norwegian Rolf Widerøe can be considered the first designer of an accelerator. In 1925, after being turned down in Karlsruhe, he obtained permission from the Technical University of Aachen to build a new type of accelerator, an induction accelerator called betatron, during his doctoral thesis. Early on, he envisaged using a transformer whose secondary coil would be replaced by electrons circulating in a glass tube in the shape of a slightly flattened torus. In this vacuum tube, the electrons would very quickly reach speeds close to that of light, as a magnetic field would hold them in the torus. However, he could only stabilize the electron beam in the plane of the orbit, and thus was unable to put his idea into practice. Faced with this failure, he set about implementing the method suggested two years earlier by Gustav Ising. This involved applying repeated electrical impulses to the particles, their speed increasing with each stimulus. He successfully built the first linear accelerator (linac) using a radiofrequency electric field. He published this work in *Archiv für Electrotechnik* (Wideröe 1928). This article paved the way for Rutherford to apply the principle of resonance to the cyclotron. Norwegian by birth, Widerøe continued to work in Germany during the Second World War. After being fined by

Norway for this collaboration, he emigrated to Switzerland in 1946. He published over 180 papers and submitted over 200 application patents.

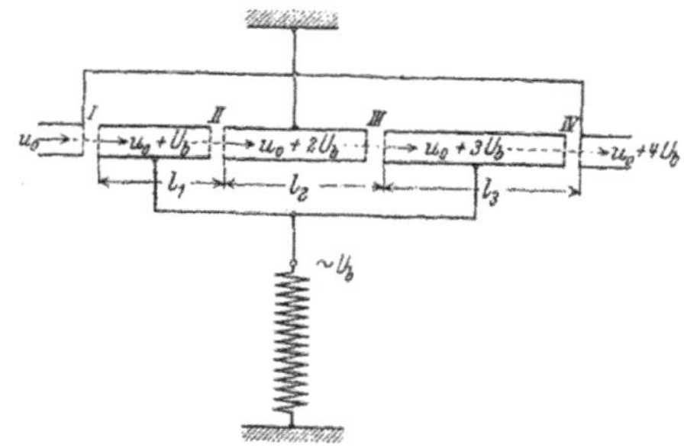

Image 1. Principle of voltage transformation with potential transmitters.

Figure 11.1. *Principle of voltage transformation with potential fields. Figure taken from Wideröe (1928)*

In 1932, building on Widerøe's work on the circular accelerator, Siemens–Schuckertwerke employee Max Steenbeck developed and patented a new magnetic field configuration that stabilized electrons on their circular trajectory. Electrons were accelerated up to 1.8 MeV. Over a period of 30 years, Siemens built 137 betatrons and sold them to research centers and radiotherapy departments around the world (Usch and Bautz 2005).

M. Steenbeck also worked on the concept of a cyclotron, but was unable to publish his discovery, let alone build the device. At the end of the war, Steenberk was arrested by the Soviet military, for whom he worked on uranium enrichment by ultracentrifugation. He returned to East Germany in 1956 as Professor of Plasma Physics at the University of Jena.

In France, Jean Thibaud, having learned of Widerøe's thesis, built a linear accelerator that functioned perfectly (Thibaud 1933). However, to achieve significant accelerations, he would have to build a device that went beyond the limits of his laboratory, in this case, Maurice de Broglie's laboratory.

At the end of 1930, he designed and built a circular accelerator. He presented a paper on the subject at the International Electrical Congress in Paris in 1932, but his opportunism and compromise with the Vichy regime and the Occupation Forces during the Second World War, followed by a plagiarism case in the 1950s, led to his work being forgotten.

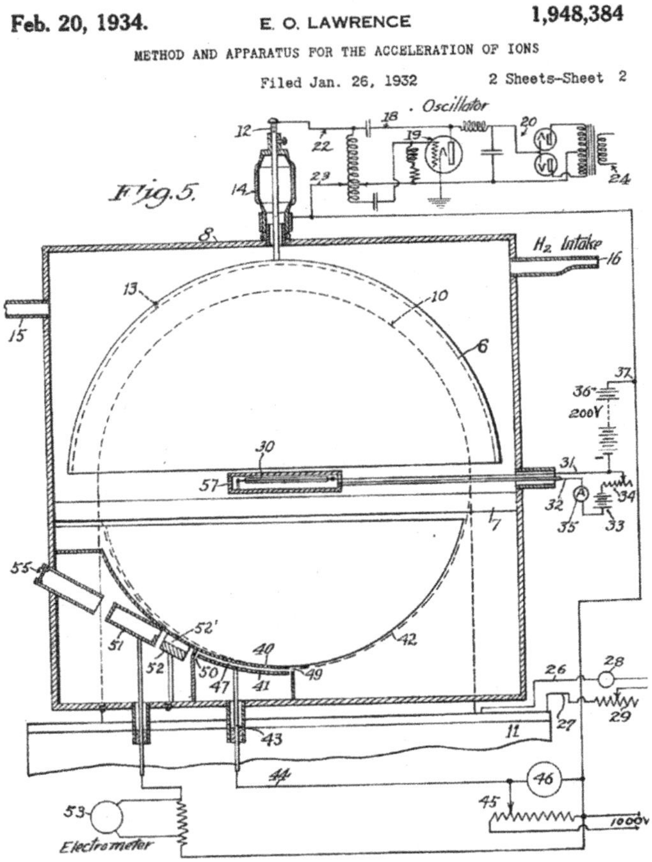

Figure 11.2. *Diagram of a cyclotron from Lawrence's patent on the invention of the cyclotron (Lawrence 1934)*

Professor at the University of California at Berkeley, Ernest O. Lawrence, who had also read Widerøe's article on the linear accelerator, imagined applying the same principle to a circular accelerator. He entrusted the project to two of his students: Nels Edlefsen, who built two rudimentary models, and Stanley Livingston, who continued the development, culminating in the first working cyclotron in the fall of 1930. This small 5-inch (127 mm) cyclotron accelerated protons up to 80,000 V with 1000 V applied to the accelerating electrodes (Lawrence and Livingston 1932). This first cyclotron is said to have cost just US$25. Lawrence then built a series of increasingly powerful cyclotrons: 1.25 MeV H with the 11-inch cyclotron, 3 MeV H and 5 MeV D with the 27-inch cyclotron (1932), 8 MeV D and 16 MeV α with the 37-inch cyclotron (1937), 25 MeV H and 50 MeV α with the

60-inch cyclotron (1939) and 200 MeV H and 400 MeV α with the 184-inch synchrocyclotron (1946) (Lawrence et al. 1939; Brobeck et al. 1947; Lawrence 1951). The magnets of the latter two weighed 220 tons and 4,000 tons, respectively. In 1934, Lawrence patented the method and devices for accelerating ions (Figure 11.2) (Lawrence 1934).

Hungarian physicist Leó Szilárd filed two patents, one on a linear accelerator in 1928 and the second on cyclotrons in 1929, while working at Berlin's Humboldt University. During his time in Berlin (1920–1933), he registered 31 patents (Telegdi 2000).

In Japan, groups led by Yoisho Nishina and Shoji Nishikawa joined forces in 1935 to build the first Japanese cyclotron. With a diameter of 26 inches (66 cm) and a weight of 23 tons, this cyclotron accelerated 3 MeV deuterons in April 1937. The cyclotron was used for nuclear physics experiments and biomedical studies. Just after completion of this small cyclotron.

Nishina began construction of a larger cyclotron at the same time as Lawrence built his 60-inch cyclotron. As the magnets for the two projects were identical, Lawrence commissioned identical magnets for the Japanese group. The poles of this 200-ton magnet were machined in Japan. It initially accelerated protons to 9 MeV, but with very low intensity. Having discovered that their rf resonator and pumping system were completely different from those of the Americans, the Japanese obtained all the plans courtesy of Lawrence. The 9 MeV proton beam reached 4 μA in February 1944.

In 1938, a 24-inch cyclotron was commissioned at Osaka University, delivering a 20 μA beam of 5 MeV protons. Kyoto University began construction of a cyclotron in 1940, but was unable to complete it.

Much of this was achieved during wartime (1937: start of the Sino-Japanese war; 1941: attack on Pearl Harbor; August 6 and 9, 1945: bombing of Hiroshima and Nagasaki). In November 1945, the occupying forces destroyed the two RIKEN cyclotrons by dumping them in Tokyo Bay, followed by those in Osaka and Kyoto.

Subsequently, in the 1950s, Japan once again acquired cyclotrons: a 26-inch at RIKEN, a 40-inch at Kyoto, a 44-inch at Osaka, a 60-inch at INS Tokyo, etc. The first was used to study the effects of irradiation on highly polymerized compounds.

Most of the information on the history of Japanese cyclotrons comes from the transcript of a speech by Hirao (Institute for Nuclear Studies, Tokyo) during the banquet at the *"Cyclotrons and their Applications"* conference in Tokyo (Hirao 1986).

Europe also built cyclotrons. The first European cyclotron was built by Igor Kourtchatov for the Radium Institute in Leningrad. It went into operation in 1937. In Germany, a cyclotron was built in Heidelberg under the direction of Walther Bothe and Wolfgang Gentner, and became operational in 1943.

In France, Frédéric Joliot called on Lawrence in 1937, who sent him the plans. After being built in Zurich, the cycloron was installed in the basement of a building at the Collège de France in 1939.

The number of accelerators with radio-frequency cavities that have been (or are still being) used for materials irradiation is limited. At the risk of overlooking a major machine, we can mention the following:

– In Germany, *GSI Helmholtzzentrum für Schwerionenfirschung GmbH–Materialforschung* has a long tradition in materials research, first with the linac (unilac 3.6–11 MeV/u) and more recently at very high energy with the SIS18 synchro-synchrotron.

– In France, *GANIL-SPIRAL2 Grand Accélérateur National d'Ions Lourds*, whose materials activity began in 1982, is managed by the CIRIL platform, Centre Interdisciplinaire de Recherche avec les Ions Lourds. Three energy ranges are available for irradiation: IRRSUD 0.1–1 MeV/u, SME 4–13 MeV/u and high energy 25–95 MeV/u.

– In Russia, the *Joint Institute for Nuclear Research* in Dubna has five cyclotrons, one of which, the DC 130, is dedicated to irradiating materials with 4.5 MeV/nucleon ions.

– In Japan, cyclotrons are dedicated to the study of materials, such as the K110 AVF cyclotron at *Takasaki Advanced Radiation Research Institute* or the 50 MV cyclotron at *Tohoku University*.

– In China, the *Institute of Modern Physics* has a set of accelerators with which it is possible to carry out irradiation experiments at energies as high as 4 GeV/u.

– In Belgium, *Cyclotron Resource Center* (CRC), whose main activities focus on the testing of electronic components and the production of nanostructured membranes. *Ion beam application* (IBA), a spin-off from this Resource Center, is the world leader in cyclotrons for proton therapy treatment of cancer.

11.1.2. *Electrostatic accelerators*

In 1930, probably under the influence of Rutherford, John Douglas Cockroft, a fellow at St John College, Cambridge and Ernest Thomas Sinton Walton, an Irish student, described in their paper to the Royal Society of London (Cockcroft and Walton 1930) the advantages of an accelerator for studying the structure of nuclei: higher intensity than α sources (1 μA is equivalent to 180 g of radium) and a beam uncontaminated by β and γ. The energy of helium nuclei must be higher than that of polonium alphas (5.2 MeV), that is, a high voltage of over 2.6 MV.

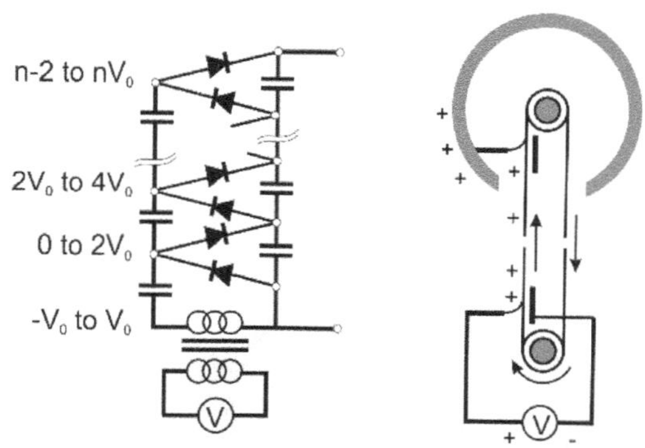

Figure 11.3. *Schematic diagram of Cockcroft–Walton and Van de Graff accelerators*

Cockroft and Walton, both thesis students at the Cavendish Laboratory, developed a generator which, supplied by a low alternating voltage, delivered a very high quasi-continuous voltage. Their creators apparently built this high-voltage generator independently of Greinacher, whom they did not mention in their 1930 article (Cockcroft and Walton 1930). Indeed, as early as 1913, Heinrich Greinacher of the Physikalisches Institut der Universität de Zürich had proposed a voltage-multiplying device: the Greinacher cascade (Greinacher 1921). The first electrostatic accelerator to be built around this technique was a 600 kV proton accelerator (Cockcroft and Walton 1932). This was also the opportunity to carry out the first nuclear reaction ($^{7}Li(p,\alpha)^{4}He$) using an accelerator.

Meanwhile, at Princeton University, Robert Jamison Van de Graaff, Karl Taylor Compton and Lester Clare Van Atta built an electrostatic generator based on the charging of a sphere by a belt carrying charges. It reached a voltage of 80 kV (Van de Graaff et al. 1933). Posterity will only remember Van de Graaff's name, but his two co-signers had successful careers. Compton[2] became president of MIT, and Van Atta became head of the Antenna Group at MIT and then at the Naval Research Lab (Washington, D.C.), ending up as Deputy Director of NASA's Electronics Research Center. Initially at the Carnegie Institute in Washington, Van de Graaff built several accelerators, including the first to be used for nuclear physics research. These machines accelerated protons to 600 keV, then 1.2 MeV. After joining the Massachussetts Institute of Technology, he set about building a much more ambitious 7 MV machine, but as this high voltage is produced in air, the distances required for electrical isolation must be very great. So, this generator, featuring two 15-foot-diameter (~4.6-m) aluminum spheres mounted on 22-foot-high (~6.7-m) insulating columns, was built in an old airship hangar (Figure 11.4). The 1.2-m wide belts were made of 0.43-mm-thick paper. In reality, this machine reached only 5.1 MV (Van Atta et al. 1936).

Figure 11.4. *The 5.1 MV electrostatic generator in the former airship hangar (Van Atta et al. 1936)*

At virtually the same time, in 1932 at RIKEN[3] in Japan, Yoisho Nishina was building a Van de Graaff generator, but it was limited to 600 kV by ambient humidity. The following year, he built a Cockcroft–Walton accelerator (Hirao 1986).

2 Not to be confused with Arthur Holly Compton (1892–1962), who discovered Compton diffusion.

3 RIKEN: short for Kokuritsu Kenkyū Kaihatsu Hōjin Rikagaku Kenkyūsho (Institute of Physical and Chemical Research).

In 1925, Ray Herb and his colleagues at the University of Wisconsin decided to build their own Van de Graaff electrostatic generator (Herb et al. 1935). But the ceiling height of the room they were allocated severely limited the maximum voltage. So they decided to operate it under pressure, like the one built by Barton et al. (1932). With the generator shown in Figure 11.5, the breakdown voltage was 190 kV at atmospheric pressure, rising to 750 kV in air at a pressure of 45 lbs[4]. In operation, the small size of the accelerator tube was another limiting factor.

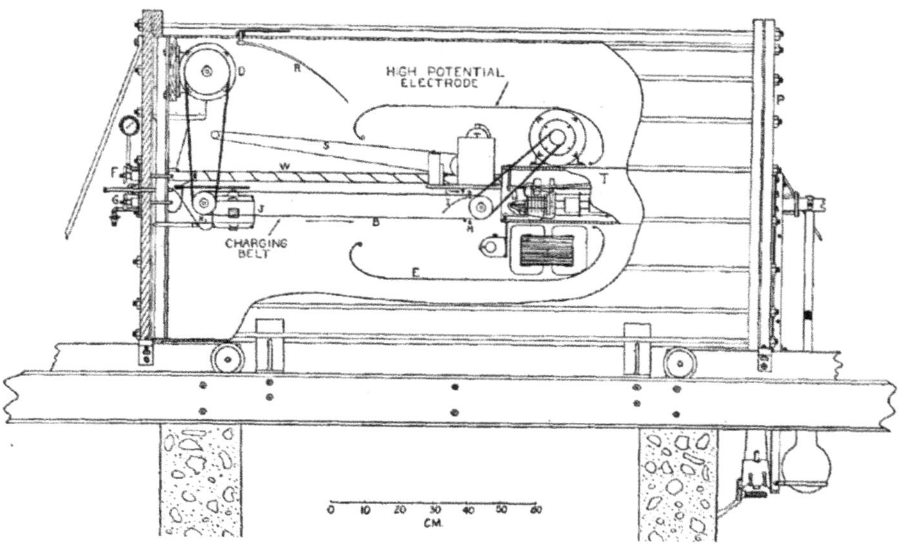

Figure 11.5. *The electrostatic generator in its steel tank (Herb et al. 1935)*

In the article "Van de Graff generator for general laboratory use", J.G. Trump and colleagues describe a compact, low-cost 500 keV Van de Graaff accelerator: "The unit described above was built at a cost of about $325, which includes all material and all shop time, but does not include the time involved in assembling and adjusting the finished parts. [...] A single belt 750-kV generator, for example, might be produced for about $425" (Trump et al. 1938). This would be equivalent to building a 750 kV accelerator for around $10,000_{2021}$ taking into account the increase in the US consumer price index between 1938 and 2021.

4 Amazing unit for pressure lbs = pounds, so it must be pound-force per square inch (psi). 45 psi = 3.1 bar.

It is worth noting that the first electron accelerator was built in 1937 to produce X-rays for medical applications at Harvard Medical School.

During the Second World War, accelerators made great strides: two machines operated at Los Alamos, one at 2 MV and the other at 4.3 MV, a vertical 4 MV machine with the introduction of series resistors to linearize the voltage gradient across the columns and tube. Later, Ray Herb introduced an important innovation. He replaced the belt with a chain whose links are isolated from each other: the Pelletron system. He founded NEC (National Electrostatics Corporation) in 1964. By 2021, NEC had manufactured more than 240 Pelletrons, including a 25 MV tandem for Oak Ridge National Laboratory.

11.1.3. *Tandem electrostatic accelerators*

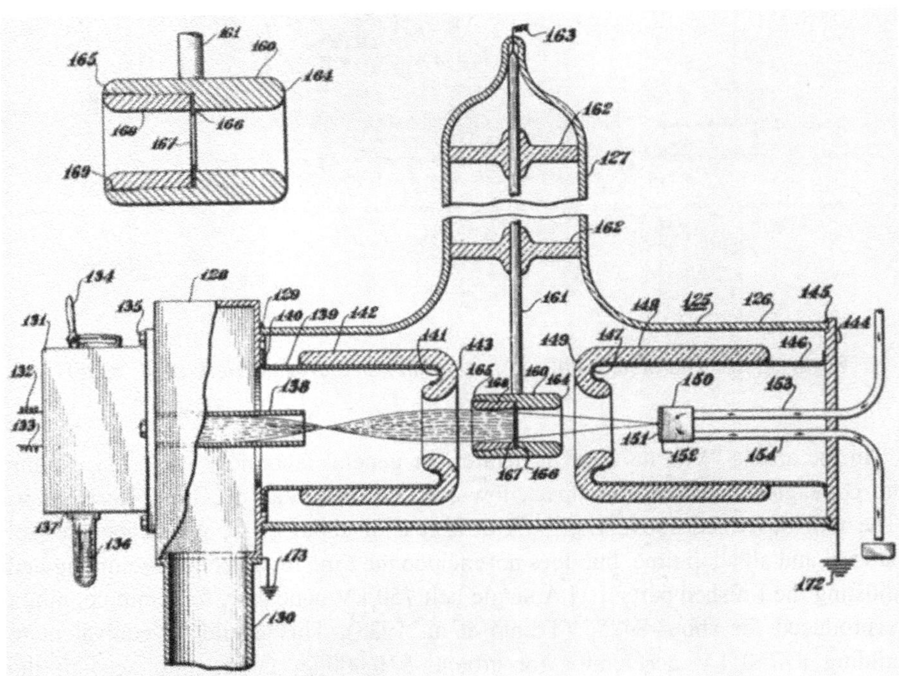

Figure 11.6. *Copy published in Rose and Wittkower (1970) of the original patent drawing on the principle of changing charge state to double ion energy (patent filed by Willard H. Bennett in 1937, Ohio State University). Negative ions are emitted to the left (138) and accelerated to the terminal (160), where they pass through a thin foil (167). Ions that have become positive are again accelerated to the target (151)*

The idea of taking advantage of a change in the charge of ions to re-accelerate them was put forward in 1932 at the American Physical Society meeting by A.J. Dempster[5] of the University of Chicago. However, the description of the process in this summary is brief. Positive protons are emitted from a positive potential (+22.5 kV). After acceleration, a fraction of the ions are neutralized in a high-pressure hydrogen cell, so they can return to their initial potential without any loss of energy. At this potential, neutral atoms can be positively reionized by collision and thus re-accelerated up to twice their original energy. A.J. Dempster wrote that this process could continue indefinitely, albeit with a loss of intensity at each pass. Nevertheless, by deflecting the ions with a magnetic field, he observed ions that had acquired an energy of 45 kV (Dempster 1932).

Seemingly independently, Willard H. Bennett (Ohio State University) came up with the idea of a tandem configuration, but with negative ions. Produced at earth potential, they were accelerated through a high-voltage tube to a terminal at positive potential. At this point, the ions passed through a thin foil in which they lost electrons, producing positive ions that could be multi-charged. They were then accelerated again in a second tube. As he stated in the introduction to an article (Bennett 1953), to test this idea, he designed a source of negative atomic hydrogen ions (Bennett and Darby 1936).

Bennett patented this accelerator design[6] but was unable to take it any further: "Conditions at Ohio State University at that time did not permit trying this idea in a high-voltage tube, and nothing further was done with this idea until the hydrogen ions part of it was reinvented recently". The idea was forgotten for 15 years, until it was revived in 1951 by Luis W. Alvarez[7]. An accelerator of this type was designed and built from a conventional Cockcroft–Walton accelerator at the University of California. It accelerated protons to 4 MeV. He stated that a sheet of 1 μg/cm^2 was achievable, with a loss of only a few hundred electron volts and little dispersion (Alvarez 1951).

High Voltage Engineering Corporation (HVEC), which marketed the first Van de Graaff accelerators, was founded in 1947 by John Trump, Denis Robinson and Robert Van de Graaff. Its European subsidiary was created in 1958 under the name HVEE (High Voltage Engineering Europa). By 1981, HVEC had installed 13 tandems over 10 MV (six in Europe and five in the USA).

5 A.J. Dempster (1886–1950) is best known for developing the first modern mass spectrometer in 1918.

6 US Patent 2,206,558.

7 Department of Physics, Radiation Laboratory, University of California.

A more detailed history of electrostatic accelerators can be found in "The Van de Graaf generator, an electrostatic machine for the 20th century"[8] (Brenni 1999) as well as in "History of the electrostatic accelerator" (McKay 2005).

11.1.4. *Pulsed electron accelerators*

Radiolysis experiments only gave access to chemical kinetics and the existence of the solvated electron when time-resolved optical absorption experiments could be carried out. These measurements are only possible if we are able to send an irradiation pulse intense enough to generate a detectable signal, the duration of which defines the temporal resolution of the measurement. In addition, the kinetics of signal fading dictate the frequency of pulse recurrence.

The first experiments were carried out thanks to the development of linear accelerators capable of producing pulses of a few microseconds with a peak current of several hundred milliamperes. The first experiment involved the detection of the benzyl radical (McCarthy and MacLachlan 1960). In this case, the 1–10 MeV electron accelerator, built by Varian Associates for the Du Pont Experimental Station (Wilmington USA), delivered pulses of adjustable duration between 1.5 and 5 µs with a repetition rate of 180 pulses/s. A similar approach was developed at the Physics Research Department of the Christie Hospital and Holt Radium Institute in Manchester with 2 µs pulses from a 4 MeV linac (Keene 1960). Other pioneers included Argonne National Laboratory, which used a 1.8 MeV linac delivering pulses of 1 µC and 2 µs duration (Hart and Boag 1962). These microsecond macropulses actually consisted of a train of sub-nanosecond pulses. The principle was greatly improved at the University of Toronto thanks to an original idea: to use the Čerenkov light produced by the electrons that trigger radiolytic processes to probe transient absorptions. Light and electron pulses have the same duration, and the probe pulse can be shifted in time by lengthening its optical path. This is a stroboscopic effect. In the case of the Toronto linac, delivering electrons of over 30 MeV, the beam was made up of 0.02 ns pulses spaced 0.35 ns apart, forming 30 ns macro-pulses. The recurrence period of these macro-pulses was 350 ns (Bronskill and Hunt 1968). However, the dose per pulse was relatively low (~0.3 nC). In Japan, at the University of Tokyo, radiolysis experiments with 20 ps pulses were carried out using two linacs, one irradiating the sample with a single piosecond pulse, the other generating an equally picosecond pulse of Čerenkov light (Kobayashi and Tabata 1985).

8 Van de Graaff is often written with a single f.

Pulses lasting a few nanoseconds (4–5 nC) have also been obtained by triggering electron emission from a Van de Graaff (Hunt and Thomas 1967).

Then came the Febetron, a pulsed electron accelerator from Field Emission Corporation based on the principle of the high-voltage generator invented in 1923 by German engineer Erwin Otto Marx. This electrical circuit generated a high-voltage pulse from a low-voltage DC source, charging capacitors in parallel and discharging them in series. For the operating principle, see the patent filed in 1974 by Field Emission Corp (Hafner 1974). Febetron pulses are in the nanosecond range, but of much higher intensity than linacs: 7,000 A for 3 ns at 600 keV for the Febetron 406.

Field Emission Corporation was founded in 1954 by Walter Dyke as part of the Linfield Research Institute in McMinnville, Oregon. Its main commercial application was an X-ray flash system. In 1973, it merged with another company also working in field emission to form field electron and ion (FEI), well known to microscopists.

The latest improvement has been the development of linear accelerators with a cathode photo-triggered by a pico- or even femto-second laser, generating electron pulses of between 100 fs and 20 ps, depending on the machine. The laser beam is also used to measure transient absorptions. In 2020, there were six such facilities, including four in Japan (University of Tokyo, Sumitomo Heavy Industries, Waseda University and Osaka University), one in the United Sates (Brookhaven National Laboratory) and one in France (Université Paris-Saclay) (Yang et al. 2020)

11.2. Nuclear reactors

The discovery of induced fission in 1938 quickly gave rise to the idea of chain reactions and energy production. Thus, in early May 1939, the Caisse Nationale de la Recherche Scientifique (French National Scientific Research Fund), at the instigation of Frédéric Joliot-Curie, Hans von Halban and Lew Kowarski, filed three patents for applications of the energy released by nuclear fission. The first, entitled "Device for producing energy", describes the operating principle of a nuclear reactor (Caisse Nationale de la Recherche Scientifique 1939a). The second, submitted the following day, was aimed at stabilizing the energy production system by regularly interrupting the chain reaction (Caisse Nationale de la Recherche Scientifique 1939b). The third, whose title is not explicit, "Improvements to explosive charges", describes the creation of an atomic weapon (Caisse Nationale de la Recherche Scientifique 1939c)[9].

The industrial history of nuclear energy began with the first stone, or rather the first graphite brick, laid for the construction of the Fermi pile, which produced its first chain reaction on December 2, 1942 in Chicago. It was a major step toward the mastery of nuclear energy, even if, made up of 6 tons of uranium metal, 50 tons of uranium oxide and 400 tons of graphite, it only produced around 100 W for a few days. The pile, known as CP-1 for Critical Pile 1, had to be shut down after just a few days' operation, as it endangered operators without biological protection. For those who want to know more about the Chicago pile, the report "The first reactor, 40th anniversary" gives an accurate account of its first divergence (Allardice et al. 1982). This chain reaction was only achieved through extraordinary work between 1939 and 1942. During this period, no fewer than 38 pile configurations were built. Some, devoid of uranium, were used to measure the neutron parameters of graphite, while others were used to test different forms of fuel (U_3O_8, UO_2 , U metal), with the quantity of uranium ranging from a few kilograms to 8 tons (Reed 2020). None of them came close to critical mass, but they did provide sufficient data to enable CP-1 to be correctly sized. Until the end of the war, seven other piles were built at Argonne, Oak Ridge, Hanford and Los Alamos. All were dedicated to plutonium production. The 250 MW Hanford pile could produce around 180 g of plutonium per day at full power[10].

According to Reed (Alma College, Michigan), between 1939 and the end of 1946, some 90 piles were developed in six countries. The vast majority of these, around 60, were located in the United States, but around 20 were built in Germany and a few more in France, Great Britain, Canada and Russia (Reed 2020).

In the UK, experiments with a ton of uranium oxide were conducted by George P. Thomson. He concluded that a chain reaction was not possible with oxide and water or parafin (Clark 1961). In Europe, however, it was in the UK, near Harwell at the Atomic Energy Research Establishment, that the first nuclear reactor, named

9 Once the war was over, although a partial agreement was reached with the British, the French claim was rejected by the Americans on the grounds of insufficient description of the devices envisaged. Furthermore, since 1946, the Atomic Energy Act has prohibited the United States from patenting fissile materials, or even exchanging information on the subject. However, in 1968, French anteriority in nuclear technology discoveries was officially recognized, with a $35,000 "compensation" for the inventors (two of whom are now deceased).

10 An electronuclear reactor produces around 1 g of ^{239}Pu per thermal MW per day.

GLEEP for Graphite Low Energy Experimental Pile, was built. It diverged in 1947 and was decommissioned in 1990. Then, in 1948 came BEPO (British Experimental Pile 0), a 6 MW reactor in which it was possible to irradiate materials cooled to −196°C with a flux of 10^{18} fast neutrons/cm^2 and transfer them cold to laboratory characterization systems (see, for example, Kinchin and Thompson (1958)).

Germany was also active, with the construction of 19 piles in various locations: Hamburg (1), Berlin (8), Leipzig (4), Heidelberg (2), Gottow[11] (3) and Haigerloch[12] (1) (Walker 1993; Reed 2020). The first, built in May 1940, contained 200 tonnes of uranium oxide surrounded by dry ice, but no neutron multiplication was detected. The last, made of uranium cubes suspended in heavy water, was destroyed by a bombing raid in April 1945.

Figure 11.7. *ZOE atomic pile at the Centre d'Études Nucléaires de Fontenay-aux-Roses. Photo credit: CEA Historical Archives*

11 Gottow is about 50 km south of Berlin.

12 Haigerloch is about 60 km south of Stuttgart.

In France, under the impetus of Frédéric Joliot, Hans Halban, Lew Kowarski and Francis Perrin, six piles were built in Paris between March 1939 and January 1940, the first of which contained 250 kg of uranium oxide immersed in water and surrounded by parafin. Neutron production was detected (Von Halban et al. 1939).

Figure 11.8. *Open pool of the 7 MW TRITON reactor at CEA Fontenay-aux-Roses. The reactor operated from 1951 to 1982. The equipment in the upper right foreground is a liquid hydrogen cryostat, enabling irradiation at 21 K in the reactor core (photo CEA Fontenay-aux-Roses). For a color version of this figure, see www.iste.co.uk/bouffard/irradiation.zip*

In the 1960s–1980s, as knowledge of the effects of irradiation on metallic materials became more refined, low-temperature irradiation became increasingly important. Low-temperature irradiation was possible in reactors at Oak-Ridge (4.2 K), Argonne (4.2 K), Garching (4.2 K), Karlsruhe (11 K), Grenoble (27 K) and Fontenay-aux-Roses (20 K). Documentation from the Union of Soviet Socialist Republics is conspicuously lacking for this period, but it is fair to assume that such equipment was being developed there. Indeed, a first experimental natural uranium and graphite-moderated reactor diverged in 1946, followed two years later by an industrial-scale reactor. The first 5 MW power reactor was commissioned in June 1954 at Obninsk. These achievements were the brainchild of Igor Vassilievitch Kourtchatov, a former pupil of Abram Fiodorovitch Ioffé, the true father of Soviet atomic energy. He died in 1960, before the MIR reactor, dedicated to metallurgical studies, went into service in 1966. This information on nuclear power in the USSR is taken from Pierre Zemskoff's master's thesis (Zemskoff 1980).

Figure 11.9. *Low-temperature (10K) infrared irradiation and absorption measurement device installed on the medium-energy beam at GANIL (Caen). Photo: Serge Bouffard. For a color version of this figure, see www.iste.co.uk/bouffard/irradiation.zip*

In France, the first experimental pile diverged on December 15, 1948, and was named ZOE for Zero Energy, Uranium Oxide and Heavy Water (Figure 11.7).

From the ZOE pile to the Jules Horowitz reactor (under construction), France has built more than 30 research reactors.

In some reactors, samples are introduced into the core via a "rabbit" system. The rabbit is a small cylindrical box which is transported inside a pneumatic tube by gas pressure to the center of the pile. The time required to reach the center of the stack from the outside is just a few seconds, so exposure can be precisely controlled.

EL3	Osiris	Mélusine	Siloé	Siloette	Triton	EL2	EL4	Rapsody	Phoenix	Orphée	RJH
2.8 MW	70 MWt	18 MW	35 MWt	1 W	6.5 MW	600 kW	0.1 W	17 MWt	250MWe	70 MW	100 MW
1952	1956	1957	1957	1958	1959	1959	1959	1959	1961	1966	2025

Table 11.1. *France's main research reactors. Dates correspond to their first divergence*

11.3. Recent developments

The current trend is to install instruments on irradiation equipment in order to follow in situ and even in operando changes in material properties under irradiation. These include X-ray diffractometers, UV-visible, infrared or RAMAN spectrometers, mass spectrometry of gases emitted during irradiation, electrical measurements and even transmission electron microscopy. The International Atomic Energy Agency (IAEA) has identified eight facilities coupled to a transmission electron microscope (IAEA 2018).

The other trend is to reproduce the complex field of industrial irradiation in the laboratory. To achieve this, two or even three accelerators are coupled together. The IAEA has identified 12 multi-beam facilities (IAEA 2018).

11.4. Chapter 11 references

11.4.1. *Biographies of some of the chapter's personalities*

Numerous personalities have contributed to the development of irradiation tools. Below are brief biographies of five of them, chosen in an almost arbitrary manner.

John Douglas Cockcroft (1897, Todmorden, England–1967, Cambridge, England) won the Nobel Prize in Physics with Ernest Thomas Sinton Walton in 1951 for their work on the artificial disintegration of the atomic nucleus. After training in electrical engineering, he continued his post-graduate studies at Cavendish Laboratory under the supervision of Rutherford. With Walton and Oliphant, he built an accelerator that came to be known as the Cockcroft–Walton generator. After the war, Cockcroft became Director of the Atomic Energy Research Establishment (AERE) at Harwell, where Western Europe's first nuclear reactor was built. His colleague Walton made his career at Trinity College, Dublin.

Ernest Orlando Lawrence (1901, Canton, South Dakota, USA–1958, Palo Alto, California, USA) won the Nobel Prize in Physics (1939) for his invention and development of the cyclotron. In 1925, he obtained his PhD from Yale University, where he continued as an assistant professor. In 1928, he became Associate Professor at UC Berkeley. His first cyclotron, built in 1929–1930, was only 10 cm in diameter, with subsequent ones becoming larger and larger: 27, 37, 60 and 184 inches. For the Manhattan Project, he helped build an isotope separator capable of separating uranium isotopes. After the war, he campaigned for the development of Big Science. He founded the Berkeley Laboratory at the University of California, which he headed until his death. With Edward Teller, he was also behind the Livermore facilities. After his death, these laboratories were named Lawrence Berkeley National Laboratory (LBNL) and Lawrence Livermore National Laboratory (LLNL).

Robert Jemison Van de Graaff (1901, Tuscaloosa, Alabama, USA–1967, Boston, USA) was an American physicist best known for designing and building the eponymous electrostatic machine. In 1923, he obtained a Master of Science degree in mechanical engineering from the University of Alabama. Having attended Marie Curie's lectures at the Sorbonne in 1926, he returned to Oxford University to defend his doctorate under the supervision of Towsend (1928).

Enrico Fermi (1901, Rome, Italy–1954, Chicago, USA), a gifted student, became a professor at Rome's La Sapienza University at the age of 25. In 1939, he emigrated to the United States, where he worked with Leó Szilárd at the University of Chicago on the development of an atomic pile, which diverged on December 2, 1942. He then joined the Los Alamos National Laboratory (LANL) to work on the Manhattan Project. After the war, he accepted a professorship at the Institute for Nuclear Physics at the University of Chicago. He died prematurely at the age of 53.

Figure 11.10. *Enrico Fermi and Ernest O. Lawrence (right) in 1930*

Rolf Widerøe (1902, Oslo, Norway–1996, Obersiggenthal, Switzerland) left Norway in 1920 to study electrical engineering in Karlsruhe. It was here that he came up with the concept of electromagnetic induction to accelerate electrons. His thesis involved building an experimental betatron. Since it was impossible to stabilize the beam, he decided to build a linear accelerator using a radio-frequency electric field. The article describing this instrument inspired Lawrence to design the cyclotron. During the war, he continued to work in Germany, and was fined by the Norwegian government. He published over 180 articles in scientific and technical journals and filed over 200 patents.

11.5 References

Allardice, C., Trapnell, E.R., Fermi, E., Fermi, L., Williams, R.C. (1982). The First Reactor, 40th Anniversary. Report, U.S. Department of Energy, DOE/NE-0046.

Alvarez, L.W. (1951). Energy doubling in dc accelerators. *Review of Scientific Instruments*, 22(9), 705–706.

Barton, H.A., Mueller, D.W., Van Atta, L.C. (1932). A compact high potential electrostatic generator. *180th Meeting of the American Physical Society*. Physical Review, Chicago.

Bennett, W.H. (1953). Nuclear reactions between heavy nuclei. *Review of Scientific Instruments*, 24(10), 915–916.

Bennett, W.H. and Darby, P.F. (1936). Negative atomic hydrogen ions. *Physical Review*, 49(1), 97–99.

Brenni, P. (1999). The Van de Graaff generator An electrostatic machine for the 20th century. *Bulletin of the Scientific Instrument Society*, 63, 6–13.

Brobeck, W.M., Lawrence, E.O., MacKenzie, K.R., McMillan, E.M., Serber, R., Sewell, D.C., Simpson, K.M., Thornton, R.L. (1947). Initial performance of the 184-inch cyclotron of the University of California. *Physical Review*, 71(7), 449–450.

Bronskill, M.J. and Hunt, J.W. (1968). Pulse-radiolysis system for the observation of short-lived transients. *Journal of Physical Chemistry*, 72(11), 3762–3766.

Caisse Nationale de la Recherche Scientifique (1939a). Dispositif de production d'énergie. US Patent, FR976541A.

Caisse Nationale de la Recherche Scientifique (1939b). Procédé de stabilisation d'un dispositif de production d'énergie. US Patent, FR976542A.

Caisse Nationale de la Recherche Scientifique (1939c). Perfectionnements aux charges exposives. US Patent, FR971324.

Clark, R.W. (1961). *The Birth of the Bomb: The Untold Story of Britain's Part in the Weapon that Changed the World.* Phoenix House, London.

Cockcroft, J.D. and Walton, E.T.S. (1930). Experiments with high velocity positive ions. *Proceedings of the Royal Society of London, Series A Mathematical and Physical Sciences*, 129(811), 477–489.

Cockcroft, J.D. and Walton, E.T.S. (1932). Experiments with high velocity positive ions. (I). Further developments in the method of obtaining high velocity positive ions. *Proceedings of the Royal Society of London, Series A Mathematical and Physical Sciences*, 136(830), 619–630.

Crookes, W. (1879). On the illumination of lines of molecular pressure, and the trajectory of molecules. *Philosophical Transactions of the Royal Society of London*, 170, 135–164.

De Guericke, O. (1672). *Experimenta Nova (ut vocantur) Magdeburgica de vacuo spatio.* Amstelodami, Joannem Janssonium, Waesberge.

Dempster, A.J. (1932). The production of high-velocity protons by repeated accelerations. *Physical Review*, 42, 901–902.

Greinacher, H. (1921). Über eine Methode, Wechelstrom mittels elektrischer Ventile und Kondensatoren in hochgespannten Gleichstrom umzuwandeln (About a method of converting alternating current into high-voltage direct current by means of electric valves and capacitors). *Zeitschrift für Physik,* 4(2), 195–205.

Hafner, M.P. (1974). Marx surge pulser having stray capacitance which is high for input stages and low for output stages. US Patent, 3,783,289.

Hart, E.J. and Boag, J.W. (1962). Absorption spectrum of the hydrated electron in water and in aqueous solutions. *Journal of the American Chemical Society*, 84(21), 4090–4095.

Herb, R.G., Parkinson, D.B., Kerst, D.W. (1935). A Van de Graaff electrostatic generator operating under high air pressure. *Review of Scientific Instruments*, 6(9), 261–265.

Hirao, Y. (1986). The history of cyclotrons in Japan (banquet speech). In *11th International Conference on Cyclotrons and their Applications*, Tokyo.

Hunt, J.W. and Thomas J.K. (1967). Pulse radiolysis studies using nanosecond electron pulses: Observation of hydrated electrons. *Radiation Research*, 32(1), 149–163.

IAEA (2018). Accelerator simulation and theoretical modelling of radiation effects in structural materials. Publication, IAEA Nuclear Energy Series, No. NF-T-2.2.

Keene, J.P. (1960). Kinetics of radiation-induced chemical reactions. *Nature*, 188(4753), 843–844.

Kinchin, G.H. and Thompson, M.W. (1958). Irradiation damage and recovery in molybdenum and tungsten. *Journal of Nuclear Energy*, 6(4), 275–284.

Kobayashi, H. and Tabata, Y. (1985). A 20 ps time resolved pulse radiolysis using two linacs. *Nuclear Instruments & Methods in Physics Research Section B: Beam Interactions with Materials and Atoms*, 10–11, 1004–1006.

Lawrence, E.O. (1934). Method and apparatus for the acceleration of ions. US Patent, 1948384.

Lawrence, E.O. (1951). The evolution of the cyclotron. Nobel Lecture, Stockholm.

Lawrence, E.O. and Livingston, M.S. (1932). The production of high speed light ions without the use of high voltages. *Physical Review*, 40(1), 19–35.

Lawrence, E.O., Alvarez, L.W., Brobeck, W.M., Cooksey, D., Corson, D.R., McMillan, E.M., Salisbury, W.W., Thornton, R.L. (1939). Initial performance of the 60-inch cyclotron of the William H. Crocker Radiation Laboratory, University of California. *Physical Review*, 56(1), 124.

McCarthy, R.L. and MacLachlan, A. (1960). Transient benzyl radical reactions produced by high-energy radiation. *Transactions of the Faraday Society*, 56(0), 1187–1200.

McKay, J. (2005). History of the electrostatic accelerator. In *Electrostatic Accelerators: Fundamentals and Applications*, Hellborg, R. (ed.). Springer, Berlin, Heidelberg.

Reed, B.C. (2020). Piles of piles: An inter-country comparison of nuclear pile development during World War II. Report, Alma College, Michigan.

Rose, P.H. and Wittkower, A.B. (1970). Tandem Van de Graaff accelerators. *Scientific American*, 223(2), 24–33.

Telegdi, V.L. (2000). Szilard as inventor: Accelerators and more. *Physics Today*, 53, 25–28.

Thibaud, J. (1933). Production d'ions positifs de vitesse élevée par accélérations multiples. *Comptes rendus hebdomadaire des scéances de l'Académie des sciences*, 188, 1394–1396.

Trump, J.G., Merrill, F.H., Safford, F.J. (1938). Van de Graaff generator for general laboratory use. *Review of Scientific Instruments*, 9(12), 398–403.

Usch, U. and Bautz, W. (2005). Die Entwicklung von Elektronen-Kreisbeschleunigern bei Siemens in Berlin und Erlangen (The development of electron circular accelerators at Siemens in Berlin and Erlangen). *Zeitschrift für Medizinische Physik*, 15(2), 87–101.

Van Atta, L.C., Northrup, D.L., Van Atta, C.M., Van de Graaff, R.J. (1936). The design, operation, and performance of the round hill electrostatic generator. *Physical Review*, 49(10), 761–776.

Van de Graaff, R.J., Compton, K.T., Van Atta, L.C. (1933). The electrostatic production of high voltage for nuclear investigations. *Physical Review*, 43, 149–164.

Von Halban, H., Joliot, F., Kowarski, L. (1939). Liberation of neutrons in the nuclear explosion of uranium. *Nature*, 143(3620), 470–471.

Walckiers de St Amand, C.L. (1784). La machine électrique nouvellement inventée. Extract, des registres de l'Académie Royale des Sciences du 17 mars, 1784, 29.

Walker, M. (1993). *German National Socialism and the Quest for Nuclear Power, 1939-49*. Cambridge University Press, Cambridge.

Wideröe, R. (1928). Über ein neues Prinzip zur Herstellung hoher Spannungen (About a new principle for producing high voltages). *Archiv für Elektrotechnik*, 21(4), 387–406.

Yang, J., Kan, K., Gohdo, M., Yoshida, Y. (2020). Femtoseconde pulse radiolysis. In *Novel Imaging and Spectroscopy*, Yang, J. (ed.). IntechOpen, London.

Zemskoff, P. (1980). L'électronucléaire en U.R.S.S. Bilan & perspectives. Master's Thesis, Université Paris VIII, Section d'Etudes Slaves.

12

Irradiation Applications

Among the applications, I will not mention those that flourished at the beginning of the 20th century, but instead take advantage of the enthusiasm of men and women for the discoveries made at the end of the previous century. Many of these pseudo-applications were charlatanism or swindles, but some of them were proposed by sincere people. This is probably the case of Émile Armet de Lisle. This industrialist was so passionate about radium that, in 1903, he set up a factory in Nogent, France, to manufacture radium and strontium salts. Then, in 1906, he set up the Laboratoire Biologique du Radium (Radium Biological Laboratory) in a building on the rue d'Artois in Paris. He went so far as to create the Banque du Radium (Radium Bank) to finance projects using this element. Before the First World War, with pharmacist Alexandre Antonin Jaboin, he marketed exclusively radium-based pharmaceutical products, with two objectives in mind: to combat cancer and, through the use of infinitesimal doses, to stimulate cellular functions. After the war, the price of radium remained high, around 1,000 F_{1920} for 1 mg, or ~ 2,300 $€_{2023}$/mg, which fortunately limited its spread. Although its toxicity was known at the time, there were no regulations limiting its handling and distribution (at least in France). With the development of a radium-rich mine in Haut-Katanga (now a province of the Democratic Republic of Congo), radium supply became easier and the cost halved. As a result, the market for radium-based products grew spectacularly. In the pharmaceutical field, microcurie therapy began spreading outside medical practices, with the advantage of consuming little or no radioactive material. Perfumers were among those taking advantage of this craze to market products that were supposedly scientifically based. Laboratories flourished were as follows: Rhemda, Radiumthérapie, Laboradium, Microradium, Radium Chema-France, etc., and beauty creams such as Alpha-Radium or Tho-Radia. The latter cream reinforced the scientific aspect by using the Curie surname, "formulated by Dr Alfred Curie" (Figure 12.1). Alfred Curie, who had no family connection with the Curies, sold his name to the SECOR company marketing Tho-Radia. In the 1930s, the recognition of

A Historical Approach to Materials Under Irradiation,
Serge BOUFFARD.

the consequences of working with radioactive materials as an occupational disease and the inclusion of radio-elements in the table of toxic substances, making it necessary to label beauty creams “poison”, put an end to these dubious applications. This information on the first uses of radium and thorium is taken from the work of Lefebvre and Raynal (2011, 2012).

Figure 12.1. *Lid of a can of Tho-radia powder containing 0.01 µg radium bromide and 0.1 g thorium sulfide. For a color version of this figure, see www.iste.co.uk/bouffard/irradiation.zip*

In a lecture given to the Nordic Society for Radiation Research and Technology in Helsinski (Finland) in 1975, Joseph Silverman of the Laboratory for Radiation and Polymer Science at the University of Maryland reported on radiation applications in industry (Silverman 1975) starting with what he described as the oldest application: *the radiosterilization of single-use biological and medical supplies*. Its success stemmed from ease of packaging and use, despite high cost due largely to unreliable accelerators. He then described the following:

– *Radiation cross-linked polymers* for the manufacture of electric cable insulation, heat-shrinkable materials, etc.

– *Permanent-pleat fabrics* for garments that retain their shape after manufacture (e.g. uniforms): cotton and polyester fabrics treated by radio-induced grafting. In the 1970s, Milliken produced 50 million meters of these fabrics. It was marketed in the United States under the name Visa.

– *Irradiated surface coatings* to improve surface resistance to aggression. These were found in automobiles in the United States. Ford Motor Company coated dashboards with an electrostatic spray, then irradiated them with an electron beam. The same process was applied to wood panels to harden them.

– *Wood and plastic compounds* for materials stronger than wood but retaining its appearance. In the 1970s, this process was only used to manufacture parquet strips. Arco Chemical had a capacity of 330,000 square meters per year.

– *Irradiation of polymers* to degrade them before recycling. This process was considered by several industries. To obtain polyethylene oxide of different molecular weights, Union Carbide Corporation proposed cutting the chains of a high-molecular-weight product by a gamma-induced chain reaction in the presence of oxygen. In a more conventional approach, Columbia Research Corporation degraded polytetrafluoroethylene (PTFE) molding powders by electron irradiation in air. This degradation produced lubricants.

– *Irradiation to initiate chemical synthesis*: this method remained a marginal activity, with the exception of Dow Chemical Company's annual production of around 1,000 tonnes of ethyl bromide.

Focusing on organic matter, this list is far from exhaustive, since it does not include all medical applications, semiconductor implants, nanostructured materials, etc.

12.1. Medical applications

The medical applications of radiation are certainly the most numerous in terms of their diversity and frequency of use. In France, for example, 30.5 million X-ray images and 5.7 million X-ray scans were carried out in the private medical sector in 2020 (Social Security data).

12.1.1. *Radiography*

The quickest transition from discovery to application is certainly that of X-rays. Just two weeks after the discovery of X-rays was announced, Otto Walkoff, a dentist in Braunsschweig, Germany, produced the first dental radiography. To do so, he placed a photographic plate between his teeth and tongue, which he exposed to X-rays for 25 min. The following year, together with Friedrich Giesel, he set up the first dental X-ray practice. The popularity of this technique, enabling us to see the bones inside the body, was exceptional. On January 13, 1896, two months after the discovery, a demonstration was organized in Berlin for Kaiser Wilhelm II. In Paris, in the spring of 1896, Dufayel's department stores offered two attractions: a radioscopy of the hand and a film show.

As early as 1921, in his book *X-rays and Radium in the Treatment of Diseases of the Skin*, MacKee (1921) describes on page 19 the evils of this general enthusiasm:

> The history of roentgen therapy can be divided roughly into three eras – optimistic, pessimistic and realistic; this is particularly true of this country. At first enthusiasm and carelessness overcame caution. Many physicians installed apparatus and attempted to employ the X-rays for practical therapeutic purposes, without making a study of the subject. Even the scientific and conscientious workers did not at first realize that they were dealing with an exceedingly dangerous agent. It was natural, therefore, that many patients received serious injuries. Not only were patients injured, but operators, by repeatedly testing the penetrability of the rays by inserting their hand between the tube and a fluorescing screen, developed an erythema which in many instances led to serious sequelae. These facts, together with the discovery by Brown and Osgood that sterility was produced by the X-rays naturally caused operators to be a little more cautious. However, optimism reigned until about 1906. During those years the rays, to a large extent, were empirically used and they were tried out on nearly every chronic disease. The literature was misleading as it was full of case reports of wonderful cures, the occasional paper from the pen of a good man being ignored or overlooked by the average X-ray operator of the period and in spite of repeated warnings from capable men the "radiomaniacs" held the reins.

Despite these initial setbacks, the contribution of X-rays to health is enormous. The reduction in radiation dose from X-rays has benefited from risk awareness and technological progress, while at the same time providing images of much better quality. The main stages in this development can be summarized as follows:

– At the end of the 19th century and the beginning of the 20th century, the first X-rays were taken using rudimentary devices that often generated high doses of radiation. Practitioners were not fully aware of the risks associated with radiation exposure.

– In the first half of the 20th century, technical improvements made it possible to reduce unwanted doses, notably by introducing filters and collimators to limit the exposed area.

– Between the 1950s and 1970s, dose reduction came from the use of more sensitive films, while maintaining diagnostic image quality. In addition, the optimization of imaging techniques led to better use of X-rays, reducing unnecessary exposure.

– From the 1980s to the present day, the introduction of digital technologies has enabled major advances in both image quality and dose minimization. Soft-tissue radiography and computed tomography have particularly benefited.

12.1.2. *Radiotherapies*

While at the beginning of the 20th century, the use of X-rays in radiology was justified by the novelty of images of inside the body, and their use in radiotherapy had no scientific or observational justification. Nevertheless, it was only a few weeks after the announcement of their discovery that the first attempt at treatment with X-rays was made. The names of four pioneers, who could best be described as adventurers, stand out: Emil H. Grubbé, Leonhard Voigt, Victor Despeignes and Leopold Freund.

In the United States, 37 years after the discovery of X-rays, Grubbé (1933) claimed in an article published in *Radiology* that he should be credited with being the first to have (1) been exposed to X-rays to the point of developing dermatitis, (2) used X-rays therapeutically on pathological lesions and (3) used lead foil as a means of protection against the undesirable effects of X-rays. In fact, at the end of January 1896, when he was only a medical student, he treated two patients, one with recurrent carcinoma of the breast, the other with lupus vulgaris. Both patients died within a month of exposure to X-rays. No written record remains of these two treatments, which appear to have been carried out outside protocol.

Leonhard Voigt informed the Hamburg Medical Society in February 1896 that he had treated a patient with inoperable nasopharyngeal cancer. The treatment consisted of 80 30-min sessions (two sessions per day). As in the previous case, there is no clinical report, nor any clear justification for the treatment.

Victor Despeignes, a French doctor and dean of Lyon's Faculty of Medicine, carried out the first attempt at X-ray treatment on a patient with stomach cancer. In an 1896 article, he reported the results: the one-week treatment reduced pain and shrank the size of the tumor, but the patient did eventually die.

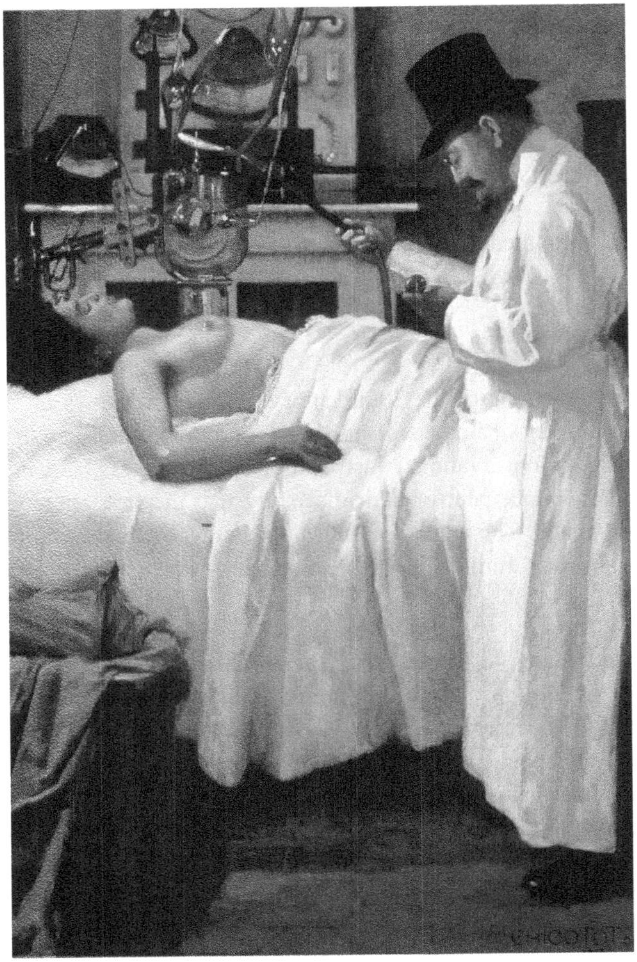

Figure 12.2. *Painting by Georges Chicotot, 1917, depicting the first trials of cancer treatment using X-rays (Musée de l'Assistance Publique – Hôpitaux de Paris). For a color version of this figure, see www.iste.co.uk/bouffard/irradiation.zip*

Leopold Freund was another pioneer of radiotherapy. When he performed his first treatment, Freund was only 28 years old and had been practicing as a dermatologist in Vienna for a year. Amid the proliferation of poorly justified treatment trials, he accurately described and motivated the treatment he applied to a five-year-old girl child suffering from a hairy nevus. By the end of the treatment, the hair had fallen out. The hair never returned. Her case is well documented, and she was followed for 75 years after her exposure to X-rays. In 1904, Freund published a

538-page book on radiotherapy for practitioners (Freund 1904). The medical chapters are entitled “Treatment with high-frequency currents”, “Treatment with X-rays”, “Becquerel rays” and “Treatment with heavy and light rays”.

Dosimetry for radiotherapy was poorly mastered until the introduction of hot-cathode Coolidge tubes and, for some, electron focusing using a Wehnelt[1], which improved the spatial distribution of the beam. The switch from film dosimeters to thermoluminescent dosimeters, and then to electronic dosimeters, also brought significant gains.

12.1.3. *Nuclear medicine*

Medical imaging and radiotherapy using low-concentration radioactive ions are the two components of nuclear medicine. Nuclear medicine has many origins. George Hevesy[2] (1923) was the first to use radioactive tracers to study the absorption and diffusion of lead in plants. To this end, the presence of ^{210}Pb, a radioactive isotope of lead, was measured in the ashes of various parts of the plant. After plants, he turned his attention to the circulation of bismuth and lead in a living organism. A solution of bismuth salt mixed with a solution of radium E, a radioactive isotope of bismuth ^{210}Bi, was injected into the muscles of a rabbit. After 15 days, the rabbit was sacrificed and the activity of the various parts was measured. A slightly different procedure was used for lead (Christiansen et al. 1924a, 1924b). Hevesy also made a totally anecdotal but unethical use of radiotracers. He suspected his landlady of serving recycled meat scraps, which she denied. A few days later, after adding a radioactive substance to the leftovers, he used an electroscope to show that the meal was radioactive, a dubious practice that is difficult to apply today.

Following Pierre and Marie Curie’s discovery of artificial radioactivity in 1936, John Lawrence used cyclotrons, his brother Ernest’s invention, to create radioactive isotopes for medical purposes. He was the first to use ^{131}I, with a half-life of 8.02 days, to diagnose and treat thyroid problems, and ^{32}P to treat leukemia (Lawrence 1940). He is often referred to as the “father of nuclear medicine”. In the 1950s, the introduction of radionuclide generator technology revolutionized radiopharmacy. These generators produce radio-pharmaceuticals from radioactive precursors that decay into substances usable in diagnostic and therapeutic

1 These improvements were introduced by William Coolidge in 1913 while working for General Electric. He later became director of the General Electric Research Laboratory and vice-president of the company.

2 He was awarded the Nobel Prize in Chemistry in 1943 for the discovery of isotopes.

procedures. This has considerably improved the availability and ease of use of radiopharmaceuticals.

Since then, nuclear medicine has made huge strides in effectiveness with alpha radio-immunotherapy, in which the alpha emitter is transported by an antibody that recognizes the cancer cells.

12.1.4. *Radiosterilization*

In the field of health, improved asepsis conditions have been a major step forward. Originally, objects were sterilized using biocides such as heat or chemical agents (phenol, alcohol, etc.). Objects were processed in bulk. Between sterilization and use, therefore, there was handling and storage, creating a risk of contamination. One of the advantages of sterilization using ionizing radiation is that it can be used to treat items packaged separately in airtight containers. Another advantage is the ability to sterilize objects containing materials sensitive to heat or chemical agents. On an industrial scale, the first plants went into operation in the early 1960s in the United States. The number of gamma-ray sterilization plants rose from 3 in 1960 to 62 in 1975, 65% of them in Europe, according to the IAEA.

Marketing of radiation-sterilized biomedical supplies began in 1949, when Johnson & Johnson acquired the Ethicom company. Ethicom still exists today, marketing absorbable sutures. To increase the speed of resorption, sutures can be irradiated. This irradiation splits the molecular chains, reducing the thread's tensile strength and accelerating the hydrolysis process.

12.2. Food processing

In 1896, the year after the discovery of X-rays, Franz Minck (1896a, 1896b) of Munich published two articles suggesting that such radiation kills bacteria and could be used therapeutically. In 1904, Prescott, studying the bactericidal effects of radium radiation on two parasites, came to negative conclusions: "any advantages derived from the therapeutic use of radium must be explained in some other way than the direct weakening or destruction of the microorganisms of decease" (Prescott 1904).

But such was the craze for new radiation at the very beginning of the 20th century that on January 28, 1905, a miller, Joseph Appleby, and a chemist,

Arthur John Banks of Lancaster County, filed a patent[3] entitled "Improvements in or relating to the treatment of foodstuffs, more especially cereals and their products" (Appleby and Banks 1906). The theme of the patent was:

> The invention consists essentially in exposing the substance to be treated to the action of the emanation from radium or radio-active substances or to the action of the α (alpha) β (beta) or γ (gamma) rays derived from radium or the radio-active substances.
>
> Treatment by exposure, under proper conditions to such rays, very materially reduces the time required to mature cereals and their products and food stuffs amenable to such action.

They planned to place the substance to be treated in a stream of air passing through a container containing radium bromide. The most frequently cited example of a possible candidate is flour. The drawing of the irradiation chamber is very rudimentary, which is not the case for the patent by David Collins Gillett (Tampa, Florida), in which the device consisting of a set of X-ray generators is described in detail (Figure 12.3) (Gillett 1918).

In 1926, Coolidge of the General Electric Company's research laboratory developed a tube capable of producing several milliamperes of cathode rays at 350,000 V (Coolidge 1926). In a second article, he and Moore described the effects of this radiation on a number of materials and organic substances. They indicated that milk and butter rapidly acquires a rancid taste and a strong odor, but gave no indication of dose (Coolidge and Moore 1926).

Two decades later, Huber (1948) of Electronized Chemicals Corporation demonstrated that, when exposed in sealed containers to cathode ray pulses of very short duration, many varieties of food could be preserved for long periods without alteration of taste or appearance. In the United States in the late 1950s, food irradiation experiments used spent fuel rods as the source of irradiation. These were abandoned because of dosimetry problems, in favor of cobalt-60 sources. After supporting research in this field, the US Army concentrated its efforts on radiation-sterilized meat products. In Europe, national research programs were launched in the United Kingdom, Belgium, France, the Netherlands, Poland, the Soviet Union and the Federal Republic of Germany.

However, it was not until ionization sources were able to deliver much higher particle fluxes that food preservation by ionization was put into industrial use. It was

3 Accepted January 4, 1906.

not until 1957, in Stuttgart, Germany, that the first commercial application saw the light of day. In that year, an industrialist began irradiating spices with a Van de Graaff electron accelerator, with the aim of improving their hygienic quality. But in 1959, a food law prohibited ionizing radiation treatment. The plant was dismantled and returned to ethylene oxide fumigation[4]. In Canada, in 1965, a plant equipped with cobalt-60 sources began treating potatoes to inhibit sprouting. It closed after just one season due to financial difficulties.

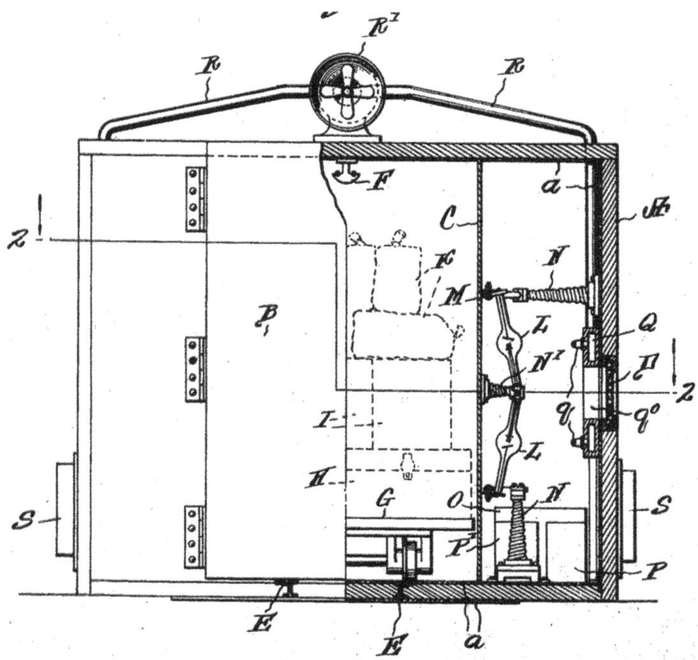

Figure 12.3. *Half-section view of D.A. Gillett's apparatus for preserving organic materials with X-rays. The material to be treated is on cart G, and the X-ray generators on L (Gillett 1918)*

At the time of the first International Symposium of Food Irradiation organized in 1966 in Karlsruhe, Germany, by the International Atomic Energy Agency (IAEA), only three countries (Canada, the United States and the Soviet Union) had given the green light to the consumption of five irradiated foods. By 1989, according to the World Health Organization, 33 countries had authorized ionizing food treatments.

4 Ethylene oxide, with the formula C_2H_4O, is a highly flammable cyclic ether, toxic to living organisms and carcinogenic. The European Commission banned it in food and food additives only on August 11, 2022.

A detailed history of food irradiation can be found in Josephson (1983) and Diehl (1995, 2002).

12.3. Polymer irradiation applications

In 1967, Charlesby (1967) published an article entitled "Industrial polymer irradiation" in the *Journal of the British Nuclear Energy Society*, in which he explained why the current economic climate is favorable to polymer irradiation applications. The main advantage lies in the control of parameters, which is easier with irradiation, and in the ability to initiate solid-phase reactions in conditions where chemistry is ineffective. However, despite the increasing sophistication of irradiation facilities, the cost remains higher than that of chemical processes, unless there is no reasonable alternative. Economic efficiency is increased if the product has a high intrinsic value, such as sterilized pharmaceuticals or reactions with very high radiolytic yields.

Around 1960, the General Electric Company was the first to market an irradiated polymer in the form of rolls of electrical insulators made from irradiated polyethylene. Its trade name was Irrathene. This product is now manufactured by Von Roll in the United States, and marketed under the brand names "Thick flexible film in heat-shrink polyethylene Irrathene 201, 202 and 210". Around the same time, W.R. GRACE and Company marketed their Cryovac Type L heat-shrink film for meat and poultry protection. Their "Irradiation treatment of polyethylene" patent is dated October 1958 (Rainer et al. 1958).

In 1973, the Western Electric Company began manufacturing radio-crosslinked telephone wire, initially at a rate of 2,700 km/year and then 8,000 km/year. Better electrical characteristics and improved resistance to fire, abrasion and solvents made it possible to simplify the wire's manufacture and therefore its size, which is particularly important for its use in telephone exchanges.

Irradiation cross-linking and polymer chain scission are not the only industrial uses of radiation. The grafting of polymerizable monomers can, after irradiation, form side chains grafted onto the main chain. For example, grafting acrylic acid onto polyethylene turns this material into an excellent adhesive on metals. This product was marketed by Total-Chimie (Chapiro 1988).

Degradation of polymers by irradiation has not been exploited for reasons of cost and the tonnages to be processed. The aim would be to facilitate the recycling of polymeric waste. For example, after degradation, polytetrafluoroethylene or Teflon can be incorporated into lubricants for heat engines.

12.4. Semiconductor doping

The implantation of dopants in semiconductors is certainly the industrial activity that mobilizes the largest number of accelerators. The IAEA estimates that there are 12,000 implanters worldwide. Implantation doping makes it possible to control the location of the dopants. Transmutation doping, on the other hand, enables the entire silicon ingot to be homogeneously doped. Doping is achieved by neutron capture as follows: $^{30}Si + n \rightarrow {}^{31}Si \rightarrow {}^{31}P + \beta$.

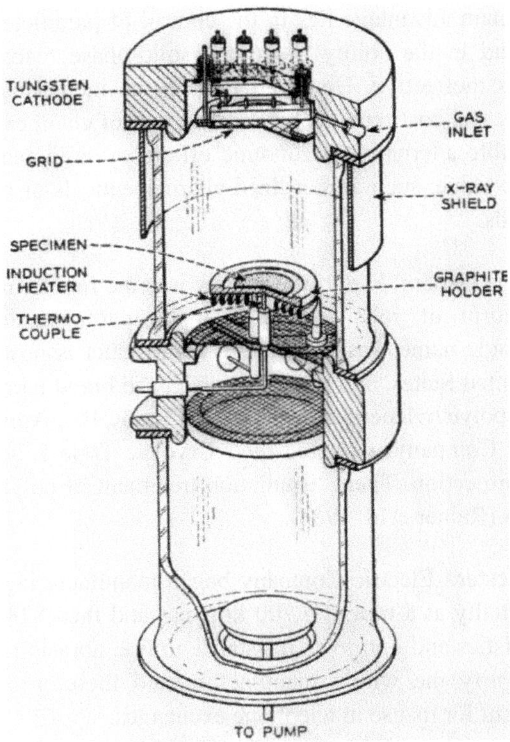

Figure 12.4. *Irradiation system developed at Bell Telephone Laboratories. The gas injected into the cell (gas inlet) is ionized by the electrons emitted by the tungsten cathode. The silicon sample is at negative potential. Its maximum diameter is 1.5 inches. Ions are accelerated between the grid and the sample holder (from 100 eV to 30 keV). The temperature can be adjusted from 20°C to 400°C (Ohl 1952)*

We saw in Chapter 7 that the first irradiation of a semiconductor dates back to 1950 (Ohl 1952). The author of this chapter demonstrates that the electrical properties of the silicon surface can be controlled by bombardment with positive ions.

12.4.1. *Doping by implantation*

Of course, it was not until 1947, when William Shockley, John Bardeen and Walter Brattain at Bell Telephone Laboratories made the first junction, that semiconductor irradiation was applied. They were awarded the Nobel Prize for Physics in 1956. As early as 1949, on April 27, 1949 to be exact, Shockley filed a patent application entitled "semiconductor translating device" (accepted on January 19, 1954), in which the realization of p- or n-type conductivity zones is produced by bombardment with nuclear particles in a semiconductor material of the opposite conductivity type (Shockley 1949). Figure 12.5 shows an example of the creation of an n-type zone within a p-type zone.

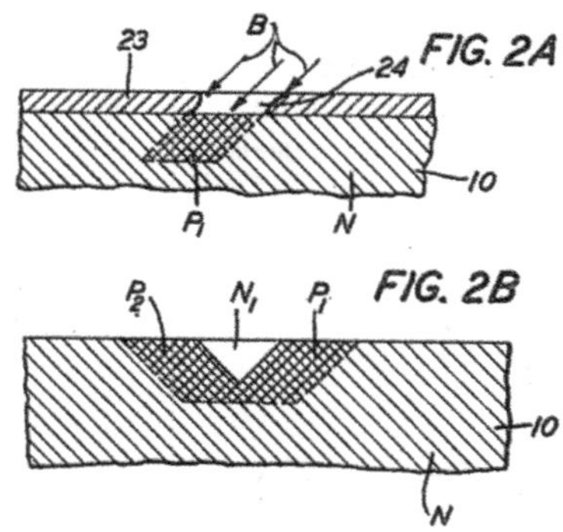

Figure 12.5. *Schematic representation of a p-n junction in a semiconductor (Shockley 1949)*

In the fall of 1957, six printed circuit board manufacturers got together to form the Institute of Printing Circuits (IPC), with the aim of creating industry standards and supporting industry progress. In 1958, the Institute published the book *How to Design and Specify Printed Circuits*, which went on to sell over 25,000 copies.

The International Conference on Electromagnetic Isotope Separators, held in Aarhus, Denmark, in 1965, was the first conference at which ion implanters were recognized as tools for device manufacture[5]. Three papers dealt with the applications of implantations. The first application, presented by Stanford and Aarhus

5 It was also at this conference that the channeling process was discussed for the first time.

Universities, used the Aarhus accelerator and isotope separator to implant thulium and neodymium in silicon, AsGa, CdS and ZnS. Preliminary results of resistivity, Hall effect, mobility and photoluminescence were given (Gibbons et al. 1965). Industries were involved in the other two applications, Sprague Electric Company for junction production by implantation of phosphorus or iodine in silicon (Manchester et al. 1965) and Ion Physics Corporation (IPC), which described processes developed to manufacture semiconductor devices by implantation (King et al. 1965). In four and a half years, the following devices were produced on the basis of p-n junctions: solar cells, radiation detectors, unipolar field-effect transistors and bipolar transistors. Preliminary results on wide bandgap materials were presented.

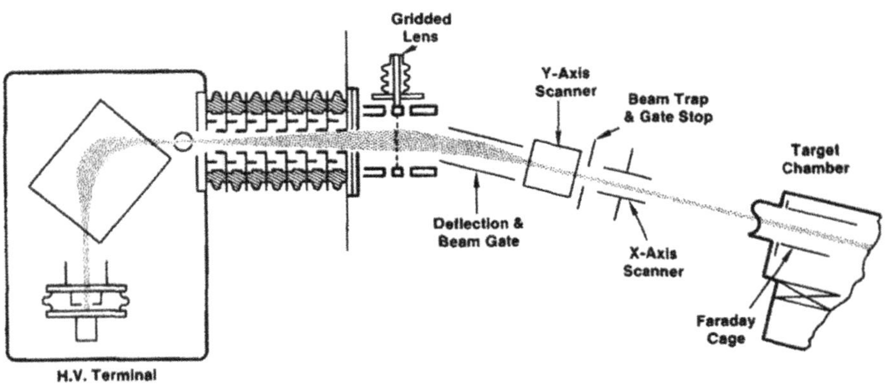

Figure 12.6. *Optical devices of the first 150 keV Extrion implanters (after Rose 1985)*

The main advantage of this technology is the ability to control and reproduce the number of atoms of dopant impurities introduced into the semiconductor, and to control the depth and spatial location of these impurities. The rapid development of this industry has led to a growing demand for ion implanters.

The first company to offer these accelerators was High Voltage Engineering Corporation (HVEC) in the United States. HVEC was founded in 1947. Among its founding members was Van de Graaff. This company had several subsidiaries, including IPC, created in 1970, and spun off several start-ups, such as General Ionex in 1971, which was finally acquired by Varian SEA in 1998. Also in the United States, Nova Corporation began manufacturing high-current implanters in 1978. Eaton Corporation first bought Accelerators Incorporated in 1978, and then Nova Corporation in 1980, becoming the first company to offer a complete catalog of medium- and high-current machines.

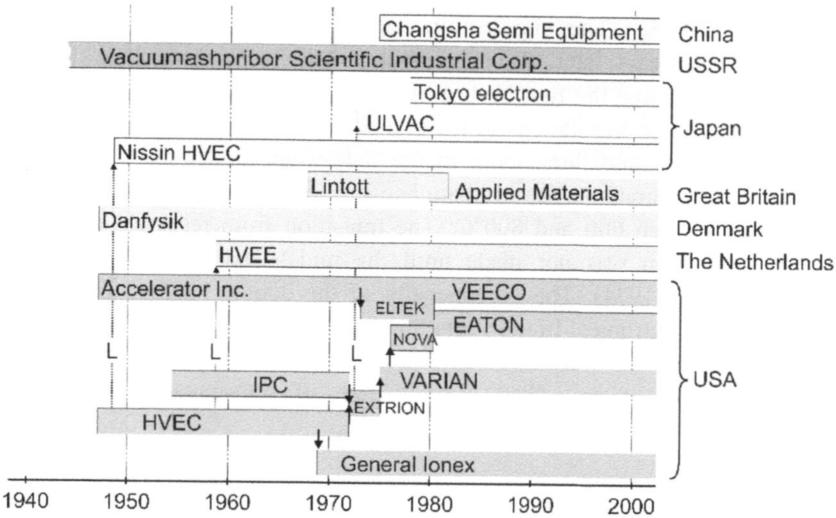

Figure 12.7. *Leading implanter manufacturers. Arrows marked with an L indicate the start of licensed activity. This figure is based on data from Rose (1985) and Yarling (2000)*

High Voltage Engineering Europa (HVEE) is Europe's oldest accelerator manufacturer. It began operations in the Netherlands in 1959, under license from HVEC for accelerator technology. In the same year, Danfysik (in Denmark) began industrial production of implanters. In the UK, Lintott Ltd was acquired by Applied Materials in 1980, and in France, Thomson-CSF manufactured implanters based on technology developed by the Bernas laboratory at Orsay. In Japan, three manufacturers started out under American license. In chronological order, these were Nissin Electric Company Ltd in 1949, Ulvac under license from Varian Associates in 1978 and Tokyo Electron Ltd in 1978. Data are less clear for China and Russia, but it seems that the implantation activity began very early in the USSR, probably before 1950. Most of this information comes from (Yarling 2000).

The first technological developments in ion implantation for transistor realization were the subject of a paper in a conference of the Institute of Electrical and Electronics Engineers (Fair 1998).

12.4.2. *Transmutation doping*

The possibility of doping silicon with phosphorus through transmutation of ^{30}Si to ^{31}P by neutron capture followed by β decay of ^{31}Si was first proposed by

Lark-Horovitz[6] in 1951. The long neutron range ensured unequalled homogeneity of the doping. Ten years later, Tanenbaum and Mills (1961) of Bell Telephone Laboratories carried out the first neutron-induced transmutation doping experiment. They performed three irradiations in the graphite-moderated reactor at Brookhaven National Laboratory and three more in the light-water-moderated reactor at Oak Ridge National Laboratory. The irradiations were followed by annealing at various temperatures between 600 and 800°C. The transition from research laboratory to industrial production was not made until the mid-1970s by SIEMENS AG in Munich (Schnöller 1974). The homogeneity of the doping resulted in significant gains for power electronics. In the years that followed, many reactors opened up to this activity, which has been the subject of an IAEA report including a list of reactors that have or have had a production of transmutation-doped silicon (IAEA 2012).

12.5. Radiation resistance of electronic components

The radiation resistance of electronic components is not really an application of irradiation. It does, however, use a significant amount of beam time to understand the processes leading to component failure, or simply to test these components. In fact, electronic components are subjected to natural radiation (cosmic or telluric radiation) and can be subjected to artificial radiation from particle accelerators, nuclear reactors or nuclear weapons. All types of radiation can temporarily disrupt them by modifying a bit, or damage them permanently by creating a short-circuit, for example. Satellites not protected by the Earth's magnetic field and atmosphere are particularly vulnerable. Perhaps the first failure attributable to cosmic rays was that of Telstar 1 in November 1962, four months after it was placed in orbit.

Space applications were therefore the driving force behind the development of research into radiation sensitivity in the 1960s, with particular attention paid to the choice of electronic systems used for the Apollo program. Military applications also played a crucial role in the development of early research into radiation resistance. Electronic communication, navigation and control systems for missiles and aircraft had to be capable of operating in hostile environments.

Growing understanding of the effects of radiation led, in the 1970s, to the formulation of standards and specifications for electronic components used in radiative environments. But the evolution of technologies, notably the miniaturization of microelectronics increasing sensitivity to radiation, requires

6 Lark-Horowitz was then Head of the Physics Department at Purdue University in West Lafayette, Indiana.

continuous adaptation of test methods and design strategies to maintain reliability in the face of radiation.

From the 1990s onwards, electronic components specifically designed to withstand radiation were developed.

12.6. Ion track technology

Above a certain energy deposition threshold, the modifications induced by fast heavy ions are cylinders whose disorder can go as far as amorphization. Young (1958) first observed the impact of fission fragments on a material, in this case lithium fluoride. Chemical etching of the irradiated crystal revealed the presence of modified zones observable in optical microscopy. The first direct observation of tracks of fission fragments was made by transmission electron microscopy in muscovite mica at Harwell by Silk and Barnes (1959). To do this, they stacked alternating sheets of mica and aluminum onto which a micrometer layer of uranium had been deposited. These multilayers were then irradiated for a few minutes in Harwell's BEPO reactor to fission uranium atoms. With a resolution of around 50 Å (described as high resolution), they provided little information about the tracks, apart from whether they were continuous or discontinuous.

Price and Walker of the General Electric Research Laboratory soon observed tracks in natural micas. These tracks were produced over a very long period by the spontaneous fission of uranium. When the mica came into contact with a uranium mineral such as uraninite, from the density of the tracks, they deduced the length of time the mica had stored tracks, 10^6 years in the case of the mica in this study (Price and Walker 1962b). In another article, again in 1962, these two authors point out that tracks in synthetic mica are clear and stable, whereas tracks in natural micas fade during observation and are difficult to measure. This gradual degradation can be eliminated by chemically attacking the tracks with fluoridic acid. Tracks are thus irreversibly etched (Price and Walker 1962a). These were the first chemical attacks on tracks of fast heavy ions, a prerequisite for the many current and potential applications of the revealed tracks.

The first applications of ion tracks date back to the 1950s and 1960s. At the end of 1962, Fleisher joined the Walter-Price team, where he demonstrated that the track revelation procedure developed for mica could be used in a wide range of materials, including plastics such as polycarbonate (Fleischer and Price 1963). The oldest of these applications is certainly their use as particle detectors. Anecdotally, one helmet from the Apollo 8 mission and three from the Apollo 12 mission[7] were used as dosimeters for heavy particles. These helmets were made of Lexan polycarbonate, a polymer that records heavy ion tracks whose electronic stopping power exceeds that of neon at ~ 7 MeV/nucleon. As the flux is quasi-isotropic, it is given in tracks/(cm².s.sr). The highest value is 1.71×10^{-7} track/cm^2/s^1/sr^1. They deduced that the fraction of killed cells can reach 0.12% in the cerebral cortex (Comstock et al. 1971). A major advance in dosimetry came with the arrival of a standardized material: CR39 (allyl diglycol carbonate), proposed by Cartwright, Shirk and Price of the University of California, Berkeley. In their paper, they point out that the main qualities of CR39 are that it is commercially available, optically clear and amorphous. In addition, its response to irradiation is homogeneous and highly sensitive (Cartwright et al. 1978).

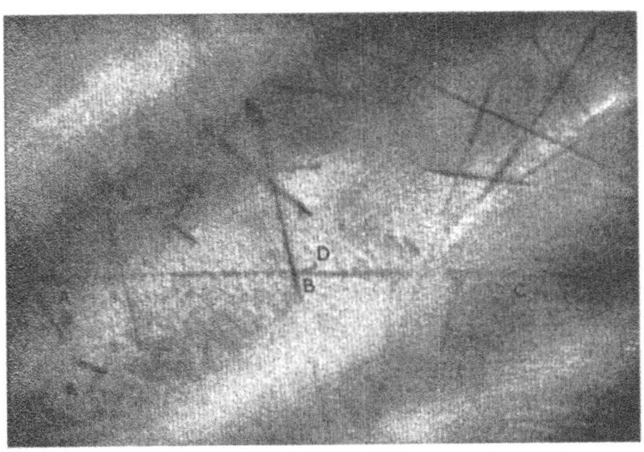

Figure 12.8. *Fission fragment tracks in mica (Silk and Barnes 1959)*

The idea of using the etched tracks to create highly selective filters matured quite rapidly. It was the subject of a patent filed on February 28, 1962 by Price and Walker on behalf of the General Electric Company (Price and Walker 1967b). The method used to produce the molecular sieve, using the term used in the patent title, is described. The pores were made by chemically etching the tracks left by the fission fragments of uranium-235, the fissions being induced by neutrons from a reactor located in Vallecibo, California. General Electric Company marketed its molecular sieve under the name *Nuclepore*. Production of the Nuclepore filter was entrusted to a new company, Nuclepore Corporation of Pleasaton. This company was subsequently sold to several investors, including Normura Oversea Enterprise

7 Apollo 8 spent 24 h in orbit around the moon, while two Apollo 12 astronauts spent 31 h on the moon. A helmet carried in a balloon at an altitude of 41,000 m was used as a reference.

(Japan). An important development was the switch from reactor irradiation to the use of sources of ^{252}Cf, an isotope that fissions spontaneously. On September 11, 1963, they filed another patent for a dosimeter based on etched track technology (Price and Walker 1967a).

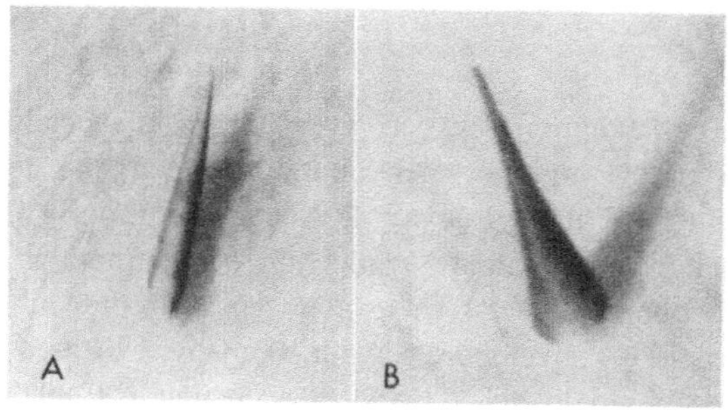

Figure 12.9. *Examples of cosmic ray tracks inside an Apollo space helmet. (A) Track of a particle entering the helmet. (B) Final track of a particle that had passed through the helmet and stopped. Tracks were 500 and 700 μm long, respectively (Comstock et al. 1971)*

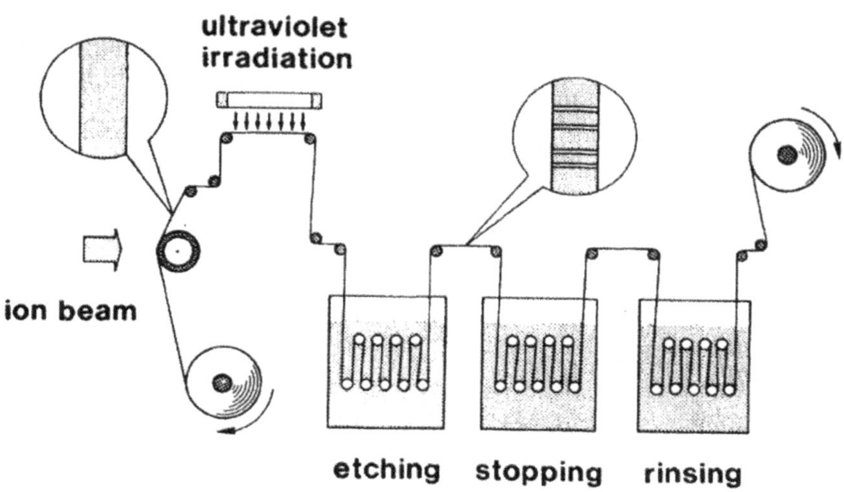

Figure 12.10. *Diagram of a heavy-ion nuclear track filter production line. A wide heavy-ion accelerator irradiates a film of polymer, which travels at a speed of several meters per second. An ultraviolet irradiation station enhances the contrast between the part to be revealed and the rest of the material. The foil then passes through the etching bath, before the stop bath stops the etching at the desired size (Fischer and Spohr 1983)*

DeBlois and Bean (1970), still from the same laboratory, developed a method for counting submicron particles and determining their size. Using an optical microscope, they sampled an area containing just a few pores, attached it to a disk and sealed all but one of the holes with epoxy glue. The particles driven by the electric field applied in the pore were characterized by the voltage pulse they generated.

In Europe, the collaboration between Yves Jongen, a specialist in ion sources, and Roger Legras of the high-polymer laboratory led to the development of an industrial activity in 1983 at the Louvain la Neuve cyclotron in Belgium. In 1985, they filed a French patent entitled "Production of perforations in a solid sheet material by particle bombardment". This was extended to the United States in 1990 (Legras and Jongen 1990). Cyclopore S.A., a company dedicated to manufacturing and marketing filters based on etched tracks, was created in 1989. Their particularity was to irradiate with relatively light ions (Ar), and to increase the contrast between the attack velocities of the material and the track, they applied a UV treatment. Today, the company it4ip (Ion track Technology 4 Innovative Products), founded in 2006, continues this activity.

In Russia, there is also a historic production of membranes at the Joint Institute for Nuclear Research, Dubna, with the IC-100 cyclotron from 1985, then with this renovated cyclotron capable of accelerating heavier ions.

In 2004–2006, the DC-60 cyclotron entered service at the Gumilev Eurasian National University Interdisciplinary Research Center (Astana, Kazakhstan), where it was used for solid-state physics experiments and the production of etched track membranes.

This industrial production has been supported by dynamic upstream research. Ion track technology is not limited to the production of pores with controlled diameters, down to one pore per membrane. The functionalization of pore surfaces also gives them chemical selectivity. The etched tracks can be used as matrices for the electrodeposition of nanowires.

Pioneers in the field are Fleisher, Price and Walker of the General Electric Company and, unranked, with the risk of forgetting important players: Reimar Spohr of Gesellschaft für Schwerionenforschung, Darmstad, Étienne Ferain of Université Catholique de Louvain, Ottignies-Louvain-la-Neuve and Natacha Betz of Commissariat à l'Énergie Atomique, Saclay.

12.7. Cultural and historical heritage materials

Particle beams are ideal for use in museums to authenticate the origin and date of historical and cultural heritage objects, determine the techniques and materials used in their manufacture and assess their state of preservation. The nondestructive nature[8] of particle beam analysis techniques is their main advantage, along with the ability to "see" into the thickness of the object. In addition to X-ray beams, ions are used for ion beam analysis (IBA). IBA covers a range of techniques: RBS (Rutherford backscattering spectroscopy), PIXE (particule-induced X-ray emission), PIGE (particule induced gamma-ray emission) and NRA (nuclear reaction analysis), to which we can add AMS[9] (accelearator mass spectrometry).

In June 2012, at the *Multidisciplinary Applications of Nuclear Physics with Ion Beams* conference, Jean-Claude Dran and Thomas Calligaro of C2RMF (Centre de Recherche et de Restauration des Musées de France), in their presentation, began the use of IBA techniques in the early 1970s (Dran and Calligaro 2013). In his short history, Alessandro Zucchiatti gives the same origin (Zucchiatti 2019). It was at the *Chemical Analysis by Charged Particle Bombardment* conference (Namur, Belgium, September 1971) that, for the first time, three papers were devoted to the analysis of cultural heritage objects.

Gordon and Kraner (1972) (Brookhaven National Laboratory) describe the development of a system for analyzing trace elements using X-ray fluorescence induced by high-energy protons or alphas (now known as PIXE). One application concerns the analysis of iron-based decorative pigments in archaeological pottery shards. In the same year, a team from the Institute of Nuclear Sciences (New Zealand) analyzed obsidian rocks of archeological interest using 3 MeV inelastic proton scattering (now called PIGE), with the aim of determining trade routes associated with pre-European Maori culture (Coote et al. 1972).

The possibility of Rutherford retrodiffusion spectroscopy was presented at the American Physical Society meeting in Stanford in December 1949 by Rubin and Rasmussen (1950) of the California Institute of Technology. But it was not until 1981 that an application of this technique to heritage objects was published.

---

8 With a few precautions to limit the creation of defects (e.g. color changes) induced by irradiation.

9 AMS is not really an IBA technique, but the techniques of ion acceleration and detection are similar. The difference is that the sample to be analyzed is the source of the ions.

The analysis, combining RBS, PIXE and NRA, was carried out on a 2,200-year-old arrowhead found in the mausoleum of the emperor Qin Shihuang in Shanxi province. RBS spectroscopy was performed with ^{4}He accelerated by a locally designed 3 MeV Van de Graff at Fudan University (Chen et al. 1981). The arrow was made of 82% copper, 17% tin and 0.5% lead, that is, bronze. This experiment was carried out in a vacuum, and it was not until two years later that Doyle (1983) of Sandia National Laboratory performed the first experiment with a beam extracted in air. The exit window consisted of a 10 μm-thick sheet of beryllium. This breakthrough was essential, as it made it possible to analyze objects that were bulky or could not be placed in vacuum. The first application with an extracted beam was a study of ink penetration in ancient manuscript parchment (Giuntini and Mandò 1994). This work, which preceded the study of the ink in the folios of Volume 72 of Galileo's *Manoscritti galileiani*, aimed to reestablish the chronology of this document (Del Carmine et al. 1996).

The first publication of an ion-beam analysis work in an archaeological journal *Archeometry* was published in 1976. The objects analyzed were cuneiform pendants with yellow areas suspected of containing gold (Ahlberg et al. 1976a, 1976b).

A far cry from IBA, but still involving heritage objects, irradiation can be used to disinfect these objects. This was the case in 1977 for the mummy of Ramses II, which was severely damaged by parasites.

Disinfection was carried out on May 9, 1977, using gamma radiation from ^{60}Co at the Nucléart irradiator in Grenoble, France, for 12 h and 40 min, reaching a dose of 18 kGy. Less famous mummies have also undergone the same treatment.

Old wood damaged by insects and/or too much traffic can be hardened by transforming it into a wood–polymer composite. For example, a parquet floor dating from 1787 in the Hôtel de Lesdiguières in Grenoble was impregnated with a monomer, methyl methacrylate, which was polymerized by gamma irradiation (20 kGy). Similar operations can be carried out on wood that has been soaked in water for a long time. After replacing the water in the wood with polyethylene glycol, the polymer can be cured by irradiation. This process was used to save a Gallo-Roman barge found in the port of Arles (Musée Départemental Arles Antique).

12.8. References

Ahlberg, M., Akselsson, R., Forkman, B. (1976a). Part II analysis of wedge-shaped pendants by proton-induced X-ray emission spectroscopy (PIXE). *Archaeometry*, 18(1), 42–49.

Ahlberg, M., Akselsson, R., Forkman, B., Rausing, G. (1976b). Gold traces on wedged-shaped artefacts from the late neolithic of Southern Scandinavia analysed by proton induced X-ray emission spectroscopy. *Archaeometry*, 18(1), 39–42.

Appleby, J. and Banks, A.J. (1906). Improvements in or relating to the treatment of foodstuffs, more especially cereals and their products. British Patent, GBD190501609 19050126

Cartwright, B.G., Shirk, E.K., Price, P.B. (1978). A nuclear-track-recording polymer of unique sensitivity and resolution. *Nuclear Instruments and Methods*, 153(2), 457–460.

Chapiro, A. (1988). L'utilisation des rayonnements dans l'industrie. *Les cahiers du MURS*, 11, 69–78.

Charlesby, A. (1967). Industrial polymer irradiation. *Journal of the British Nuclear Energy Society*, 6, 301–304.

Chen, H.-S., Chen, J.-X., Ren, C.-G., Xu, Z.-W., Yang, F.-C., Zhao, G.-Q., Zhuo, Z.-Y. (1981). Ion beam analysis in archaeology. *Nuclear Instruments and Methods in Physics Research*, 191(1), 391–396.

Christiansen, I.A., Hevesy, G., Lomholt, S. (1924a). Recherches, par une méthode radiochimique, sur la circulation du bismuth dans l'organisme. *Comptes rendus hebdomadaires des séances de l'Académie des sciences*, 178, 1324–1326.

Christiansen, I.A., Hevesy, G., Lomholt, S. (1924b). Recherches, par une méthode radiochimique, sur la circulation du plomb dans l'organisme. *Comptes rendus hebdomadaires des séances de l'Académie des sciences*, 179, 291–293.

Comstock, G.M., Fleischer, R.L., Giard, W.R., Hart, H.R., Nichols, G.E., Price, P.B. (1971). Cosmic-ray tracks in plastics: The Apollo helmet dosimetry experiment. *Science*, 172(3979), 154–157.

Coolidge, W.D. (1926). The production of high-voltage cathode rays outside of the generating tube. *Journal of the Franklin Institute*, 202(6), 693–721.

Coolidge, W.D. and Moore, C.N. (1926). Some experiments with high-voltage cathode rays outside of the generating tube. *Journal of the Franklin Institute*, 202(6), 722–736.

Coote, G.E., Whitehead, N.E., McCallum, G.J. (1972). A rapid method of obsidian characterisation by inelastic scattering of protons. *Journal of Radioanalytical Chemistry*, 12(1), 491–496.

DeBlois, R.W. and Bean, C.P. (1970). Counting and sizing of submicron particles by the resistive pulse technique. *Review of Scientific Instruments*, 41(7), 909–916.

Del Carmine, P., Giuntini, L., Hooper, W., Lucarelli, F., Mandò, P.A. (1996). Further results from PIXE analysis of inks in Galileo's notes on motion. *Nuclear Instruments and Methods in Physics Research Section B: Beam Interactions with Materials and Atoms*, 113(1), 354–358.

Diehl, J.F. (1995). How it all began. In *Safety of Irradiated Foods*, 2nd edition. Marcel Dekker Inc., New York.

Diehl, J.F. (2002). Food irradiation – Past, present and future. *Radiation Physics and Chemistry*, 63(3), 211–215.

Doyle, B.L. (1983). Non-vacuum Rutherford backscattering spectrometry. *Nuclear Instruments and Methods in Physics Research*, 218(1), 29–32.

Dran, J.-C. and Calligaro, T. (2013). Ion beam analysis in cultural heritage studies: Milestones and perspectives. In *Multidisciplinary Applications of Nuclear Physics with Ion Beams (Ion Beams'12) AIP Conference Proceedings*, Legnaro-Padova.

Fair, R.B. (1998). History of some early developments in ion-implantation technology leading to silicon transistor manufacturing. *Proceedings of the IEEE*, 86, 111–137.

Fischer, B.E. and Spohr, R. (1983). Production and use of nuclear tracks: Imprinting structure on solids. *Reviews of Modern Physics*, 55(4), 907–948.

Fleischer, R.L. and Price, P.B. (1963). Tracks of charged particles in high polymers. *Science*, 140(3572), 1221–1222.

Freund, L. (1904). *Element of General Radiotherapy for Pratitioners*. Rebman Company, New York.

Gibbons, J.F., Moll, J.L., Meyer, N.I. (1965). The doping of semiconductors by ion bombardment. *Nuclear Instruments and Methods*, 38, 165–168.

Gillett, D.C. (1918). Apparatus for preserving organic materials by the use of X-rays. US Patent, 1,275,417A.

Giuntini, L. and Mandò, P.A. (1994). External beam RBS in an unenclosed helium environment. *Nuclear Instruments and Methods in Physics Research Section B: Beam Interactions with Materials and Atoms*, 85(1), 744–748.

Gordon, B.M. and Kraner, H.W. (1972). On the development of a system for trace element analysis in the environment by charged particle X-ray fluorescence. *Journal of Radioanalytical Chemistry*, 12(2), 181–188.

Grubbé, E.H. (1933). Priority in the therapeutic use of X-rays. *Radiology*, 21(2), 156–162.

Hevesy, G. (1923). The absorption and translocation of lead by plants: A contribution to the application of the method of radioactive indicators in the investigation of the change of substance in plants. *Biochemical Journal*, 17(4–5), 439–445.

Huber, W. (1948). Electronic preservation of food. *Electronics*, 21, 74–79.

IAEA (2012). Neutron transmutation doping of silicon at research reactors. Report, IAEA-TECDOC-1681, International Atomic Energy Agency, Vienna.

Josephson, E.S. (1983). An historical review of food irradiation. *Journal of Food Safety*, 5(4), 161–189.

King, W.J., Burrill, J.T., Harrison, S., Martin, F., Kellett, C. (1965). Experience in fabricating semiconductor devices using ion implantation techniques. *Nuclear Instruments & Methods*, 38, 178–179.

Lark-Horowitz, K. (1951). Nucleon bombarded semiconductors. In *Reading Conference on Semiconductor Materials*, Butterworth, London.

Lawrence, J.H. (1940). Nuclear physics and therapy: Preliminary report on a new method for the treatment of leukemia and polycythemia. *Radiology*, 35(1), 51–60.

Lefebvre, T. and Raynal, C. (2011). Du radium dans les pharmacies ! Première partie : les usages pharmaceutiques du radium avant la Première Guerre mondiale. *Revue d'histoire de la pharmacie*, 431–446.

Legras, R. and Jongen, Y. (1990). Method of producing perforations in a solid materials in sheet form, an irradiation device for working the method, and the resulting perforated materials. US Patent, 4 956 219.

MacKee, G.M. (1921). *X-rays and Radium in the Treatment of Diseases of the Skin*. Lea & Febiger, Philadelphia.

Manchester, K.E., Sibley, C.B., Alton, G. (1965). Doping of silicon by ion implantation. *Nuclear Instruments and Methods*, 38, 169–174.

Minck, F. (1896a). Zur Frage Uber die Einwirkung der Röntgen'schen Strahlen auf Bacterien und ihre eventuelle therapeutische Verwendbarkeit. *Münchener Medicinische Wochenschrift*, 43(5), 101–102.

Minck, F. (1896b). Zur Frage Uber die Einflusswirkung der Röntgen'schen Strahlen auf Bacterien. *Münchener Medicinische Wochenschrift*, 43(9), 202.

Ohl, R.S. (1952). Properties of ionic bombarded silicon. *The Bell System Technical Journal*, 31(1), 104–121.

Prescott, S.C. (1904). The effect of radium rays on the colon bacillus, the diphtheria bacillus and yeast. *Science*, 20(503), 246–248.

Price, P.B. and Walker, R.M. (1962a). Electron microscope observation of etched tracks from spallation recoils in mica. *Physical Review Letters*, 8(5), 217–219.

Price, P.B. and Walker, R.M. (1962b). Observation of fossil particle tracks in natural micas. *Nature*, 196, 732.

Price, P.B. and Walker, R.M. (1967a). High level radiation dosimeter having a sheet which is permeable to damage track producing particles. US Patent, 3 335 278.

Price, P.B. and Walker, R.M. (1967b). Molecular sieves and method for producing same. US Patent, 3 303 085.

Rainer, W.C., Redding, E.M., Hitov, J.J., Sloan, A.W., Stewart, W.D. (1958). Irradiation treatment of polyethylene. US Patent, 2,855,517.

Raynal, C. and Lefebvre, T. (2012). Du radium dans les pharmacies ! Seconde partie : les usages pharmaceutiques du radium entre les deux guerres. *Revue d'histoire de la pharmacie*, 373, 73–86.

Rose, P.H. (1985). A history of commercial implantation. *Nuclear Instruments and Methods in Physics Research Section B: Beam Interactions with Materials and Atoms*, 6(1), 1–8.

Rubin, S. and Rasmussen, W.K. (1950). Chemical analysis by proton scattering. *Physical Review*, 78, 83.

Schnöller, M. (1974). Breakdown behavior of rectifiers and thyristors made from striation-free silicon. *IEEE Transactions on Electron Devices*, 21(5), 313–314.

Shockley, W. (1949). Semiconductor translating device. US Patent, 2 666 814.

Silk, E.C.H. and Barnes, R.S. (1959). Examination of fission fragment tracks with an electron microscope. *Philosophical Magazine*, 4, 970–972.

Silverman, J. (1975). The applications of radiation technology in industrial processes: Current and future perspectives. *What Can Radioisotopes Do for Man?*, 17–1, 15–22 [Online]. Available at: https://www.iaea.org/sites/default/files/17105391522.pdf.

Tanenbaum, M. and Mills, A.D. (1961). Preparation of uniform resistivity n-type silicon by nuclear transmutation. *Journal of the Electrochemical Society*, 108(2), 171.

Yarling, C.B. (2000). History of industrial and commercial ion implantation 1906–1978. *Journal of Vacuum Science & Technology A*, 18(4), 1746–1750.

Young, D.A. (1958). Etching of radiation damage in lithium fluoride. *Nature*, 182(4632), 375–377.

Zucchiatti, A. (2019). Ion beam analysis for the study of our cultural heritage. A short history and its milestones. *Nuclear Instruments and Methods in Physics Research Section B: Beam Interactions with Materials and Atoms*, 452, 48–54.

Conclusions

C.1. An active community

Over time, knowledge of materials under irradiation has become increasingly precise. It is therefore legitimate to question the relevance of continuing to invest in research into materials under irradiation. An initial answer can be provided by estimating the dynamism of this activity. The evolution of the proportion of publications on irradiation in relation to the total number of publications can be a criterion of the vitality of this field. As a first approximation, the number of publications on the irradiation of materials can be obtained by searching with the keywords "irradiation AND material AND (ion OR electron OR neutron)". Several sources indicate the total annual number of publications. These include the *Ministère de l'Enseignement Supérieur et de la Recherche* (French Ministry of Higher Education and Research), the *Haut Conseil de l'Evaluation de la Recherche et de l'Enseignement Supérieur* (French High Council for the Evaluation of Research and Higher Education), UNESCO and Web of Sciences. Figure C.1 shows that over the past 30 years, the number of publications on irradiation has increased by 7% per year. This growth shows that the irradiation field is still active. This increase is partly due to the arrival of new players such as China and India, and the evolution of research toward industrial materials and irradiation conditions close to those in use. The number of articles on the irradiation of materials with a Chinese address has risen from one Chinese article for every one from the United States before the 2000s, to 30 times more Chinese articles than American in 2023 (Figure C.2).

Over the past 20 years, new experimental and theoretical approaches have enabled us to better describe certain phenomena and provide new data on defects. For example, in cubic-centered iron, interstitials were systematically described as having a dumbbell structure. A team from the Physical Metallurgy Research Department at CEA Saclay have discovered that, within a displacement cascade, interstitials can organize directly into nanometric crystalline precipitates (Laves C15 phase) (Marinica et al. 2012).

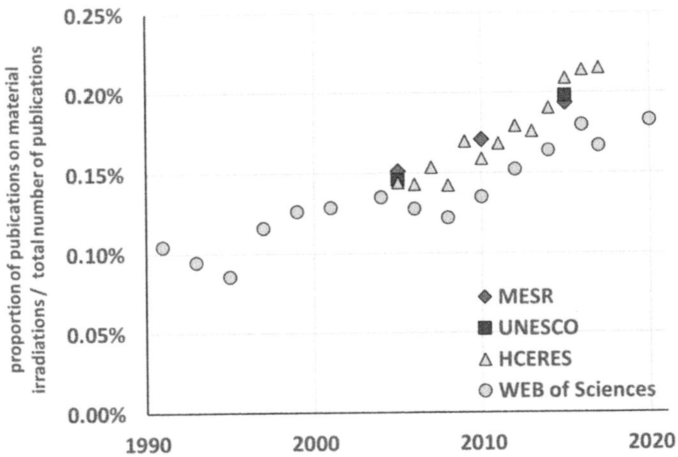

Figure C.1. *Number of publications with the terms "irradiation AND material AND (ion OR electron OR neutron)" (source Web of Sciences) normalized to the total number of publications from four sources (MESR, UNESCO, HCERES and Web of Sciences). For a color version of this figure, see www.iste.co.uk/bouffard/irradiation.zip*

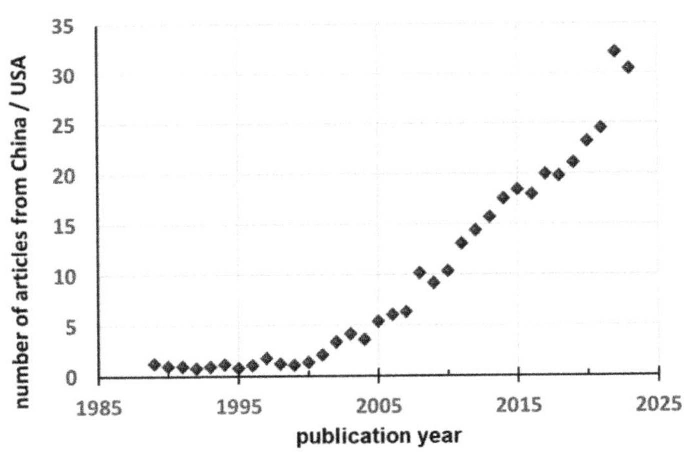

Figure C.2. *Number of publications containing the terms "irradiation AND material AND (ion OR electron OR neutron)" with an address in China normalized by the same number with an address in the United States (source Web of Sciences)*

The development of a new instrument often opens up new possibilities for study. For example, atom probe tomography has made it possible to "see" Cottrell atmospheres (Blavette et al. 1999). The concept of Cottrell atmospheres was introduced in 1949 to explain the role of impurities in the plastic deformation of metals (Cottrell and Bilby 1949).

Point or extended defects introduced into a material in a controlled manner by irradiation are part of the toolbox available to solid-state physicists for understanding the properties of solids. Defects have been, for example, used as a probe to study the superconducting order parameter in iron-based superconductors (Cho et al. 2018).

C.2. Future prospects

Although the main mechanisms governing the behavior of materials under irradiation are now known, there are still many points that have not been fully clarified.

– Basic knowledge of defects in ion-covalent materials is still patchy, except for nuclear fuel. Displacement thresholds are often determined only for the species easiest to displace, with considerable uncertainty; the configurations of elementary defects and small clusters are known for only a few materials; migration mechanisms under and outside irradiation and diffusion coefficients have rarely been studied; the same applies to defect-impurity interactions.

– The radiolysis of polymers is known in much less detail than the radiolysis of water. Adapting pulsed radiolysis experiments to the specific conditions of polymers would enable us to better describe the mechanisms of solid-phase radiolysis. This may require working on model polymers with optical absorption adapted to the experiment.

– Elastic collisions are decoupled from electronic excitations. In particular, the interaction potential is the same before and after the minimum approach distance, whereas there is a high probability that the charge states of the target atom and projectile have changed during the collision. It would be interesting to evaluate the effect of this elastic–inelastic coupling.

– The temperature rise in a displacement cascade is generally attributed to the numerous sub-threshold shocks, but in high-energy cascades, a fraction of the energy is dissipated in the form of ionizations and electronic excitations. The fate of this energy is generally not taken into account. In insulators, this energy contributes to a local rise in temperature. Can this brief rise in temperature promote the diffusion of species and modify the initial configuration of defects?

– The mechanism of track formation by intense electronic excitation is still described phenomenologically. The most widely used model is the thermal spike model, which lends itself to good agreement between the calculation of a molten zone and the observation of a transformed zone. Its coupling with a molecular dynamics simulation is a further step toward understanding, even if there is still much room for improvement.

– Industrial materials are generally much more complex than those used for basic mechanism studies. Is this separation of parameters relevant? The same question remains for the dose rate increase used to simulate material aging in the fuel cycle: simulating 60 years of reactor vessel ageing (flux × ~750,000) or 500 years of nuclear glass ageing (flux × ~3 × 10^6) in a few days of irradiation.

These examples show that, beyond the necessary studies of real materials subjected to irradiation, many questions remain as to the mechanisms governing irradiation-induced changes. We can even dream of the ideal equipment for studying the development of heavy ion tracks and, why not, a cascade of displacements: an accelerator delivering sub-femtosecond pulses synchronized with laser pulses of the same duration, or even shorter, whose absorption or fluorescence is sensitive to the presence of disorder. Laser acceleration may one day make this dream come true.

Research into materials under irradiation has benefited from strong demand from industry, of course from the nuclear energy sector, but also from the electronics and, to a lesser extent, medical sectors. It has also benefited from the development of nearby research fields. These include, for example, nuclear physics, which has provided the means for irradiation, and the physical chemistry of materials, for which it uses numerous analytical tools (X-ray and neutron diffraction, electron microscopy, optical spectroscopy, etc.).

– Is the environment conducive to ambitious projects?

For a number of years now, risk-taking in the experimental approach has been on the decline. Exaggerating slightly, it could be said that in the past, the researcher designed the device that would contribute to solving their problem. Today, many of us choose a subject in relation to the performance of the recently acquired device. In France, this phenomenon is exacerbated by calls for projects, which favor investment over operations, and remove around 10–15% of research staff to respond to calls for projects and, if successful, to manage them and justify their execution. This loss in person–years is accentuated by the change in the ratio of technical staff to researchers, which has fallen sharply.

Let us hope that all the parties involved in research will soon realise that scientific production would be more effective if it were based on a contract of trust between laboratories and funders, setting out objectives and the associated budget. The evaluation at the end of the contract should focus mainly on the scientific aspect.

C.3. References

Blavette, D., Cadel, E., Fraczkiewicz, A., Menand, A. (1999). Three-dimensional atomic-scale imaging of impurity segregation to line defects. *Science*, 286(5448), 2317–2319.

Cho, K., Konczykowski, M., Teknowijoyo, S., Tanatar, M.A., Prozorov, R. (2018). Using electron irradiation to probe iron-based superconductors. *Superconductor Science and Technology*, 31(6), 064002.

Cottrell, A.H. and Bilby, B.A. (1949). Dislocation theory of yielding and strain ageing of iron. *Proceedings of the Physical Society, Section A*, 62(1), 49.

Marinica, M.C., Willaime, F., Crocombette, J.P. (2012). Irradiation-induced formation of nanocrystallites with C15 laves phase structure in bcc iron. *Physical Review Letters*, 108(2), 025501.

Index

A, B

Allen, 227–230, 234, 238
alloy(s), 6, 98, 118, 127, 128, 139, 146, 147, 149–151, 153, 202
amorphous, 6, 11–14, 76, 95, 183, 184, 284
annealing, 14, 117, 142–144, 147, 156, 157, 159–167, 187, 188, 198, 282
Argonne, 78, 123, 153, 184, 202, 212, 231–233, 238, 239, 254, 256, 258
atomic pile, 214, 215, 257, 261
Becquerel, 3, 29–33, 36, 38, 47, 50, 54, 55, 79, 91, 94, 147, 207, 209, 224, 273
Bernoulli, 10, 24, 25
Bethe, 61–63, 68, 69, 79, 81
Biersack, 68, 69, 131
Bohr, 57, 60–62, 64, 66, 72, 79, 81, 115, 117, 167, 238
Bourgoin, 120, 121, 188
Bragg, 4, 34, 51–53, 55, 57, 79, 80, 158, 196
Brookhaven, 4, 107, 120, 122, 125, 217, 238, 255, 282, 287
Burton, 107, 108, 216, 228, 230, 237, 238

C, D

cascade(s), 117–119, 125, 130, 153, 154, 184, 249, 293, 295, 296
cathode, 15, 25, 26, 28, 29, 31–33, 38, 40, 57, 58, 79, 92, 93, 109, 128, 129, 177, 207–209, 243, 255, 273, 275, 278
Chapiro, 210, 216, 219, 277
Charlesby, 213, 214, 216, 217, 219, 277
Chicago, 4, 179, 228, 229, 244, 253, 256, 261
Cockroft, 19, 184, 249
cold working, 96, 139, 141–144, 160
copper, 6, 17, 72, 74–76, 96, 117, 120, 122–125, 130, 141–144, 147–149, 151–154, 158, 160–165, 167, 288

Corbett, 120, 122, 151, 162–165, 167, 185, 188, 190, 191
Crawford, 178, 180
Cronstedt, 8–10
Crookes, 4, 26, 27, 30, 35, 38, 40, 243
cross-linking, 90, 208, 209, 211, 213–216, 218, 219, 277
Cu_3Au, 127, 128, 146, 149–151, 154
Curie, 4, 32, 36, 50, 94, 101, 223–226, 255, 261, 267, 273
cyclotron, 64, 69, 147, 160, 176, 244–248, 261, 262, 273, 286
Davies, 71, 74, 75, 78
Debierne, 36, 224, 226, 227
defect(s), 1, 6, 13, 31, 78, 89, 92, 95, 96, 98, 100, 102, 107, 109, 114, 117, 119–121, 123–125, 132, 133, 139–141, 143, 145–148, 151–160, 162, 163, 165–168, 175, 180, 181, 183–190, 196–202, 213, 216, 243, 287, 293, 295
degradation, 210, 211, 213, 214, 216, 217, 269, 277, 283
diffusion, 26, 57, 64, 96–98, 126, 139–143, 149, 153, 157, 160, 164, 187, 188, 217, 234, 235, 250, 273, 295
  retro, 287
dislocation, 110, 111, 139, 143–145, 167, 168, 187, 199, 201
disorder, 4–6, 12, 14, 76, 89, 96, 115, 118, 126–128, 149–151, 153, 180, 183, 195, 283, 296
displacement, 69, 73, 93, 95, 99, 109–111, 114–123, 125, 127, 128, 130, 143, 151–153, 163, 179, 181, 183, 184, 187, 200, 201, 293, 295, 296
Dole, 214, 215, 219

E, F

electron(s), 4, 19, 24, 26, 28, 29, 32, 38, 51, 57, 61–66, 69, 79–81, 92, 94, 110–113, 115, 117, 120–123, 127, 129, 130, 139, 143, 151–153, 155, 157–159, 161–164, 168, 176, 178–187, 190, 198–201, 209, 210, 212, 216, 218, 228–236, 243–245, 252–255, 260, 262, 269, 273, 276, 278, 283, 293, 294, 296
Faraday, 17, 25, 38, 175
Fermi, 62, 63, 65, 67, 71, 115, 256, 261, 262
Flory, 211, 218, 219
Frenkel, 4, 97–100, 102, 103, 119, 124, 125, 128, 141, 153, 159, 166
Fricke, 227, 238

G, H

gadolinite, 12–14, 38
Geiger, 56–60, 79
Geissler, 25, 26, 38, 40
germanium, 74, 116, 120–122, 157, 176, 178, 180, 181, 183, 184, 186
gold, 6, 40, 58, 60, 76, 77, 96, 122, 130, 141, 144, 147, 149, 153, 161, 167, 288
Goldstein, 4, 26, 28, 31, 38, 40, 92–94, 109
Guericke, 16, 18, 19, 38, 243
Harwell, 4, 76, 110, 114, 117, 118, 130, 133, 134, 151, 199, 201, 202, 213, 214, 219, 256, 261, 283
Hauksbee, 15–17, 21, 22, 25, 38, 39
Huntington, 98, 120–122, 133, 142, 167

I, J

implantation, 69, 78, 175–177, 188, 189, 278–281
interstitial, 98–100, 111–113, 115, 119–121, 125, 126, 128, 133, 141, 142, 144, 147, 149, 153, 154, 156–166, 181, 185, 186, 188, 200, 202, 293
ionization, 47–52, 54, 56, 57, 66, 79, 80, 109–113, 115, 117, 188, 196, 208, 210, 215, 216, 219, 225, 234–238, 275, 295
Joly, 34, 196

K, L

Kinchin, 1, 117–119, 133, 154, 257
Koehler, 115, 118, 127, 146, 162, 168, 184
Lawrence, 51, 64, 80, 239, 246–248, 261, 262, 273
linac, 231, 244, 248, 254, 255
Lind, 57, 94, 100, 101, 195, 208, 209, 219, 225
Lindhard, 62, 63, 66, 67, 69, 79, 81, 115, 119
Los Alamos, 70, 71, 239, 252, 256, 261

M, N

Magee, 230, 234, 235, 239
medical, 37, 47, 216, 227, 252, 267–269, 271, 273, 296
  bio, 133, 247, 274
metamict, 14, 95, 195, 197, 202
mica, 34, 57, 131, 196, 199, 207, 283, 284
microscope, 59, 123, 129, 155–157, 168, 186, 199–201, 260, 286
Monte Carlo, 71, 72, 79, 117, 118, 125, 235, 236
neutron(s), 70, 71, 107, 115, 116, 127, 130, 131, 139, 140, 146, 149–151, 153, 154, 157, 176, 178, 180, 183–186, 197, 202, 203, 210, 211, 244, 256–258, 278, 281, 282, 284, 293, 294, 296
Nobel, 4, 27, 29, 38, 48, 51, 79–81, 96, 101, 102, 132, 176, 181, 190, 211, 261, 273, 279
Norgett, 118, 119

O, P

Oak Ridge, 4, 72, 100, 107, 115, 118, 125, 127, 146, 149, 151, 176, 178, 180, 202, 209, 212, 252, 256, 258, 282
octane, 208
oxidation, 213, 215, 217, 238
Pease, 1, 117–119, 133, 154
pleochroic, 34, 35, 196
Plücker, 26, 38
polymer, 69, 114, 209–219, 268, 269, 277, 284–286, 288, 295
Pooley, 112–114
Przibram, 4, 33, 110, 195

Q, R

quartz, 6, 10, 127, 130, 185, 198, 202, 232
radical(s), 210, 213, 215, 227–229, 233–235, 238, 254
radioactivity, 1, 4, 5, 14, 29, 34, 36, 38, 74, 79, 80, 91, 225, 226, 243, 273
radiography, 269, 271
radiotherapy, 37, 216, 245, 271–273

radium, 31–36, 50–57, 80, 94, 95, 101, 139, 195, 207–209, 211, 223–225, 229, 231, 243, 248, 249, 254, 267, 268, 270, 273–275
reactor(s), 2, 5, 107, 132, 133, 139, 140, 143, 146, 149–151, 167, 176, 186, 197, 198, 202, 209, 210, 212, 228, 244, 255–261, 282–284, 296
Robinson, 72, 74, 75, 79, 118, 119, 153, 253
Röntgen, 3, 19, 27, 28, 31, 32, 35, 38, 47, 48, 79, 91, 94, 212, 217, 225
rubber, 17, 47, 207, 209, 210, 218
Ruhmkorff, 18, 19, 26, 38
Rutherford, 4, 34, 35, 48, 49, 53, 54, 56–60, 64, 66, 79, 80, 102, 115, 116, 195, 209, 244, 249, 261, 287

S, T

Schottky, 100, 102
Seitz, 98, 107, 110, 111, 115–118, 127, 133, 142, 144, 146, 150, 165, 168, 184
silicon, 12, 78, 120, 122, 177, 179, 181, 183, 185–189, 278–281
simulation, 68, 70, 75, 95, 115, 118, 119, 124, 140, 153, 235, 296
Smart Cut, 188, 189
sputtering, 74, 76, 128–130
SRIM, 51, 53, 72, 73, 116, 117, 120, 179
Szilárd, 132, 247, 261
thermal spike, 126–128, 130, 180, 202, 296
threshold, 73, 115–118, 120–124, 151–153, 181–183, 200–202, 283, 295
track(ing), 71, 74, 131, 197, 199, 200, 219, 235, 283–286, 296

V, W

vacancy, 98, 99, 110–113, 119–121, 125, 128, 133, 141, 142, 144, 147, 153–161, 163–166, 181, 185, 187, 198, 200
Van Atta, 19, 249, 250
Van de Graaff, 16, 17, 19, 38, 212, 243, 249–251, 253–255, 276, 280
Varley, 110–112, 114, 134
Walton, 19, 48, 184, 249, 250, 253, 261
water, 6, 19, 24, 47, 48, 69, 92, 101, 107, 197, 202, 209, 212, 219, 223–234, 236, 256–259, 282, 288, 295
Widerøe, 244, 245
Wigner, 107, 132, 133

X, Z

X-rays, 1, 4, 14, 19, 24, 27–29, 32, 35, 36, 38, 47, 51, 80, 91, 110, 111, 139, 198, 207, 227, 228, 243, 252, 269–272, 274, 276
Ziegler, 51, 53, 68, 69, 115, 120
zircon(s), 13, 196, 197, 202

Printed and bound by CPI Group (UK) Ltd, Croydon, CR0 4YY
13/09/2026
14961073-0003